GENETIC CROSSROADS

GENETIC CROSSROADS

The Middle East and the Science of Human Heredity

Elise K. Burton

STANFORD UNIVERSITY PRESS
Stanford, California

Stanford University Press
Stanford, California

© 2021 by the Board of Trustees of the Leland Stanford Junior University.
All rights reserved.

No part of this book may be reproduced or transmitted in any form or by any means, electronic or mechanical, including photocopying and recording, or in any information storage or retrieval system without the prior written permission of Stanford University Press.

Printed in the United States of America on acid-free, archival-quality paper

Library of Congress Cataloging-in-Publication Data

Names: Burton, Elise K., author.
Title: Genetic crossroads : the Middle East and the science of human heredity / Elise K. Burton.
Description: Stanford, California : Stanford University Press, 2021. | Includes bibliographical references and index.
Identifiers: LCCN 2020021120 (print) | LCCN 2020021121 (ebook) | ISBN 9781503611917 (cloth) | ISBN 9781503614567 (paperback) | ISBN 9781503614574 (ebook)
Subjects: LCSH: Human genetics—Political aspects—Middle East—History. | Nationalism—Middle East—History. | Nationalism and science—Middle East—History.
Classification: LCC QH428.2.M628 B87 2021 (print) | LCC QH428.2.M628 (ebook) | DDC 616/.0420956—dc23
LC record available at https://lccn.loc.gov/2020021120
LC ebook record available at https://lccn.loc.gov/2020021121

Cover design: Rob Ehle

Cover map: "Persia, Afghanistan and Beluchistan Showing Lord Ronaldshay's Route," from *Sport and Politics Under an Eastern Sky* by the Earl of Ronaldshay, William Blackwood and Sons, 1902. Perry-Castañeda Library Map Collection, University of Texas.

Typeset by Motto Publishing Services in 10.5 on 14.5 Brill

To my family

Contents

List of Abbreviations ix
Note on Transliteration xi
Acknowledgments xiii

Introduction
An Uneasy Inheritance   1

PART I. RACE AND NATION
1   Drastic Measurements   29
2   Truth Serum   67

PART II. MEDICINE AS ANTHROPOLOGY
3   The Traffic in Blood   101
4   Sickling Sociologies   128
5   Genes Against Beans   154

PART III. COLONIAL AND ETHNIC VIOLENCE
6   Collection Agents   183
7   Domesticating Diversity   213

Conclusion
Genomes Without Borders?   243

Notes   263
Bibliography   307
Index   353

Abbreviations

ABGL	Anthropological Blood Grouping Laboratory (Beirut)
AGHP	Anatolian Genetic History Project
AUB	American University of Beirut
BGRL	Blood Group Reference Laboratory (London)
CHP	Cumhuriyet Halk Partisi (Republican People's Party, of Turkey)
CISNU	Confederation of Iranian Students—National Union
DDT	dichlorodiphenyltrichloroethane (a pesticide)
DNA	deoxyribose nucleic acid
DTCF	Dil Tarih-Coğrafya Fakültesi (Language and History-Geography Faculty at Ankara University)
FAO	Food and Agriculture Organization
G6PD(d)	Glucose-6-phosphate dehydrogenase (deficiency)
GP	Genographic Project
HbS	hemoglobin S
HGDP	Human Genome Diversity Project
HLA	human leukocyte antigen

IBP	International Biological Program
ICSU	International Council of Scientific Unions
IGP	Iranian Genome Project
IHGP	Iranian Human Genome Project
IHMGB	Iranian Human Mutation Gene Bank
INBTS	Iranian National Blood Transfusion Service
JDC	Jewish Joint Distribution Committee
mtDNA	mitochondrial DNA
NIH	National Institutes of Health (USA)
PKK	Partiya Karkerên Kurdistanê (Kurdistan Workers' Party)
RAF	Royal Air Force (United Kingdom)
SAVAK	Sāzimān-i Iṭṭilāʿāt va Amniyat-i Kishvar (National Organization for Security and Intelligence, of Iran)
SPGL	Serological Population Genetics Laboratory (London)
TGP	Türkiye Genom Projesi (Turkey Genome Project)
UNESCO	United Nations Educational, Scientific and Cultural Organization
UNRWA	United Nations Relief and Works Agency
WHO	World Health Organization

Note on Transliteration

Hebrew and Persian have been transliterated according to the Library of Congress systems. Arabic follows the transliteration system of the *International Journal of Middle East Studies*. Personal names are transliterated without diacritics or follow the individual's preferred spelling in Roman script.

Acknowledgments

Nearly ten years have passed since I began the research for this book. The idea was hatched while I was at the University of California, Berkeley, trying to figure out how to pursue biology and Middle Eastern studies together. Emily Gottreich wisely counseled me that I was better suited to being a historian than an anthropologist. I can't thank her enough for setting me on the right path. Likewise, I am enormously grateful to Susan Kahn for believing so fiercely in my project and seeing me through my early years at Harvard. This research required a fleet of supportive faculty and staff from the Center for Middle Eastern Studies, the History of Science Department, and the Program for Studies of Women, Gender, and Sexuality. Afsaneh Najmabadi, Sarah Richardson, and Janet Browne regularly went beyond the call of duty to make sure this project, and my career, could succeed. Cemal Kafadar, Sheila Jasanoff, Steve Caton, and the late Roger Owen also offered crucial advice at different stages of the project.

Generous funding for my research was provided by the Social Science Research Council, the Council on Library and Information Resources, the American Institute of Iranian Studies, the Harvard Graduate School of Arts and Sciences, and the Harvard Center for Middle Eastern Studies. Support for language training in Persian and Turkish was also provided by the Institute of Turkish Studies and the Critical Language Scholarship Program.

Accessing material from Iran presented constant challenges. I am forever indebted to Ehsan Amini, who provided incredible research assistance from Tehran. I thank him for relentlessly pursuing all leads to track down missing documents and for interviewing the geneticist Dariush Farhud. Mostafa Karimkhanezand went to great lengths to help me acquire a copy of the Iranian map pictured in Figure 11. Shervin Malekzadeh very kindly granted permission to use his photography in Figure 10. I am also grateful to friends and colleagues who helped me obtain copies of materials in far-flung libraries, namely Jenny Tan, Alex Warburton, Rachel Schine, and Jim Ryan.

Many colleagues during my time at Harvard, Cambridge, and around the world provided advice and support. In the early stages of my research on Turkey, I met with Sanem Güvenç-Salgırlı, Murat Ergin, and Nazan Maksudyan in Istanbul. I thank them for their advice and encouragement. In 2015, I enjoyed several months at the Ankara hostel of the American Research Institute in Turkey, where I was inspired by chats with Cheryl Anderson and Anat Goldman. I also thank Murat Gülsaçan for sharing his discovery of a rare Turkish high school biology textbook from 1939. In Israel, the hospitality and generosity of Michal Hasson and Ohad Shamir made Jerusalem and Rehovot feel like home. Other colleagues and friends at Harvard who offered an exceptional degree of companionship, advice, support, and feedback on my work include Ceyhun Arslan, Mou Banerjee, Arbella Bet-Shlimon, Shireen Hamza, Neelam Khoja, Bethany Kibler, Andrew McDowell, Ian McGonigle, Sreemati Mitter, Eli Nelson, Shaun Nichols, Mircea Raianu, Kathryn Schwartz, Alex Shams, Sarah Shehabuddin, Arslan Tazeem, and Rustin Zarkar.

A junior research fellowship from Newnham College at the University of Cambridge made it possible for me to write this book in its current form. I am grateful for the support and encouragement offered by Newnham fellows and affiliates Kate Fleet, Manali Desai, Yael Navaro, Sheila Watts, Janine Maegraith, Gillian Sutherland, and Gabriela Ramos. I greatly appreciated the companionship of Newnham's other research fellows, especially Asiya Islam, Mezna Qato, and Cécile Bushidi. I also benefited extensively from colleagues in Cambridge departments and faculties, especially the History and Philosophy of Science Department, where Mary Brazelton, Helen Curry, and Nick Hopwood were founts of advice. It was a special pleasure to work on this book alongside Jenny Bangham as she finished her own. Our mutual interests in Arthur Mourant and the other fascinating historical figures in British blood group genetics made for fruitful and delightful conversations. Charu Singh, Sarah Abel, and Sertaç Sehlikoğlu served as incredibly important sounding boards in the development of this book. I also enjoyed the guidance and feedback of Hana Sleiman, Chris Wilson, Assef Ashraf, Helen Pfeifer, Arthur Asseraf, Andrew Arsan, and Saul Dubow in the Faculty of History, and Khaled Fahmy in the Faculty of Asian and Middle Eastern Studies.

Special thanks go to Projit Mukharji for being a particularly strong supporter of this book; I greatly value all of his insights and feedback on my work. Other important interlocutors and sources of encouragement include Warwick Anderson, Soha Bayoumi, Claire Beaudevin, Ruha Benjamin, Luc Berlivet, Edna Bonhomme, Soraya de Chadarevian, Iris Clever, Nathaniel Comfort, Rosanna Dent, Claude-Olivier Doron, Vivette García-Deister, Snait Gissis, Jaehwan Hyun, Ceyda Karamursel, Nurit Kirsh, Liat Kozma, Tanya Lawrence, Ilana Löwy, Taylor M. Moore, Eve Troutt Powell, Joanna Radin, Ahmed Ragab, Jenny Reardon, Sherene Seikaly, and Edna Suárez-Díaz.

Like all historians, I depend on the work of many archivists and librarians, some of whom I never have the chance to meet face-to-face or learn their names. This book uses materials from nearly thirty archives and libraries in nine countries. I'm very grateful for the patience, kindness, and generosity I experienced from the staff in most of these locations, such as the staff at the Milli Kütüphane in Ankara who made me a copy of the DVD recording of Muzaffer Aksoy's mysterious oral history interview. In particular, I must thank the managers of small local collections who enabled me to access rare and uncatalogued material, including Claire Wallace and Steven Kruse at the Whipple Museum in Cambridge; Jeffrey Monseau, manager of the archives and special collections at Springfield College in Massachusetts; Christian George, director of the archives of Johannes Gutenberg University in Mainz; and Ofer Tzemach and Michael Vinegrad, who administer the central archives of the Hebrew University of Jerusalem.

Where possible, I took the opportunity to interview several geneticists and physicians whose work is related to the material discussed in this book. I thank Adel Aulaqi, Habib Fakhrai, and Marc Haber for their time and willingness to discuss difficult topics with me.

I am grateful to the editors Kate Wahl and Sandra Korn, as well as the anonymous manuscript reviewers they recruited, whose feedback has been so instrumental to improving this book and making it a reality.

I owe the greatest debt to my parents and my husband. My fascination with the Middle East comes from my mother. I learned to love biology from my father. Both parents taught me to value education and travel. For this, and their unwavering faith in my career, I thank them. Finally, to Ari, my partner in every sense of the word, who has never known me without this project: thank you for sustaining me, tolerating an unasked-for education in genetics, and sharing the ups and downs of the entire journey.

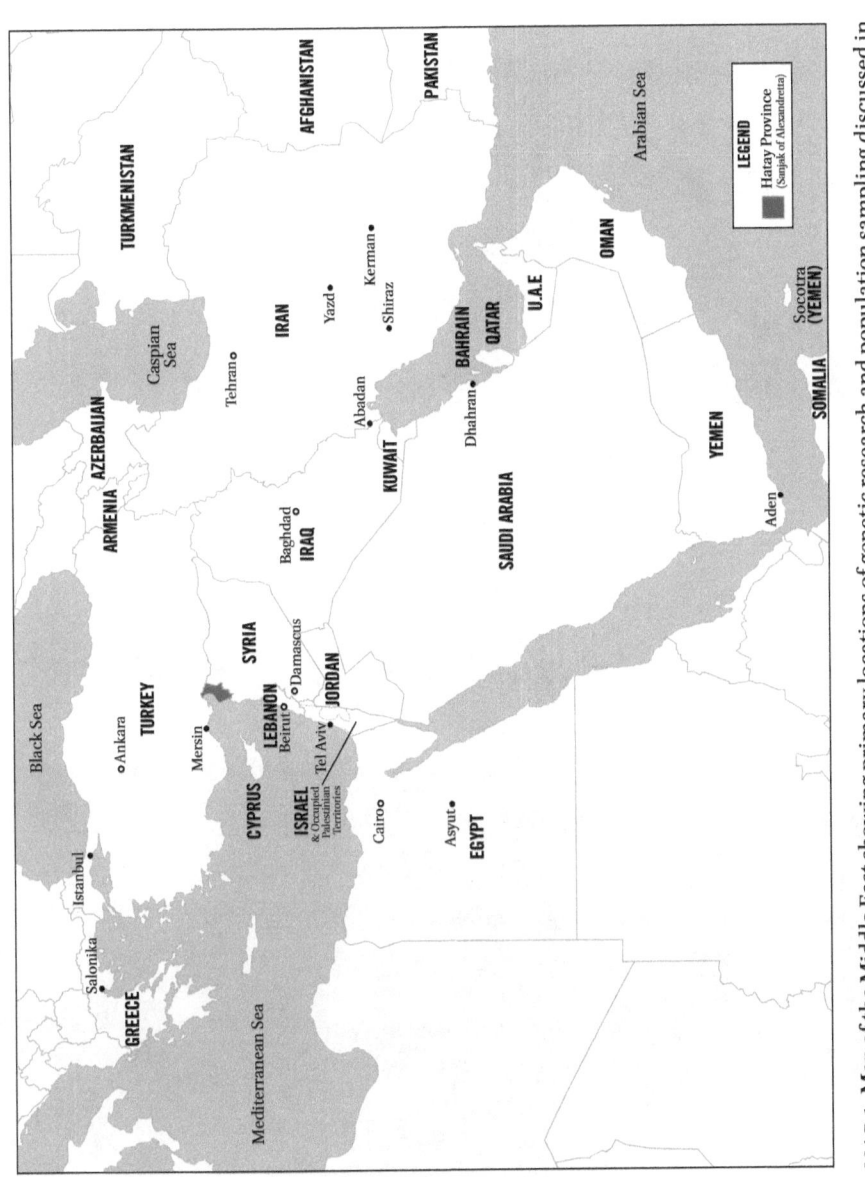

MAP 1. Map of the Middle East showing primary locations of genetic research and population sampling discussed in the book.

GENETIC CROSSROADS

Introduction
AN UNEASY INHERITANCE

IN AUTUMN 2015, DARIUSH DANESHVAR FARHUD, A TEHRAN UNIversity professor and self-proclaimed "father of Iranian genetics," sat down for an interview with Iran's Channel 3 talk show *Imzā*. In his old age, Farhud has an unassuming figure that belies his long career holding leadership positions in the Iranian biomedical community. His status further stems, in part, from the prestige of many connections abroad: he has collaborated extensively with geneticists from across Europe and North America, published regularly in international scientific journals, and served on World Health Organization committees for bioethics and medical genetics. On the walls of his private genetic counseling clinic near busy Valiasr Square (and on his clinic's website), he proudly displays his foreign diplomas and awards.

Over the course of an hour, *Imzā* host Ehsan Karami asked Farhud about his family background, school years in Tehran, university education in West Germany, and subsequent career as a prominent medical and anthropological geneticist since returning to Iran in 1972. The interview, which aired on December 4, features several exchanges that might seem jarring to non-Iranian viewers. For example, Farhud explained that he had refused to leave Iran during and after the 1979 revolution because of his deeply felt nationalism (*īrān-dūstī*) and patriotism (*mīhan-parastī*), proclaiming, "I was born here, and I'll die here." Farhud suggested that his sentiments were "genetically" inherited, since his parents "were the same way." Karami asked dubiously, "Do you mean that all patriotic people in the world have patriotic offspring?" Farhud clarified with other traits that he considered to result from both genetic and (familial) environmental contributions, such as talents in calligraphy and music. Later, Karami asked Farhud's opinion on the concept of "gene drain" (*farār-i zhinhā*), an extrapolation of "brain drain." Farhud claimed, "I never want to talk about politics," but he explained that when young people migrate, "they take their genes with them," causing not only a brain drain in developing regions but also the loss of those young people's potential children. Through this process, "some countries of the world" profit not only from the natural resources of developing countries but also from their "human resources."

A conversation about Farhud's research similarly turned toward questions of nationality and Iran's place in the world. Farhud credited himself with initiating population genetics in Iran, namely comparative studies of the blood groups, serum proteins, hemoglobins, and physical traits of the country's many ethnic groups to understand their differences from one another. For example, he explained, there are differences between Turkic speakers (*turk'hā*) and Baloch tribes. By determining the genetic structure of different Iranian ethnicities (*sākhtār-i zhinitīk-i qawmiyyāt-i īrān*), his work had enabled studies on ethnic differences in disease frequencies. Finally, to conclude the episode, Karami asked how Iranian genes were different from others. For example, were Iranians smarter than Koreans, Japanese, and Americans? Farhud affirmed that "genetically," Iranians were more intelligent than many other peoples (*mardum*), but not "the smartest in the world," and that they suffered from a customary lack of perseverance and work ethic. Farhud ended his interview with an entreaty to Iranian youth to cultivate a stronger sense of patriotism and honor their identity (*huviyyat*), citing the alleged saying of the Prophet Muhammad that "love of homeland is part of faith" (*ḥubb al-waṭan min al-īmān*).

Farhud's self-narrative throws into sharp relief many questions about the social implications of genetic research in Iran. What is the relationship between his identities as a professional geneticist and an ardent nationalist? Why does he slip so easily between speaking about the genetic differences between Iranian ethnicities and declaring intelligence to be an essential trait of all Iranians? His gestures toward the biological determinism of intelligence and other behavioral traits are hardly unique among prominent geneticists, including those in the West.[1] But his speculations about unitary Iranian national traits seemingly contradict the premise of his anthropological research, which highlights the country's ethnic and genetic diversity. Addressing these questions requires an assessment of Farhud's professional life, in which he acts simultaneously as a representative of transnational research in population genetics to his fellow Iranian citizens, and also as a representative of his country within a global scientific infrastructure that has often neglected the contributions of Middle Eastern workers.

The discussion about gene drain invokes scientific and social discourses that ascribe value to genes rather than to individual people. Although Farhud and Karami lament "gene migration" as the depletion of a national resource, in prior decades geneticists from around the world flocked to Iran and the rest of the Middle East specifically to study genetic evidence of past human migrations. The belief that the region had been a historical crossroads of genes effectively transformed the Middle East into a global crossroads of scientists. Yet the practices through which this genetic evidence was collected and interpreted have often been contentious. Given Iran's close association with the concept of brain drain since the 1960s, Farhud's comments about gene drain reflect ongoing geopolitical anxieties about the exploitation of developing regions by "other countries"—that is, the West or the Global North. Meanwhile, his exhortations to other Iranians to cultivate an allegedly innate patriotism highlight the relationship between scientific nationalism and the transformation of human genetic data into socially meaningful identities.[2] To parse out where Farhud's views fit into globally standardized practices of biomedical research and where they expose specific contingencies of pursuing such research in Iran or the broader Middle East, I begin with a brief overview of the entanglements between genetics and social knowledge.

HUMAN GENETICS AND THE PRODUCTION OF ETHNICITY

Today, many explanations of genetic research turn immediately to the molecular level, describing how scientists detect the sequence of chemical units in deoxyribonucleic acid (DNA), the hereditary material found in all living things. Both professional geneticists and the lay public have largely succumbed to the "DNA mystique," casting DNA sequences as the essence of unique individual identities as well as the "immortal text" containing the history of all life on Earth.³ However, the history of "genetics" (a term first proposed in 1905) far precedes both the discovery of the DNA molecule and the invention of technology to sequence it. For much of the twentieth century, scientists studied "genes" as units of heredity without direct reference to DNA.⁴ Accordingly, the majority of the research I analyze in this book, which begins with the First World War, predates the widespread adoption of efficient DNA sequencing techniques. Rather than examining variations in the molecule itself, geneticists working on humans observed the physiological manifestations of such variations: the size and shape of body parts, the presence of different proteins in the blood, and the inheritance patterns of certain diseases.

At present, genetic research on human subjects occurs within a number of specialized subfields, and my principal concerns with racial classification and ethnic ancestry are now commonly associated with that subfield known variously as anthropological genetics, molecular anthropology, genetic anthropology, or genetic history. However, during most of the period covered by this book, even the broader category of genetics was not a well-defined field of inquiry, and many of its practitioners identified as medical professionals. Indeed, many of the individuals I write about worked for hospitals or blood banks and received their research funding from national and international health-care agencies. Accordingly, I approach the history of human genetics not as a unified discipline with a clear trajectory but rather as a conglomeration of knowledge about human history and prehistory predicated upon tracing the relationships of individuals to groups and of groups to each other. I use "geneticists" as a shorthand for the scientists and physicians trained in physical anthropology, evolutionary biology, biochemistry, and medical pathology (especially hematology and serology, the study of blood and serum, respectively) who studied inherited traits to answer questions about human history.

These geneticists, in turn, did not apply consistent labels to the field they were building. In 1965, the prominent British hematologist Arthur Ernest Mourant, the director of the Blood Group Reference Laboratory in London, who worked with many Middle Eastern geneticists, mused at length over the name of his proposed new research unit: the Serological Population Genetics Laboratory. "As regards the first part of the name, designating the subject of research, it would appear that the [Medical Research] Council have raised no objections. I take it that the word 'Serological' sufficiently distinguishes it from the Population Genetics Research Unit. [. . .] One other possibility would be 'Serological Anthropology.'"[5] The lack of nominal conventions for this enterprise, suspended between medicine and anthropology, conceals certain remarkable consistencies that unite fin de siècle skull measurements, Cold War surveys of blood-group frequencies, and the oft-maligned Human Genome Diversity Project of the 1990s. These methodologies all demanded a specific kind of human group—a study population—in order to yield useful, generalizable scientific knowledge. Because this knowledge is often articulated in the form of statistical averages, geneticists must establish consistent criteria regarding the boundaries of their study population. In order to ensure that their data collection accurately reflects the range of hereditary traits in the group, geneticists must constantly decide which individuals "belong" to the population and which should be excluded.

This adjudication of population boundaries is rarely straightforward. It involves many social and cultural assumptions about kinship, marriage, and reproduction. For example, scholars have long assumed that certain kinds of human groups are more likely to be endogamous—that is, to marry only among themselves, and therefore retain a high fidelity to their ancestral cultural and biological traits. These include communities isolated by geography, surrounded by ocean, desert, and high mountains, or those isolated by sociocultural factors like religion and language. Geneticists tend to regard such endogamous communities as living representatives of the past, and therefore particularly valuable for studies of human evolution. For over a century, geneticists have justified the urgency of their research on remote tribal communities using narratives of the "vanishing indigene," whose biology must be preserved through "salvage genetics."[6] These narratives argued that the unique genetic composition of such communities must be documented before it inevitably disappeared due to state-imposed projects of "modernization" that aimed to disrupt traditional group identities and ways of life. This disruption was expected to cause, in scientific terms, "admixture"—the intermarriage of previously isolated social groups with other populations, and therefore the loss of their characteristic set of genetic traits.

Admixture has negative connotations for ethnic and nationalist ideologies that conceptualize their authenticity in terms of a purity of ancestry.

National histories seek to project the existence of a qualitatively distinct population ("the nation") back in time to a fixed temporal and geographical origin point, and valorize the living people who have faithfully adhered to the marriage preferences of their ancestors. At the same time, the limitations of scientific methodology meant that geneticists, too, framed admixture as problematic. The medical and evolutionary information sought by geneticists—e.g., the inheritance patterns of known disorders, the relative significance of natural selection and genetic drift, and the biological relationships between living and ancient human populations—were all simply more difficult to study in admixed populations, at least until the advent of efficient computer-based data analysis beginning in the 1970s. As philosophers Lisa Gannett and James Griesemer have explained, the problems were especially acute for the reconstruction of ancient human migrations:

> We cannot speak of group origins or unique common ancestors without well-delineated "primordial" groups locatable in space and time. We cannot speak of dates and routes of group migrations without assuming the constancy and integrity of these groups over space and time. We cannot speak of genealogical relationships other than between "qualitatively distinct" groups for which there are sorting criteria for inclusion and exclusion. We cannot speak of the admixture of groups without some modified sense of "purity" in terms of the relative homogeneity and heterogeneity of "qualitatively distinct" "gene pools" with characteristic compositions. The bounding of genes—and the people who possess and pass on these genes—across space and time is a necessary a priori assumption for all such narratives or explanations.[7]

In practice, this meant that geneticists "required a clearly demarcable, empirically manageable endogamous population," i.e., groups that could be reliably identified as "reproductively isolated" and therefore "evolutionarily coherent." An effect of this requirement was that "non-biological knowledge entered the research design," as "linguists, ethnographers, historians, sociologists and others, as well as myths and claims of collective identity" provided geneticists with the necessary evidence to identify ideal communities for genetic research.[8] The biological data collected from such research, in turn, cannot be interpreted without reference to non-biological knowledge and is therefore not a truly independent source of information about human history.

The production of ethnic categories through this process of epistemological layering and cycling through social and intellectual networks is therefore a fundamental component of all genetic research concerning human subjects. This process resembles a feedback loop, in which "folk concepts" of race and ethnicity feed into the assumptions and interpretations of scientific research, whose practitioners in turn feed their work back into the original popular discourse.[9] Through this process, genetic researchers in the Middle East effectively transformed religious, linguistic, and other social identities into ethnicities: they contrasted allegedly endogamous communities like Zoroastrians, Armenians, Jews, and Bedouin tribes with heavily admixed populations of Persians, Turks, and Arabs. Interactions between geneticists and their research subjects reified and reinforced communal identities through a hyperbolized sense of a group's historical isolation from others, via socially or geographically enforced endogamy. Emphasizing this isolation became configured as positive and desirable for researchers and community members alike, since strict practices of endogamy were believed to preserve a community's authenticity through an unbroken and undiluted genetic relationship to its ancestors. Minority groups were not passive in this process, and some pioneering ethnographic studies show how communities like Samaritans and Alawites have responded to their genetic objectification.[10]

Meanwhile, geneticists also use national labels to identify populations (e.g., "Iranian" or "Turkish"), despite the fact that such labels do not constitute biologically meaningful categories. Here, the production of ethnicity collides with the logistical pragmatism of "methodological nationalism"—that is, an unquestioned acceptance of the nation-state as a natural unit of analysis.[11] Whether used with a conscious ideological aim to identify the biological unity of a "nation" or simply to solve a practical need to identify research subjects with an ostensibly neutral geographic label, geneticists compare these admixed or agglomerated national populations to test hypotheses about national origins and trace historical migrations through the course of human evolution. Advances in technology and methodology promise to reach further and further back in time with greater resolution. However, our concepts and terminology for geography and human groups are inevitably rooted in the present and are therefore burdened by larger uncertainties about how living populations are related to historic and prehistoric human groups. The same questions about national ancestry, with the same underlying assumptions, are thus tested again and again with newer and newer methods, but the results can never disprove the political logic of the contemporary moment. Attempts to reconstruct national histories through genetic data "can never live up to their positivist commitments" because "biological evidence is incapable of undermining the original hypothesis" produced by other sources of knowledge.[12]

I must clarify here that these methodological problems do not mean that the study of human genetics is somehow pseudoscientific or that it can tell us nothing useful about human evolution or historical population movements. Indeed, the very robustness of modern genetic technology makes possible the emergent sociopolitical phenomenon scholars have called "bionationalism," "genetic nationalism," or "genomic nationalism." This phenomenon represents a shift in how individuals imagine themselves as part of a national or ethnic group—as part of a community so large they will never personally meet most of its members. Whereas Benedict Anderson conceptualized the nation as an imagined community produced by language and print capitalism, new scholarship argues that genetic science has refocused national imaginations toward biological kinship.[13] In other words, public awareness of genetic research is "reworking ethnic identities as imagined genetic communities, that is, communities in which the language, concepts and techniques of modern genetic medicine play their part in shaping identity."[14] For example, in South Korea, Taiwan, and Japan, imagining genetic communities has amounted to the conflation of nation-state identity with essentialized majoritarian ethnic categories.[15] But this phenomenon is not inevitable, because genetic research can be easily appropriated for "very different political, biological, activist, national, and transnational ends."[16] Even without relying on non-biological knowledge, genetic data is inherently relational: it becomes meaningful only through comparison between individuals, populations, or species. Furthermore, the biological genealogies linking together any one set of individuals or groups can vary widely depending on sample sizes, the number and types of genetic markers analyzed, and the statistical formulas and outgroups used to assess genetic similarities.

To account for these everyday methodological issues faced by professional geneticists, I propose a notion of genetic nationalism that incorporates Prasenjit Duara's insights on the fundamentally relational nature of national identity. Duara argues that the nation exists as "a provisional *relationship*, a historical configuration in which the national 'self' is defined in relation to the Other. Depending on the nature and scale of the oppositional term, the national self contains various smaller 'others'—historical others that have effected an often uneasy reconciliation among themselves and potential others that are beginning to form their differences."[17] The "imagined endogamous communities" of socially isolated minority groups—imagined together by geneticists and research subjects—have similarly provisional relationships to admixed, majoritarian national populations, depending on the scale of genetic comparisons being made: local, transnational, continental. By emphasizing genetic nationalism as not only essentialist but also relational, my analysis delves beyond political discourses of biological identity to focus on the material and interpretive practices of professional scientific communities in the Middle Eastern context. Through these practices, local processes of ethnic category-making are profoundly interlinked with the methodological nationalism used by international projects to compare human genetic traits.

STATES AND SCIENCE IN THE MIDDLE EAST

Every aspect of human genetic research has been, and continues to be, affected and shaped by contemporary geopolitics. Scientists face various legal and logistical restrictions in their collection of biological material (e.g., blood, saliva, and other tissue samples) and anthropometric measurements, requiring national, local, and individual permissions for their research. The site conditions under which this raw material is collected vary widely, ranging from hospital-based blood banks to remote locations far from any laboratory or clinic. In the latter scenario, biological samples must be preserved and transported to the nearest accessible testing facility, often at great financial cost to the investigators. Once they have analyzed their data, additional language and financial barriers affect researchers' ability to publish and disseminate their work. Bookending each individual scientific career are innumerable constraints on opportunities for higher education and collaboration with international colleagues, as determined by the local scale of class and social privilege and the global scale of visa regimes and (post)-colonial patronage networks. The circulation of scientific actors, equipment, specimens, and ideas are regulated and hindered by political entities asserting authority and contending for sovereignty.[18]

In the contemporary Middle East, this geopolitical context is predominantly shaped by the historical interactions between the region's empires and European powers from the eighteenth century onward. This colonial legacy in the Middle East is the root of a perceived "lag" or "lack" of scientific productivity in the region. Since the nineteenth century, much of the literature on the history of science produced by Western and Middle Eastern scholars alike has perpetuated the notion that the Arabic-speaking world made no significant scientific achievements after the Abbasid Empire's so-called golden age (circa 786 to 1258 C.E.). This apparent lack of productivity has often been attributed to cultural factors such as anti-intellectual strains of Islamic theology or the Mongol conquests of the thirteenth century. Only in recent years have historians begun to systematically dismantle this thesis of scientific decline, demonstrating continuous scientific activity under early modern Muslim empires and showing how political and economic conditions rather than cultural or religious constraints marginalized Middle Eastern scientific actors in favor of European ones.[19] While India and Algeria faced military invasion and direct colonial administration under Britain and France, the core regions of the Ottoman Empire and Qajar Iran retained political sovereignty. However, the latter became "semi-colonial" polities under the pressure of European economic imperialism and Russian military aggression during the nineteenth century. These factors, more than activity by anti-modern reactionaries, hindered the development of industrial, educational, and scientific institutions in the Middle East.

Historian Cyrus Schayegh argues that this semi-colonial position rendered Iran even more peripheral to global scientific networks than directly colonized regions, where administrative and economic necessity drove intensive data accumulation and methodological innovation. Although late Qajar and early Pahlavi Iran did not experience the "comprehensive scientific colonialism" of India, modern science in Iran emerged through colonial patterns of knowledge transfer. These patterns promoted applied science and medicine over basic research, privileged Western educators and scientists over local ones, and provided a foundational discourse for professional class formation.[20] The legacy of this semi-colonial experience continues to shape Iranian science today. Even after the Islamic Republic strove to shake off Iran's dependence on Western scientists and technicians, its integration into the global scientific community was hindered by economic problems and sanctions that limited the financial and logistical support for basic research and accelerated the country's brain drain, as well as by difficulties in publishing and disseminating scientific knowledge abroad due to language and visa barriers.[21] Sociologists studying the formation of the Iranian scientific community have observed a sort of "inferiority complex" among Iranian scientists who believe they will never be able to keep pace with Western scientific achievement due to their fewer material and professional resources.[22] Through these twentieth-century political shifts, Iran—and, I contend, the rest of the semi-colonial and postcolonial Middle East—regarded its subordination to Western powers increasingly in terms of "the neocolonial intellectual hegemony of science and technical reason" instead of "formal political and economic imperialism."[23]

Scientific developments in the different Middle Eastern states that formed after the dissolution of the Ottoman and Qajar empires did not all occur at the same time or under the same conditions, and this book makes no pretense of comprehensively covering the entire region commonly designated "the Middle East." Certain localities and professional networks take center stage in my narrative, necessarily at the expense of others. Turkey, Iran, and the Levant are best represented, while the Arabian Peninsula, Egypt, and Iraq play secondary, but no less significant, roles. North Africa west of Egypt—despite being an important nexus for Middle Eastern medical training, genetic research, and revolutionary ideas about Arab identity—has regretfully been omitted. I hope that other scholars will take this work as a launchpad to further investigate genetic science across the region and rectify these gaps. In particular, more localized studies could potentially offer a better sense of the popular understanding and reception of genetic research, which is largely neglected by my methodological focus on academic scientists, state-certified physicians, and ruling elites.

Nevertheless, this book identifies broad patterns and trends of state formation, scientific development, and racial and ethnic identity representative of the region as a whole. One of the goals of the book is to demonstrate that science in Israel fits into these broader Middle Eastern patterns. In contrast to the rest of the region, the history of genetic research on Jews in Israel has been relatively well studied. Historians and anthropologists have critically examined how the structuring assumptions of Jewish race science in early-twentieth-century Europe and North America, and their relationship to Zionist nationalism, reverberate within the genetic studies of Jewish populations by Israeli scientists from the 1950s to the present. Because most Jewish race scientists lived and worked in "the West," they have been readily incorporated and theorized into the generally Eurocentric literature on the history of science.[24] Existing scholarship on Israeli genetics emphasizes its settler-colonial nature by identifying its scientists as Western transplants who focused on the unique research opportunities presented by its Jewish populations.[25] Indeed, Israeli geneticists have often been engaged in "self-study" through their research on Jewish populations. But they held simultaneous interests in other "Middle Eastern" populations like Samaritans, Armenians, Christian and Muslim Palestinians, and Sinai Bedouins. Accordingly, I highlight Israeli scientists' prominent role in forging Middle Eastern regional networks to study various hereditary diseases or isolated communities. Even as Western transplants, they represent overall trends in Middle Eastern scientific development and research agendas. As I discuss in several chapters, exiled German Jewish scientists had a visible presence in the transformation of medical education and practice (as well as ideas of race and identity) not only in Mandate Palestine but also in Turkey, Iran, Iraq, and other Arab states.[26] I therefore contextualize Israeli genetics within its regional geopolitics.

I begin the story of Middle Eastern human genetics in the aftermath of World War I. The interwar period forms the context for the emergence of today's Middle Eastern nation-states. Through an array of international treaties facilitated by the newly formed League of Nations, the Ottoman Empire was dissolved and much of its territory divided into protectorates, mandates, and client states of France and the United Kingdom. For example, British military control over Palestine dating to 1918 was administratively formalized as the British Mandate for Palestine, recognized by the League of Nations in 1923. The same year, after three years of armed struggle against the Allied occupation of Anatolia, the Turkish National Movement, led by military commander Mustafa Kemal (who later took the name Atatürk), proclaimed the establishment of the Republic of Turkey. Meanwhile, in Iran, another military commander named Reza Khan deposed the ruling Qajar dynasty in 1925 and established his own hereditary rule as the first shah of the Pahlavi dynasty. Both European and Middle Eastern political actors seized upon existing ideological trends and legal structures to legitimize and stabilize the abrupt changes in political administration and state boundaries. These included the academic paradigm of race science, its political counterpart of ethnic nationalism, and its social interpretations into eugenic policies and population transfers.

In this period, Middle Eastern geneticists—whether they worked in sovereign states like Turkey and Iran or under European "protection" and "tutelage" in Egypt and the Mandates— faced analogous challenges. In order to function, each of these polities aimed to establish the autochthony of its ethnic majority within arbitrary territorial boundaries settled partially by international treaties and partially by the armed conflicts those treaties had provoked. All strove to reconcile the readily apparent physical, linguistic, and cultural diversity of their citizenry with an ideology of national unity strongly informed by race science, aiming to reconfigure patterns of group identification along ethnic lines. Scientific data gathered by archaeological expeditions and public health surveys became political ammunition to control the historical narratives of ethnic and religious minorities and to delegitimize the territorial claims of competing nationalisms. The consolidation of a strong central government thus hinged on the enforcement of "internal colonialisms" through political, military, and economic institutions while sharply inhibiting the social visibility and cultural expression of minority groups.[27] Scientific discourses on race, medicine, and hygiene were explicitly deployed for the purpose of these projects of internal colonialism.[28]

Of course, the ideological and infrastructural similarities linking the region's states are not deterministic, and major domestic fluctuations in their political economies have led to divergent trajectories in the development of scientific research. Regardless, Middle Eastern scientists had much in common and must be understood not only in relation to their aspirations to Western modernity but also through their local working conditions. They studied the same hereditary disorders and overlapping population categories, which formed the logical basis of a regional network. My transnational approach thus provides a framework for analyzing the research interests and practices of these geneticists not only in terms of national self-fashioning but also in terms of their regional and global positionality.

INTEGRATING THE MIDDLE EAST INTO A GLOBAL HISTORY OF SCIENCE

Drawing primarily on the archival evidence of scientists' correspondence, oral histories, and professional publications, this book advances three distinct but closely intertwined arguments. I explain these arguments in order of expanding geographical implications, from local to global.

First, I argue that genetic research has significantly influenced the notions of race and ethnicity that have been incorporated into Middle Eastern nationalist ideologies, and that these ideologies, at the same time, have influenced how Middle Eastern scientists study genetics. In other words, I apply to the history of Middle Eastern race science an analytical framework from science and technology studies (STS) commonly called "co-production," which emphasizes that human understandings of natural and social order are mutually constitutive. Science, politics, and social norms are not independent, autonomous spheres of human knowledge and activity, and therefore their interactions cannot be understood through "unidirectional arrows" of cause and effect.[29] I prefer the alternative rendering of this framework as "co-constitution," which avoids the connotations of deliberate construction suggested by "production."[30] Analyzing the processes of Middle Eastern nation-building and scientific development as co-constitutional enables us to understand the seemingly "contradictory" nature of anticolonial nationalism, particularly its ambitions to "emancipate the colonized self" through the very "same values it claimed to oppose," including faith in scientific modernity and progress.[31]

Recent scholarship has begun to take seriously the fundamental role of transnational racial and eugenic discourses in Middle Eastern nationalist thought, social class formation, and state health policies, assessing how such discourses incorporated or departed from existing local concepts of race and ethnicity.[32] Well before the Middle Eastern encounter with race science, discrimination based on skin color, religious practices, and ancestral lineages was an entrenched feature of social life, literature, and historiography.[33] When nineteenth-century Ottoman intellectuals translated and popularized Eurocentric notions of race and nation, they mobilized a range of existing local vocabularies for human identity and communal belonging. These vocabularies, even in Turkish and Persian, were substantially derived from Arabic words whose connotations fall into two general categories: typological and genealogical. Firstly, terms meaning species, types, or classes were adapted in the nineteenth century to discuss racial classification—for example, *al-anwāʿ* or *al-aṣnāf al-bashariyya*. By the early twentieth century, the term *jins* (etymologically derived from the Greek "genus") became the predominant Arabic translation for "race." Yet it was frequently discussed alongside a second category of genealogical terms connoting ancestral roots or origins (e.g., *ʿirq, sulāla, ʿunṣur*) or familial or tribal ties (e.g., *qawm, shaʿb*). A third layer of complexity emerged from late-Ottoman Arabic-Turkish political discourses that appropriated words customarily used to demarcate religious communities like *umma* and *milla* (in Turkish, *millet*) into ethnic concepts of "people" and "nation." Just as in Europe, all of these terms were combined in inconsistent and nearly interchangeable ways such that race, ethnicity, and nation became thoroughly entangled.[34] But in doing so, some writers deliberately invoked well-known medieval Arabic scholarship and imbued it with new meaning, situating the new race science within local traditions of knowledge.[35] For example, Arab journalist Jurji Zaydan titled his 1912 book on human races in two parts: *Ṭabaqāt al-umam aw as-salā'il al-bashariyya* ("Classes of Nations, or the Races of Man"). The first part of the title invokes a text by an eleventh-century Muslim jurist of Toledo, Ṣā'id al-Andalusī. The second part of the title alludes to several nineteenth- and early-twentieth-century scientific books of anthropology (such as Joseph Deniker's 1901 *The Races of Man*), although Zaydan sourced most of his material from A. H. Keane's 1908 *The World's Peoples*, a work of popular ethnology. Zaydan contrasted the medieval Islamic text's hierarchy of civilizations according to their knowledge production against his own use of the experimental science of anthropology to rebuild a (strikingly similar) hierarchy of races according to evolutionary theory.[36] The ethnological works of Zaydan and other Middle Eastern intellectuals show that they were not simply conduits for diffusing a European race science. Rather, their translation choices represent their contribution to an "active global circulation of ideas and practices concerning race," wherein European "analytic frameworks became transformed, rendered vernacular, through local adaptation and modification."[37]

After the dissolution of the Ottoman Empire, the connotations of the various words used to discuss race diverged and stabilized under the regimes of national language academies. In contemporary standard Arabic, the term *jins* corresponds more often to sex or gender, with *'unṣur* (connoting stocks, breeds, or ancestral elements) now being the preferred term for race; *jinsiyya*, however, is still commonly used to translate "nationality" (as is *qawmiyya*). Meanwhile, contemporary Persian usage in Iran renders *qawmiyyat* as ethnicity, and both Turkish and Persian now use *milliyet/milliyat* as nationality. In post-Ottoman Turkish, the Arabic-derived *ırk* became the explicit term for race.[38] Meanwhile, the anti-Arab Iranian writers of the late Qajar and early Pahlavi periods selected the Persian word *nizhād*, meaning "descent"—still a component of many surnames, e.g., Ahmadinejad—to take on the scientized meaning of race.[39] The Israeli case is distinguished by the revival of Hebrew as a national language, wherein the biblical word *geza'* (meaning stock, stem, or root) was appropriated as "race." Prior to World War II, many Jewish intellectuals used "race" (in various languages) as a category to distinguish Jews from non-Jews as well as to distinguish Jews of different geographical origins. After the war, referring to the Jews or any subset thereof as a race was studiously avoided in Israeli political and scientific discourse in favor of *'am Yiśra'el*, the Jewish (literally, Israelite) "people" or "nation." However, in the early years of Israeli statehood, *geza'* still appeared in the Israeli press, often in a disparaging way, to highlight, e.g., the difference between Ashkenazim and new immigrants from North Africa (see Chapter 3). The official term deployed to discuss intra-Jewish difference was *'edah*, "community." Like *milla* or *millet* in Arabic or Turkish, this word initially had a meaning of religious congregation or denomination. However, since the Yishuv period, *'edah* mutated into a thoroughly racialized word as applied to the ethnonational origins of non-Ashkenazi Jewish communities, invoking their history of political and social marginalization in Israel.[40]

Not all of these social and intellectual histories of race and nation in the Middle East discuss how Arab, Turkish, and Iranian participation in local race-science research affected these shifting concepts. Analyses of Egyptian and Israeli scientists' research in racial anthropology and population genetics still tend to portray them as guided by their ideological commitments, implying a unidirectional influence of politics *on* science.[41] Studies of early Republican Turkish anthropology come the closest to articulating a co-constitutional perspective, showing how the practical tensions between the civic and ethnic "faces" of Turkish nationalism relate to simultaneous scientific projects to define Turks as a race.[42] I build on this scholarship to demonstrate that differently situated Middle Eastern people, ranging from politicians and nationalist intellectuals to religious and community leaders, regularly adapted internationally circulating biomedical discourses on race and genetics to inform and describe their concerns about group origins, territorial claims, and public health. Those same concerns shaped how geneticists selected and defined study populations, collected biological samples, and interpreted their results in the laboratory. The accumulation of genetic data became politically and socially valuable evidence to substantiate historical narratives of migration, intermarriage, and religious conversion; compare these narratives with contemporary circumstances; and advocate for political action at national and international levels.

My second argument is that Middle Eastern research subjects and scientific actors have played a particularly important role in international research developments concerning human evolution and population genetics. While the literature mentioned above shows how European race science was localized and adapted to Middle Eastern contexts to serve particular national interests, such as defining the characteristics of a nation-state's population and meeting its medical needs, it does not explore the interests of the international scientific community in studying Middle Eastern populations. I show that these interests derive from the perception of the region as a "crossroads" or "contact zone" of three major continental races (African, Asian, and European) that intermixed over thousands of years through prehistoric migrations, military invasions, the slave trade, and cultural assimilation. Yet these racial borderlands were simultaneously configured as the cradle of human civilization and as a biblical landscape that harbored hundreds of nomadic tribes, remote village communities, and small religious sects who professed strict regimes of endogamy since time immemorial. Because of these unique social and geographic circumstances, Western geneticists frequently sought out data from the Middle East as part of larger-scale efforts to trace European and global population history, identify the origins and spread of inherited diseases, and reconstruct ancient human migrations. By the mid-twentieth century, the Middle East had become a professional crossroads for medical and anthropological geneticists. Well-known figures in the North Atlantic history of genetics, including J. B. S. Haldane, William C. Boyd, Arthur E. Mourant, Hermann Lehmann, Anthony C. Allison, Arno Motulsky, Walter Bodmer, and Luigi Luca Cavalli-Sforza, all traveled to the region to conduct fieldwork, participate in conferences, and consult with local researchers.

As a result, Middle Eastern scientists and research subjects have made significant contributions to the study of medical genetics and human evolution. They had an immense degree of influence over how Western geneticists collected, used, and interpreted data from their countries, because the latter rarely possessed independent knowledge of local political and social history. In many cases, they therefore uncritically accepted and perpetuated the use of labels that connoted ambiguous civic or social rather than purely biological identities (such as "Turks" or "Bedouins") or reflected nationalist identity manipulation (such as "Eti-Turks" for Alawites in Turkey). The influence of Middle Eastern geneticists in this regard is not merely symbolic; these labels were attached to nationalist-inflected historical narratives of migration, endogamy, and admixture that promoted certain populations as "model systems" to study universal mechanisms for the inheritance of congenital disorders or the reconstruction of human evolution. The work of Western scientists, in turn, has often reinforced categories and narratives originally produced in these particular contexts of ethnic nationalism.

By highlighting this circulation of knowledge about human genetics between the West and the Middle East, I build on historian Omnia El Shakry's arguments about the rise of academic social sciences in twentieth-century Egypt. Through her investigation of Egyptian anthropological studies inspired by conflicting European and Egyptian colonial claims to the Sudan, she demonstrates that the "asymmetrical conditions of power" involved in global and local scales of knowledge production allowed Egypt to simultaneously function "as both colonizer and colonized."[43] Other scholars have observed that postcolonial critiques of anthropology and race science have tended to be Eurocentric; by framing the discipline's history exclusively in terms of Western colonial power, these critiques overlook the "internal colonialisms" of non-Western nationalist regimes. However, histories of genetics and biomedicine should not be reduced to "narratives of biocolonialism that position scientists in the Global North against subjects in the Global South."[44] Instead, I treat anthropological and medical genetics as "circuits of knowledge that secure a sovereign subject status" for certain titular Middle Eastern nationalities (such as Turks and Iranians) over the marginalized ethnic, linguistic, and religious minorities residing within the borders of their nation-states.[45]

These power differentials manifested in distinct ways, reflecting the peculiarities of national politics. Professional scientists in Turkey identified as native Turkish speakers of the Sunni Muslim majority, and they chose to elide or downplay the significance of ethnolinguistic and religious diversity to their findings on the "Turkish" population. Genetic research in Iran and the Arab states often concentrated on religious and linguistic minorities—for example, Zoroastrians, Armenians and other Christian communities, and Kurds—but most of the time, members of these communities served only as consultants rather than primary investigators. Israelis occupied a similar status as outsiders to the disenfranchised communities on whose blood they built their careers. The Israeli physicians and anthropologists involved in genetic research all had Ashkenazi backgrounds, while their research subjects were predominantly new Jewish immigrants from the Middle East and North Africa (now known collectively as Mizrahim), as well as Samaritans, Bedouins, and Armenians. Regardless, the elite professional identity of these scientists granted them the authority to manipulate their research subjects and to make claims about their biological and historical identities. As a group, Middle Eastern scientists represented state hegemony while investigating "exotic" socioculturally marginalized populations in the service of national and international agendas of biomedical research.

My final argument extrapolates from this examination of the power dynamics involved in human genetics research at the local level to assess the interaction of scientific communities at the national versus the international scale. Specifically, I explore the fluctuating positionalities occupied by Middle Eastern geneticists embedded in the networks of professional collaboration that tie together these layers of research infrastructure. In doing so, I adapt Cyrus Schayegh's notion of an "interstitial" positionality occupied by practitioners and promoters of modern science in early-twentieth-century Iran, which was "located at the overlap between a local society thrown open to foreign intervention and a world defined by the shifting hierarchies and interactions of metropolitan and colonial social classes and states."[46] I emphasized above how Middle Eastern scientists, as representatives of a technocratic elite in their home countries, marshaled their national identities to speak on behalf of their fellow citizens, even when they did not belong to the specific minority communities under investigation. In this sense, they behaved as "native informants" to their Western colleagues by gatekeeping the latter's access to and knowledge about local populations.[47] Yet even while Western geneticists depended heavily on the collaboration of these informants for the collection of biological and historical material, they often sidelined their non-Western collaborators at the final stages of interpreting and publishing data, relegating these figures to subordinate statuses as secondary authors, technicians, and field assistants. In the postwar era, this marginalization was reinforced by the emergence of international agencies, such as the World Health Organization, which financed a global biomedical infrastructure that unofficially enforced a particular scientific division of labor in which "first world" countries controlled the intellectual agenda and administrative management, while less-developed, "third world" nations contributed only "basic data" (i.e., blood samples collected from local populations).[48] As a result, the collaborative relationships between Western and non-Western scientists have been shaped by a professional as well as geopolitical hierarchy of scientific prestige.

Historians of science now acknowledge that the modern natural sciences emerged through a process of mercantile, colonial, and imperial interaction between the West and the rest of the world, dependent on the work of "go-betweens" and "intermediaries" who transported and translated ideas, objects, and technologies across the globe.[49] However, scholars of the Caribbean and South Asia have recently pointed out the inadequacy of these terms. Londa Schiebinger designates African, Amerindian, and European doctors and healers in the eighteenth-century West Indies as "knowledge brokers" whose activities could not reduce them to the status of go-betweens; rather, they were "men and women situated in the push and pull of life-and-death struggles for political, economic, cultural, and personal survival."[50] Projit Mukharji argues that the notion of go-betweens problematically implies the existence of distinct, fully formed, and internally consistent "knowledge traditions" or "learned cultures," when in fact such bodies of knowledge always contain within themselves "antagonistic and mutually irreconcilable" differences of thought and method.[51] The case of human genetics in the Cold War period of decolonization similarly raises questions about how to identify the role of non-Western scientific actors who collected blood from their own communities for the use of foreign scientists while simultaneously negotiating asymmetrical geopolitical power relations and asserting new professional and national identities. Historians have variously referred to such actors as "local experts," "local assistants," and "indigenous or subject intellectuals."[52]

In agreement with the concerns of Schiebinger and Mukharji, I find that these terms do not fully capture the dynamics of sociopolitical power experienced by Middle Eastern scientists in their interstitial role between Western geneticists and doubly marginalized research subjects. As producers of medical and anthropological knowledge, they were engaged not only in the intellectual work of ethnic nationalism but also in "nation-state science," namely "the scientific work that helped imagine a national population fitted to state borders."[53] Scholars working on other non-Western contexts, ranging from Japan and India to Mexico and Brazil, have observed similar patterns wherein local scientists simultaneously strove to wrest narrative control of the nation's biology from Western scientists and supported the transformation of that biology through nationalist social and demographic policies.[54] Furthermore, these processes occurred not through a rejection of Western models of scientific practice but through relationships of *scientific collaboration*. The Middle Eastern scientific actors I consider here do not identify themselves as go-betweens, intermediaries, local assistants, or subject intellectuals; instead, they imagine themselves as collaborators in a global scientific enterprise.

To make sense of what this means, I turn to sociologist Stephen Hilgartner's definition of scientific collaboration as "a genre of knowledge-control regime based on an agreement among specific agents to participate in some joint research project or activities."[55] Generally, the notion of scholarly collaboration has positive connotations ideally involving mutually beneficial exchanges of biological samples, data, technology, and professional advancement. However, because many collaborative relationships are formed ad hoc, they often become sites of tension due to mismatched expectations and ambiguities about the substantive boundaries and obligations of the shared work.[56] Furthermore, in postwar and postcolonial politics, "collaborator" is also a loaded term, often posed as a derogatory opposite to heroic nationalists.[57] The simultaneously positive and negative valences of the actor's category of scientific collaboration demands a reconciliation of the well-documented roles of scientists, physicians, and engineers in decolonizing nationalist movements with prevalent theories of imperial and colonial collaboration.[58] In particular, scholars must contend with the agency of Middle Eastern and other non-Western scientists as participants in the asymmetrical power structures entailed by transnational scientific collaboration. In the chapters that follow, I investigate how Middle Eastern geneticists managed collaborative relationships with their Western counterparts. At the same time, I compare these relationships among professional scientists to their interactions with the human subjects of genetic research. Many Middle Eastern scientists acknowledge promoting national interests in the global scientific community; fewer admit their complicity in reinforcing different versions of colonialism at home. I highlight the basic tensions between their (conscious or unconscious) interests in consolidating the dominant national culture to which they belonged and the quest of Western researchers, who set the agenda of the major international organizations, to locate unique, isolated populations that could yield more universal information about human evolutionary history. These tensions illuminate how national scientific communities, composed of technocratic elites, reshape or resist the standard practices and assumptions of the same international scientific community through which they claim local legitimacy.[59]

THEMES AND STRUCTURE OF THE BOOK

These arguments are threaded together by a set of themes foregrounded in many historical and sociological studies of human genetics. The first theme is the fraught relationship between "race" and "nation" (not to mention "ethnicity") as concepts for sorting and classifying human populations. Historians and STS scholars have extensively analyzed how politics and science work together to invent human groups; recent studies of racial anthropology in the nineteenth and early twentieth centuries examine the notion of "national races."[60] As for postwar human genetics, the literature on North America has focused predominantly on the role of genetics in reshaping and reifying broad racial categories. Studies of other contexts, ranging from the early Soviet Union to contemporary East Asia and Latin America, characterize genetics as an iteration of a longer process in the co-constitution of national identities and racial classification.[61] For Middle Eastern scientists, identification with a European, Caucasoid, or white race has consistently been a prerequisite for defining the genetic characteristics of the national population.

Another theme is the entanglement of medicine and physical anthropology. Physical anthropology emerged in part from the anatomical research of nineteenth-century European physicians; well into the twentieth century, many medical practitioners maintained scholarly interests in racial taxonomy. Even as physical anthropologists began to consolidate the professional and methodological boundaries of their discipline, physicians took an active role in shaping social discourses on race by applying medical technologies to anthropological questions and disseminating ideas about racial hygiene and eugenics. The rise of postwar medical genetics cannot be separated from these earlier eugenic concerns.[62] Furthermore, institutional resources designated for medical genetics and biochemistry drove the simultaneous development of anthropological genetics. Regardless, anthropologists and medical researchers in the Middle East and beyond have routinely disputed one another's ethics and methodologies.

Finally, the book consistently highlights the role of violence in genetic research. The evolutionary narratives of contemporary genetics "are not possible without histories of violence," particularly colonial violence against indigenous peoples.[63] In the Middle Eastern context, much genetic research has taken place under conditions of "humanitarian catastrophe," including active war zones and sites of genocide. Geneticists opportunistically targeted prisoners of war, refugees, vulnerable migrants, and civilians under military occupation as research subjects.[64] It is not coincidental that new genetic technologies for studying racial and ethnic difference were developed at the same time and place that the Balkan Wars, World War I, and the international legal regime established by the League of Nations set about "unmixing" the diverse populations of the Middle East through "population transfer" (i.e., ethnic cleansing) and territorial partition.[65] İpek Yosmaoğlu, a historian of the Ottoman Balkans, poignantly writes that the "blood ties" that transform people into a nation "are not in the blood they imagine they share but in the blood that spills in the name of the [national] family."[66] To this I add that the collection of blood in vials for inspection of genetic traits has gone hand in hand with violent bloodshed in constituting the modern nation.

While these themes are relevant to and present in all chapters within the book, I have organized the chapters into three parts to focus on each theme in turn. The chapters also follow an overlapping chronology tied loosely to shifts in both the political geography of the Middle East and the dominant methodological approaches within human genetic research. Part I, "Race and Nation," covers the early twentieth century until World War II and examines the role of two early methodologies of genetic research—anthropometry and sero-anthropology—in the formation of racial and national identities for new Middle Eastern states. Chapter 1 focuses on anthropometry, the measurement of human bodies and body parts, as a key methodology of European race science. I explore how anthropometry was used on

Middle Eastern peoples and human remains from archaeological sites to create hierarchical racial taxonomies within the context of European imperialism. After the First World War and the collapse of the Ottoman Empire, anthropometric research acquired further political significance as different ethnic and religious communities struggled for international recognition as national groups entitled to political sovereignty. In the Levantine mandates, anthropometric reconstructions of "ancient races" like the Phoenicians and Israelites fed into political discourses about Lebanese identity and the legitimacy of Zionism. Meanwhile, the Turkish Republic and Pahlavi Iran mobilized anthropometry in different ways to affirm the ethnic homogeneity of their nation-states, claim racial membership in European civilization, and even litigate territorial disputes. Chapter 2 analyzes the emergence of sero-anthropology, a newer method for classifying races based on the detection of inherited blood-serum antigens—that is, blood groups. This method, developed on the Ottoman-Balkan front during World War I, assumed that the proportions of ABO blood types in a given population correlated with their racial ancestry. However, advocates for sero-anthropology clashed with physical anthropologists who considered anthropometry a more reliable indicator of racial identity. I examine how this methodological dispute played out in the interwar Middle East, where physicians in Egypt, Turkey, and the mandates of Syria and Palestine took part in sero-anthropological research. Some of these researchers embraced the method as a simple way to calculate racial mixture, casting Egyptians, Turks, and Middle Eastern Jews as admixed national groups in contrast to the racial "purity" of Samaritans, Armenians, and nomadic tribes; others cast doubt on the anthropological significance of blood groups. Although anthropometry persists to this day as a method for quantifying human variation, the emerging field of human genetics narrowed its focus to molecules—first, blood proteins and enzymes, followed by DNA itself. For this reason, the remainder of the book focuses on blood.

Part II, "Medicine as Anthropology," focuses on the 1950s and 1960s and the evolving relationship between medical and anthropological genetics. In this period, infrastructures originally designed to support emergency blood transfusions and research on infectious diseases were co-opted by a transnational network of geneticists to draw sweeping anthropological claims about the origins and relationships of different ethnic groups. Chapter 3 provides a broad overview of this infrastructure, which linked together institutions like hospital blood banks, forensic laboratories, and military and corporate medical clinics across the Middle East with reference laboratories in Europe and North America. The World Health Organization granted these Western labs authority to produce standardized protocols and supplies for blood testing all over the globe. This recognition enabled British and American laboratory directors to coordinate the collection and shipment of blood on a massive scale, tapping institutional and social networks to acquire blood samples from different Middle Eastern communities and evaluate their ethnic relationships. Chapters 4 and 5 offer in-depth examinations of how medical researchers used the inheritance patterns of two disorders common in parts of the Middle East to support certain narratives about racial and national origins. Chapter 4 investigates the discovery of sickle cell disease in two marginalized Arabic-speaking communities living in Aden Colony and southern Turkey. These discoveries destabilized prevalent understandings of sickle cell disease as a marker of African ancestry and drove the production of new hypotheses about racial admixture and ancient human migrations. Chapter 5 explores the research networks forged by a similar inherited blood condition, known as favism, in Israel and Iran. Israeli physicians correlated the presence or absence of favism in different Jewish communities with the biblical account of Jewish history. In Iran, favism reinforced the notion that Zoroastrians were pure remnants of the ancient Persians. However, the conflation of sickle cell disease and favism with ethnic ancestry was challenged by geneticists who proposed that both disorders could provide a natural immunity to malaria. Medical researchers across the Middle East participated in an ensuing decades-long debate over the relative significance of natural selection and genetic drift in the spread of these disorders, basing their stances on anthropological assertions about sectarian endogamy, traditional agricultural and environmental practices, and the regional history of the African slave trade.

Part III, "Colonial and Ethnic Violence," emphasizes how war and other violent legacies of colonialism and ethnic conflict have shaped Middle Eastern genetic research since the 1960s. Chapter 6 analyzes Israeli and British efforts to study different Arabic-speaking communities in the Levant and southern Arabia amidst the ongoing Israeli-Palestinian conflict as well as the Aden Emergency, an episode reflecting the twilight of Britain's empire.

The Israeli-Arab wars of 1967 and 1973, and the British withdrawal from Aden in 1967, disrupted the blood supply chains described in Part II and turned Israel into an occupying power, granting Israeli scientists expansive new access to Arab research subjects. Decolonization and Israeli military conquest therefore transformed both the logistics of blood sampling and geneticists' perceptions about who should take leading roles in scientific collaborations between Europe and the Middle East. Chapter 7 considers the effects of civil strife on Iranian and Turkish approaches to genetic anthropology in the 1970s and 1980s. Concurrently with Western attempts to integrate human genetics with physical anthropology, Iranian and Turkish scientists worked to understand the genetic characteristics of different ethnic groups in their countries. In the context of the Iranian revolution and its violent suppression of movements for regional and ethnic autonomy, as well as the revival of militant Kurdish separatism in Turkey, these anthropologists adopted divergent strategies to recognize and study ethnic difference without threatening their governments' notions of territorial integrity.

The conclusion brings the book to a close with present-day genome sequencing projects, reflecting on the logistical and discursive relationships between transnational efforts like the Genographic Project and national-scale genomic research in Turkey and Iran. The population sampling methods and historical interpretations favored by these projects demonstrate that massive advances in genetic technology and major political and social transformations in the Middle East have not significantly changed how geneticists study the region. After a century of warring over arbitrary state borders and building ethnic and sectarian identities, genetic narratives of the past remain tightly intertwined with nationalist politics.

Part I
RACE AND NATION

Chapter 1

DRASTIC MEASUREMENTS

DURING THE LAST WEEK OF JULY 1911, MORE THAN TWO THOUsand delegates from at least fifty countries crowded into the "cavernous hall" of London's Imperial Institute, enduring stifling heat and poor acoustics, to attend the first (and last) Universal Races Congress.[1] The organizers, leading figures of the European Ethical Societies movement, had brought together a motley crew of politicians, intellectuals, lawyers, and anthropologists as "representatives of the different races [who] might meet each other face to face, and might, in friendly rivalry, further the cause of mutual trust and respect between Occident and Orient, between the so-called white peoples and the so-called coloured peoples."[2] Though it was not intended to be a strictly scientific conference, the European anthropologists involved in the planning, including Alfred Cort Haddon and Felix von Luschan, sought to conduct the Congress in "a scientific manner" incorporating the latest racial theories.[3] One of the delegates for the Ottoman Empire, the physician and parliamentarian Rıza Tevfik, prepared accordingly. He contributed a discussion paper that leveraged his own knowledge of biology to critique the purportedly scientific justifications for European imperialism in Asia:

> I cannot help drawing attention to certain scientific facts, usually wrongly interpreted, which have a close connection with the question of races. I wish to speak of craniological science . . . the formal differences in the human

skull [are] of no value except for the classification of races. While we are as yet unable to agree on the rational conditions of a natural classification even of mushrooms, it would be premature and arbitrary to divide humanity into two or four great classes, or to say that there are European heads and Asiatic heads in the moral sense, and to suppose that the Asiatic head is much inferior to the European in this respect.[4]

Many intellectuals of the late Ottoman and Qajar empires, particularly those who attended the Congress, embraced science in both its broader promise of objective, rational discourse and its specific subfields and methodologies, such as racial classification and craniometry. However, they protested, their European and American counterparts habitually "wrongly interpreted" empirical data on human physiological differences by constantly imbuing them with moral (i.e., intellectual and psychological) characteristics.[5] By doing so, Rıza Tevfık argued, many works produced by Western physicians and anthropologists on human racial classification did not satisfy even the standards of scientific rigor applied to lowly mushrooms. Citing European scientists like Herbert Spencer and Ernst Haeckel, Rıza Tevfık suggested that a truly objective approach to craniometry could ameliorate the various racial and national antagonisms the Congress had explicitly set out to resolve. Rıza Tevfık urged Western scholars to achieve this objectivity by setting aside their colonial prejudices; acknowledging that their ideas of racial superiority and inferiority, "like everything else, [are] relative"; and avoiding "dragging in certain kinds of sociological and political facts which have no place in a province that is ruled by a biological principle."[6]

However, the Congress proved an inhospitable venue for scientific defenses of Ottoman sovereignty. Like other delegates, Rıza Tevfık used race science as "part of an arsenal to criticize imperial relations."[7] But his case for the Ottomans, under the regime of the Committee of Union and Progress (the so-called Young Turks), was beset on all sides by political challenges to its own imperialism. Rıza Tevfık's claim that "unity was desired by all the races in the [Ottoman] Empire" was repeatedly undermined by the representatives of Zionist, Greek, and Armenian nationalist movements.[8] The Zionist activist Israel Zangwill, who presented a paper on the "Jewish race," decried the "wrongdoings" of the Young Turks and charged that they were making a "tactical mistake" by using force to impose their "ideal of a homogeneous

Turkey," in which all races were free and equal in rhetoric but not in practice.[9] Soon after, anthropologist Ignaz Zollschan further pressed the Zionist case, urging the Congress to support Jewish settlement on their "ancient soil of Palestine" because the Jewish race, as "Orientals accustomed to the life and views of the West," could promote interracial understanding.[10]

Unusually hot weather climaxed in violent thunderstorms on the final day of the Congress, enhancing the drama of the proceedings. In one of the closing sessions, the Greek physician Spyridon Zavitzianos, under the guise of "plead[ing] for a more exact definition of 'race' and 'nation,'" insisted that the "inhabitants of the Ottoman Empire known as Turks were not connected with the real Turks; they were the descendants of the original inhabitants of the country—the Greeks and others who were Hellenized."[11] Subsequently, Rıza Tevfik found himself confronted by Garabed and Lucy Thoumaian, Armenian exiles from the Ottoman Empire, who insisted on shaking his hand in a show of solidarity. They urged the Young Turks to "[combine] peacefully all the nationalities" under their rule and "show that civilization, like light, came from the East, by doing justice and promoting harmony throughout their empire."[12] Greek-Swiss lawyer Michel Kebedgy piled on, declaring that "the peace of the Orient could only be secured by respect for the autonomy of the various peoples who composed the Ottoman Empire."[13] Rıza Tevfik demurred, responding that it was unreasonable to expect any immediate "harmonizing" of conflicting national interests and that a necessary first step involved differentiating between "political and social or religious differences."[14]

Rıza Tevfik's experience at the Universal Races Congress illustrates how the aforementioned "biological principles" studied by race scientists could never be separated from "political facts," namely the burgeoning nationalist conflicts on the eve of the First World War. In the decade following the Congress, the embattled Ottoman Empire made reluctant concessions to the Zionists, inflicted genocide upon the Armenians, and fended off a Greek invasion of Anatolia that would culminate in the expulsion of Greek Orthodox communities from Turkey. Meanwhile, across the entire length of the Eastern Front, racial anthropologists were busily measuring and documenting the bodies of prisoners of war, refugees, and newly conquered peoples.[15] When the victorious Allies set out to redraw the maps of Europe and the Middle East, they turned to race science to grant new states to some national movements and relegate others to the status of "ethnic minorities."[16] But as Rıza Tevfık noted, the "science" itself was laden with unfounded assumptions and methodological inconsistencies. Since its beginnings in the eighteenth century, race science had been not a single academic discipline but rather an assemblage of concepts, theories, and methods concerned with the classification and evolution of human diversity.[17] This hodgepodge enabled the Allies to claim scientific bases for arbitrary political decisions.

Anthropometry, the primary focus of this chapter, is an iconic method of race science that records and compares various physical measurements of the human body. These measurements range from weight and height to small dimensions of the face and skull (Rıza Tevfık's "craniological science"), features that the measurers understood to be determined partially by environment and partially by inheritance. The phenotypic variations detected by anthropometry provided an approximation, albeit imperfect, of otherwise undetectable genetic differences. In this sense, anthropometry was the original methodology of human genetics. During their heyday (circa 1860 to 1940), anthropometry-based racial taxonomies were utilized by scholars of diverse educational backgrounds toward diverse ideological ends. The first "human geneticists" in the Middle East earned degrees and subsequently held academic posts not only in fields like medicine or anthropology but also in history and sociology. This reflects the extent to which, during this period, race science and hereditary concepts pervaded all fields of academic inquiry, including history.[18]

Anthropometry, as the methodological arbiter of racial identity, became a precursor to both defining a nation's membership and locating its historical origins—projects of immense political and social importance to all new states. Although researchers often blurred or conflated the categories of race and nation, *race* was primarily invoked in relation to an international conceptual space, wherein states could be hierarchically ordered according to the racial classifications of their inhabitants. *Nation*, meanwhile, was treated as a more historically defined category—often encompassing a distinct mixture of races—around which political sovereignty should be built. Anthropometry provided a diagnostic tool for measuring racial differences within the national population, which could be amplified or minimized at will through the selection of research subjects and presentation of data.[19]

By the beginning of the twentieth century, one particular measurement of skull shape, the cephalic index, dominated all others for classification purposes.[20] Race scientists used the cephalic index to characterize national groups, determine a nation's constituent "racial stocks," and interpret the ancestral relationships between different nations. Furthermore, the cephalic index measured upon living people had an analogous measurement (the cranial index) for human remains, offering scientists the opportunity to compare members of ancient and biblical civilizations to modern inhabitants of the same territories. A growing number of skulls excavated by Western archaeologists from sites across the Middle East prompted a full-fledged obsession with racial origins: who were the closest living descendants of the Pharaonic Egyptians, Phoenicians, Israelites, Hittites, and Aryans? In the wake of the First World War and the dissolution of the Ottoman Empire, the question was no longer merely academic. Matching cephalic and cranial indices conferred scientific legitimacy on nationalist claims to represent the rightful political heirs to these civilizations.

In this chapter, I analyze anthropometric studies in the Middle East in the first half of the twentieth century, showing how this fixation on the pre-Islamic past collided with intersectional and contested meanings of race and nation. I begin by tracing the connections between European nationalism, colonialism, and the development of racial taxonomies based on the cephalic index. Turning to the Levant in the late Ottoman and Mandate periods, I examine how the anthropometric search for living "Phoenicians" and "Israelites" developed into competing ideas about Lebanese, Arab, and Jewish racial identity. I then focus on the Turkish Republic and Pahlavi Iran, two political entities that endorsed explicit ideologies of ethnonational homogenization. The performance of anthropometric research in the context of fervent state nationalism enabled entire categories of people to be invented: "titular" national populations, e.g., Turks. This research also supported the erasure of competing national groups, e.g., Kurds and Armenians in Turkey. Descriptions and calculations of average physical traits served to break down diverse local identities in favor of a single population unified by language, religion, and above all a shared genealogical history, rooted in the ancient Hittite and Aryan civilizations.

Turkish scholars in the early Republican period recorded the physical characteristics of the national population as part of a reactionary project against previous Western characterizations of Turks as Asian ("Mongoloid") migrants to Anatolia who were racially inferior to Europeans. They mobilized anthropometry to define Turks both as a European race and as autochthonous inhabitants of Anatolia, in order to defend the sovereignty of the Turkish state. Simultaneously, state-employed physicians, anthropologists, linguists, and historians reclassified peoples with competing nationalist movements, like Kurds and Alawites, as members of the Turkish race who needed to be reacquainted with their true Turkish culture.

Because of a tradition of European linguistic scholarship that favored an ancestral relationship between Iranians and Europeans, the case of Iran is somewhat different, as highlighted at the 1911 Universal Races Congress. A representative for Qajar Persia, the constitutional activist and education reformer Hajji Mirza Yahya Dawlatabadi, did not critique European race science as Rıza Tevfık had.[21] Instead, his contributed paper left "the study of [the origin of the Iranian race] to the investigation of specialists."[22] Nevertheless, he stated confidently that Europeans were "of one and the same race with ourselves."[23] Therefore, he claimed, after "unpleasant political experiences" with European imperialists, his countrymen hoped to "cement broken ties [with] their ancient kinsfolk."[24] Like Rıza Tevfık, Dawlatabadi aimed to use racial discourse to forestall the encroachment of European powers on Qajar sovereignty. At the Congress, he emphasized that "Persia desired to be free from foreign intrigues while she was engaged in repelling internal aggression and troubles," referring to British and Russian meddling during the ongoing conflicts faced by the Constitutional movement to which Dawlatabadi belonged.[25] But Dawlatabadi faced no challenges to his reductive conflation of the ethnic, linguistic, and religious diversity in the Qajar Empire with the "Persian race," which he alleged had suffered "corrupting influences" like "fusion with other races" and invasion by "barbarous hordes."[26] Following the installation of the Pahlavi dynasty in 1925, the racial ideas outlined by Dawlatabadi found their way into governing attitudes toward science and social engineering. In Pahlavi Iran, locals assisted and participated in anthropometric studies, but the leading roles were left to European and American anthropologists. Although the Pahlavi regime was preoccupied with identifying Iran as a racially "Aryan" nation, it relied primarily on scholarship in Indo-European linguistics to support these claims in the international sphere. A more pressing problem was the imposition of this identity on the diverse population within Iran. Training Iranian scientists to perform new research therefore took a back seat to social projects like forming a centralized national education system and forcing the sedentarization of nomadic populations.

COLONIALISM, NATIONALISM, AND EUROPEAN RACE SCIENCE

The emergence of political discourses of colonialism and nationalism and the development of human taxonomy and racial classification in European science were co-constitutional processes. Anthropologists now widely accept that the history of their discipline is fundamentally intertwined with the governance of European colonial empires over vast stretches of Asia, Africa, Oceania, and the Americas.[27] However, postcolonial critiques of anthropology have yet to systematically analyze the practices of anthropological research developed by Turkish and Iranian "nationalist regimes that had their own long and evolving pre-history of colonial governance over varied language-speakers (the Ottoman and Qajar empires, for example)."[28] The development of anthropometry and human genetics in the twentieth-century Middle East accordingly reflected the concerns of nation-state ideologies that simultaneously agitated against external (Western) colonialism while supporting and justifying internal colonialism—i.e., the establishment of a hegemonic national identity. While aspects of European Orientalist scholarship did gain currency within local nationalist movements, these dual concerns demand a more complex intellectual genealogy for Middle Eastern race science. A compelling parallel can be drawn between the new nationalist regimes of the Middle East and nineteenth-century Sweden, the birthplace of human taxonomy and the principal calculation of racial anthropometry: the cephalic index.

Sociologist Greggor Mattson argues that the political transition from a seventeenth-century multiethnic Swedish empire to a nineteenth-century peninsular nation-state with a "homogenous" racial identity was achieved by the presence—not the absence—of non-Swedish minorities within the state, especially the "Lapps" (Saami). Carl Linnaeus, the eighteenth-century Swedish botanist and physician who developed the system of binomial nomenclature for biological taxonomy, suggested a five-part racial division of *Homo sapiens*: white, black, yellow, red, and "monstrosities," the category in which he placed the Lapps.[29] Linnaeus's students, the German Johann Friedrich Blumenbach and the Swede Anders Retzius, extended and refined their mentor's attempt at human racial classification with an emphasis on comparative skull morphology. In the early 1840s, Retzius created the cephalic index measurement—the ratio of the maximum width of the head divided by its maximum length—to compare the Swedes to the two largest minority groups in Sweden, the Lapps and the Finns.[30] Subsequently, he calculated the average cephalic index for nearly all the peoples of Europe.[31] While these core innovations of race science emerged in the context of a homogenizing Swedish nation-state, their acceptance and adaptation by scholars working in the vast overseas colonies of European empires enabled the creation of comprehensive theories of human evolution and racial differentiation.[32]

The new methodological tools of anthropometry were applied to existing human classification schemes derived from philology, as in the transformation of Semitic and Indo-European or "Aryan" language families into racial categories. The colonial domination of India played a key role in this invention of Aryan racialism. During the 1860s, the German-born Oxford philologist Friedrich Max Müller suggested that Europeans and Indians shared biological as well as linguistic ancestry, referring to them as "brethren" who both belonged to an "Aryan race." Müller intended his rhetoric of racial kinship to critique the violence of British colonialism in India as a sort of internecine struggle.[33] However, as the Aryan race concept gained popularity toward the end of the nineteenth century, it mutated into an ideology of white racial supremacy that justified European colonial domination over the rest of the world. Yet even as the "Aryan race" came to refer to the "civilized" white race standing at the top of any colonial hierarchy over conquered "colored" races, its scientific meaning was embroiled in nationalist chauvinism within Europe.

Among the European scientists who believed in the existence of a primordial Aryan race, no consensus was ever reached on any characteristic of its hypothesized members. The most heated debates centered on the Aryans' geographic origin, direction of migration, and physical traits, from which claims about the Aryans' representative living descendants could be made. This is where anthropometry, and particularly Retzius's cephalic index, found its most contentious application. Certain ranges of cephalic index measurements, labeled at one extreme "dolichocephalic" (longheaded) and at the other "brachycephalic" (broad-headed) were assigned to racial subtypes, which in turn became strongly associated with specific regions and nationalities (i.e., the dolichocephalic "Nordics" in Germany and Scandinavia, and the brachycephalic "Alpine" French). This racial subdivision of Europeans was projected into the prehistoric past through the measurement of cranial indices on excavated human remains. The presence of dolichocephalic or brachycephalic skulls at a given geographic location indicated the presence of a racial subtype, which was subsequently incorporated into nationally inflected historical narratives of origin, migration, and conquest. Thus, although the term "Aryan" comprised a broader Indo-European racial category, numerous competing theories emerged as to whether the Aryans had originated as a brachycephalic people in Asia that migrated westward to civilize Europe, or in Scandinavia or the Danube Valley as a dolichocephalic people that migrated south to populate the world. Depending on one's preferred hypothesis, either the Germans, the French, the British, North Indians, or Iranians could proudly claim to represent the biological and cultural traits of the original Aryans.

The Aryan issue was one of many that divided European and American race scientists, many of whom came up with their own particular systems of racial classification. They believed that races existed and that race had important historical and social implications, but there, global consensus ended. Virtually everything else about the definition and even the number of races (which, depending on one's system, ranged from three or four to more than thirty) was subject to animated academic debate. Although in professional settings these debates were always expressed in a scientific vocabulary, many race scientists readily acknowledged that the true stakes were moral and political, supporting or undermining various colonial endeavors or nationalist movements.[34]

Many Middle Eastern intellectuals first encountered European race science through their educational careers. Students attending prestigious schools and universities founded by Christian missionaries, such as Beirut's Université Saint-Joseph and Syrian Protestant College (after 1920, the American University of Beirut), Istanbul's Robert College, and Tehran's Alborz College, learned about human racial taxonomies alongside the latest developments in evolutionary biology and anthropology.[35] Instructors in and graduates of these schools played a significant role in disseminating and popularizing race science in Middle Eastern vernaculars, such as the Arabic journals *al-Muqtaṭaf* and *al-Hilāl*.[36] Ottoman and Qajar imperial governments also sent students to Europe to learn medicine, engineering, and other practical forms of knowledge relevant to the military reforms that sought to stave off the encroaching forces of European imperialism. These students soon recognized the political significance of Aryanist discourse and the various methodologies of racial classification. As race science reached its European zenith in the 1880s and 1890s, the Western-educated Ottoman and Qajar intellectual elite began to adapt and instrumentalize racial discourses to serve emerging forms of nationalism.[37] This mobilization of anthropometric race science for Middle Eastern nationalist causes intensified dramatically after the First World War, as the peoples of the region struggled to create solidarity within arbitrary new state borders and advocate for political self-determination through the diplomatic structures of the League of Nations.

In her study of interwar Egyptian anthropology, historian Omnia El Shakry distinguishes between colonial and nationalist interests in race science, arguing that the former were primarily concerned with categorizing peoples into hierarchical racial taxonomies, while the latter sought to circumscribe unique, homogeneous "collective national subjects."[38] I argue that many nationalists, through their engagement with anthropometric research, were equally preoccupied with both aims, which they pursued by claiming direct biological descent from glorified ancient civilizations. European Orientalists consistently glamorized the pre-Islamic history of the Middle East, and some of them anointed Egyptian Copts and Lebanese Maronites as living representatives of Pharaonic Egyptians and Phoenicians, offering a racial justification for the special prestige these Christian groups enjoyed under British and French imperial influence. In response, many Copts and Maronites enthusiastically incorporated pharaonicism and Phoenicianism into their ethnic identities.[39] Arab, Turkish, and Iranian nationalist thinkers therefore perceived that demonstrating a biological link between new polities and pre-Islamic civilizations could legitimate their demands for self-determination in the international sphere. Because Europeans included ancient peoples like the Phoenicians, Hittites, and Aryans in their own civilizational genealogy, these ancestry claims were calculated not only to provide national populations with shared origin myths but also to establish an equivalent racial status to Europeans in a global geopolitical hierarchy. For these ends, the cephalic index became a scientific tool to draw together ancient human remains, diverse living communities, and European whiteness into new national identities.

"LIVING MONUMENTS" OF RACIAL ORIGINS

During the late nineteenth century, the Ottoman Levant offered a special fascination for European adventurers, anatomists, and archaeologists, who were as interested in reconstructing the biblical Holy Land as in providing ethnographic information about the region's living inhabitants to support colonial interests. Their evaluation of the relationship between socially and religiously distinct Arabic-speaking communities in the region was consistently entangled with their desire to imagine the appearance and customs of Israelites, Canaanites, and Phoenicians. Dating from Sir Richard Francis Burton's collection of skulls from Palmyra and other sites in Syria in 1870, European scholars concluded that the cranial and cephalic indices of "Arabs" ranged too greatly to constitute a single race. When Austrian anthropologist Felix von Luschan measured peoples from across the Ottoman Empire in the 1880s, he commented, "The somatic difference between a Bedouin from Arabia or Mesopotamia and an 'Arab' farmer from near Beyrout is striking, and they have nothing in common except their language."[40] Rather, Arabic speakers could be meaningfully classified only when subdivided into religious sects or social categories like Bedouin, townspeople, and fellahin (peasant farmers). Burton and other British researchers from the Palestine Exploration Fund argued that only Bedouins, as immigrants to the region from the Arabian Peninsula, should be considered true Arabs; meanwhile, the fellahin descended (at least in part) from the indigenous residents of the Holy Land and might be better called "Syrians."[41]

Subsequently, the French Jesuit and American Protestant universities of Beirut became significant sites for the dissemination of racial discourses and anthropometric research. Early on, their engagement in archaeological studies of the Phoenicians helped their students conceptualize a secular Syrian identity distinct from but not opposed to Arabness. However, after the First World War, the question of Phoenician versus Arab ethnicity exacerbated underlying communal tensions between Maronite Christians, who were French protégés, and other religious groups in Mount Lebanon. Von Luschan had argued that "partly through their isolation in the mountains, partly through their not intermarrying with their Mahometan or Druse neighbours, the Maronites of today have preserved an old type in an almost marvellous purity."[42] This "old type," claimed certain Maronite elites and French educators, was the Phoenician race. The French associated Maronites with a strictly non-Arab Phoenician genealogy to justify dividing "Greater Lebanon" from the rest of the Syrian mandate. Jesuit-educated Maronite allies, such as the archaeologist Maurice Chehab, worked to popularize Phoenician racialism among their coreligionists.[43] Other communities became increasingly alienated by this discourse, and the American University of Beirut (AUB) became known for opposing this brand of Phoenicianism in favor of Arab identity. Famed AUB history professor Constantine Zurayk turned to anthropometry to debunk Phoenicianist politics in his 1939 work *National Consciousness*. He insisted that according to the scientifically accepted measures of racial classification—skin and eye color, hair color and texture, blood composition, and the length and shape of the head—Phoenicians and Arabs belonged to the same Semitic branch of the Caucasian Mediterranean race (*jins*).[44] Therefore, he contended, the political separation of Lebanon from the rest of Syria on the basis of biological difference was specious and illegitimate.

Zurayk's appeal to race science reflects an awareness of over a decade of anthropometric research and education at AUB's own medical school, which was regularly reported in the Arabic and English editions of the university's alumni magazine, *al-Kulliyyah*. The AUB School of Medicine was already renowned throughout the region, taking in students from Iran as well as across the Arab world. In the late 1920s, anatomist William M. Shanklin and pathologist Harald Krischner joined the school as it underwent a major expansion boosted by funding from the Rockefeller Foundation. During the 1929–1930 academic year, prominent Dutch neuroanatomist Cornelius Ariens Kappers arrived in Beirut to conduct fieldwork in Syria while covering Shanklin's sabbatical-year teaching. Kappers, together with AUB's seroanthropologist Leland Parr (see Chapter 2), inspired Krischner and Shanklin to launch a major anthropometry program that measured several thousand individuals across the Levant, Mesopotamia, and Persia (see Figure 1). The three men and their wives left behind a lasting legacy of Middle Eastern craniometry. They recruited AUB students of different religious and ethnic backgrounds (including Druze, Armenian, Bedouin, Persian, and Kurdish)

FIGURE 1. William Shanklin using calipers to measure the head of a villager in the Akkar mountains of Lebanon, July 1935. Courtesy of American University of Beirut Libraries.

into their work by measuring them directly, training them in anthropometric methods, and using them as liaisons to their communities to collect more data.[45] Shanklin had a particular interest in Bedouins, venturing into the deserts of Syria and Transjordan to catalog the physical traits of the Ruwalla, Aqaydat, and Mawali tribes. On these occasions he relied heavily on Tamir Nassar, a laboratory technician and Shanklin's assistant neuroanatomy instructor, "who on many occasions saved the day by patiently explaining the motives and purpose of anthropological investigation to Arabs who thought that the measuring instruments were a means of stealing the mind, or, still worse, of causing premature death."[46]

Shanklin and Kappers agreed with earlier European scholars that the Arabic-speaking communities of the Levant originated from several racial stocks and that the Bedouins and "the so-called Arabs forming the settled population" were distinct racial groups.[47] The discourse of Maronite Phoenicianism, though never referenced directly in their scientific publications, clearly influenced the questions they asked of their data, including "whether or not racial differences were correlated with religious belief" and "who are the present representatives of [the Phoenician] race?"[48] Shanklin claimed to observe racial differences of skin and eye color, facial features, and cephalic indices strictly along geographical, not religious, lines. Meanwhile, Kappers relied solely on the cephalic index (*al-dalīl al-ra'sī*) to settle the Phoenician question. He contended that "real Lebanese" (i.e., the Arabic-speaking Christian communities of Mount Lebanon), as well as Armenians, Assyrians, Alawites, and Druze, primarily descended from the ancient Indo-Aryan Hurrians or "proto-Hittites."[49] In a stark reversal of the prevalent political discourse, Kappers argued that it was *not* the Maronites, but rather the Bedouin tribes of the Syrian desert, whose head measurements "strongly suggest their relation with the Phoenicians."[50] According to AUB's anthropometrists, Lebanese Christians were indeed racially distinct from Arabs, but not because the former were Phoenicians. Instead, they attributed Phoenician ancestry to the Bedouins—allegedly the most Arab of Arabs!

In addition to Bedouins, Kappers found the cranial index of excavated Phoenician skulls remarkably similar to the head shape of the Samaritans of Nablus, a sect claiming descent from the ancient Israelite tribes of Ephraim and Manasseh, and allegedly preserving their original religious practices. These claims were strongly rejected by local rabbinical Jews, who insisted that the Samaritans had no ancestral connection to the land or tribes of Israel. By the time Kappers took notice of the Samaritans, their number had dwindled to approximately 150, and their adherence to endogamy led many scholars to assume the imminent extinction of the community. The Samaritans' status as a local curiosity had already drawn the attention of men like Henry Minor Huxley, a Harvard anthropologist. Huxley traveled to the Levant between January 1900 and June 1901 with an archaeological expedition funded by a group of New York industrialists. As the expedition's specialist in physical anthropology, he collected anthropometric measurements and ethnographic data from eleven Levantine communities.[51] In February 1901, Huxley went to Nablus to measure the Samaritan community, where he remarked on the tense social relations that seemed to preserve Samaritan endogamy:

> The Samaritans themselves claim the perfect purity of their stock. Only as a last resort would they seek wives outside their own sect; and in this case they would naturally wish to marry among the people of the most closely allied religion, the Jewish. The Jews hate and despise the Samaritans with the greatest bitterness, and would do all in their power to prevent marriages

between the two sects. Syrian Christians and Moslems would be equally averse to intermarrying with the Samaritans [. . .]. These two factors, the natural inclination of the Samaritans to marry strictly among themselves, and the difficulty of forming marriages with other sects of Syria, would combine to preserve the purity of the stock, and at the same time to promote degeneracy by close interbreeding.[52]

Accepting the narrative of total endogamy, Huxley compared the Samaritans' physical characteristics to not only his own self-collected data on local populations but also published data on Ashkenazi Jews. On the basis of cephalic index measurements and the range of hair and eye colors within the small community, Huxley concluded that the Samaritans had "preserved the ancient type in its purity; and they are to-day the sole, though degenerate, representatives of the ancient Hebrews."[53]

Huxley's anthropometric observations on the Samaritans quickly attracted the attention of Jewish race scientists in Europe and America, who were embroiled in debates about the racial status of the Jews and the relationship of living Jewish communities to their biblical ancestors—debates intimately connected to the legitimacy of Zionism.[54] Huxley's work was added to the 1906 edition of the *Jewish Encyclopedia* and was swiftly translated into German for the journal of the Berlin-based Bureau for Jewish Statistics. His substantiation of Samaritan claims to Palestinian autochthony and racial purity since the time of the Babylonian exile turned Samaritans into a scientific proxy for the ancient Israelites. For example, during Samuel Weissenberg's 1908 research trip to Palestine, he, like Huxley, accepted the basic premises of Samaritan purity and endogamy. In his pursuit of the ancient autochthonous population of Palestine, he compared the Samaritans with the Palestinian fellahin (whom he believed to be more racially pure than urbanites) and the small groups of local Jews, who were "indistinguishable from the fellahin in language, dress and customs."[55] Ultimately, he argued that the dolichocephaly of the Samaritans offered "direct proof" to support his longstanding contention that both the ancient Israelites and the ancient indigenous population of Palestine (i.e., Canaanites) were dolichocephalic.[56]

Underlying the Samaritans' identity claims, and their newfound usefulness to Jewish race science, was the community's clear awareness of its own precarious demographic and sociopolitical position within the context of

Zionist settlement in the New Yishuv. As the Polish-Jewish anthropologist Henryk Szpidbaum reported in 1927, "The Samaritans believe themselves to be a vanishing tribe [due to] the insufficient number of women. [Footnote:] In order to counter the extinction, the Samaritans try to enter into mixed marriages with Jews. For the time being there is only one such a marriage."[57] Despite the alleged desire of Samaritan men to marry "more-evolved" Jewish settler women, by the early 1930s, only two additional Samaritan-Jewish mixed marriages had occurred.[58] Meanwhile, European anthropological circles completely took for granted the eventual extinction of the Samaritans. Anthropometric studies on the community made a concerted effort to measure as many community members as possible, justified through an increasingly urgent discourse of salvage anthropology.

The collection of this data was sometimes propelled by specifically Jewish interests, as in Szpidbaum's work for the anthropology laboratory of the Polish Society for the Exploration of the Mental and Physical Condition of the Jews (*Gesellschaft zur Erforschung des psychischen und physischen Zustandes der Juden*). In the introduction to his study, he described the Samaritans as "a real curiosity as a living monument [*Denkmal*] of the biblical period. This tribe can be traced back 2800 years, during which it should be noted that the Samaritans have never left their country of Palestine. Detailed knowledge of this tribe will hopefully help to solve many difficult problems concerning the anthropology of the former inhabitants of Canaan and partially [also of] today's Jews."[59] In keeping with these aims, Szpidbaum compared his measurements on ninety-four Samaritan individuals primarily to data on Polish Ashkenazim and "*spaniolisch*" Sephardim, seeking a "racial diagnosis" that would locate Samaritans among Jewish groups. Ultimately, he concluded, the Samaritans showed a "striking resemblance in almost all characteristics with the Sephardic Jews."[60] Kappers agreed with this assessment, contending that the cephalic indices of both Samaritans and Sephardic Jews revealed Phoenician-Canaanite ancestry.[61]

In contrast, Italian statistician and sociologist Corrado Gini's analyses of Samaritan demography in the 1930s reflected a more abstract interest in the Samaritans as a "degenerating" people. The Italian Committee for the Study of Population Problems, represented in the field by the anthropologist Giuseppe Genna, funded Gini's expedition to Palestine to collect data on the Samaritans. This research, which recorded ethnographic and demographic information as well as the anthropometric measurements of 171 of the community's 213 members, aimed to observe characteristics that could be correlated with the Samaritans' demographic decline (although the community had actually grown over the preceding two decades). Gini expressed skepticism about the earlier Samaritan research because of its overriding interest in the community as surviving remnants of a biblical Judaean *Urtypus*. He rejected the idea that the Samaritans represented the authentic, if degenerated, characteristics of the ancient Israelites. Although the Italian study did not indulge these Jewish research interests, several Jewish physicians and scholars of the New Yishuv participated in the work: physician Reuben Katznelson, the main local organizer of the expedition; Dena Joseph, an anthropometrist trained at the University of Chicago; and Jerusalem physicians Shimon Shimony and Jakub Cohen, who acted as interpreters. The future second president of Israel, Yitzhak Ben-Zvi, served as a historical consultant (see Figure 2). He had been studying the community since 1908, and he had even played an active part in convincing Samaritan families to allow their sons to marry Jewish women.[62] Gini's reliance on Ben-Zvi predisposed the Italians to expect a positive relationship between the Samaritans and any future Zionist state. In his concluding comments on the social changes taking place in the Samaritan community under the influence of Zionist colonization, including Samaritan-Jewish intermarriage, he predicted their "gradual fusion with the Jewish race."[63]

Indeed, once elected to the first Knesset of the Israeli state in 1949, Ben-Zvi became a prominent advocate for the Samaritan community. Largely due to his intervention, the 1950 Law of Return—which stipulated that only Jews could immigrate to Israel and obtain citizenship—was interpreted to apply also to Samaritans crossing the Green Line from Nablus.[64] In 1953, while president, Ben-Zvi instigated the creation of a Samaritan neighborhood within the city of Holon, near Tel Aviv, and encouraged more Samaritan families to leave Jordanian territory for Israel.[65] Ultimately, Ben-Zvi's actions did not result in a fusion of Samaritans with Jews, as Gini had predicted. Rather, they capitalized on new political, social, and ideological conditions for Samaritan survival as religiously distinct, but ethnically related, kindred of the Jewish people. Secular Zionism's ethnonational conceptualization of Jewish identity, combined with the anthropometric studies that lent scientific credibility to Samaritan claims of indigeneity to biblical

FIGURE 2. Yitzhak Ben-Zvi (center) mediates a feud between Samaritans Yitzhak ben Amram (of Nablus, a former Samaritan high priest) and Nur Marhiv (of Jaffa). Nablus, April 1934. Courtesy of President Ben-Zvi collection, Yad Ben-Zvi Photo Archives, Jerusalem.

Palestine, worked in concert to transform the traditionally antagonistic relationship between Jews and Samaritans into a strategic partnership, which will be explored in detail in Chapter 6.

The search for living examples of "ancient races" took different forms and suited different aims elsewhere in the Middle East, reflecting the uneven development of scientific infrastructure and the role of colonial interests in directing research agendas. Whereas the Levantine mandates formed in the wake of the First World War were directly administered by the British and French, Turkey and Iran had more ambivalent relationships to European colonialism. Although never overtly colonized, the Turkish Republic and its nationalist ideology were molded in response to the European dismemberment of the Ottoman Empire and the war to maintain an independent Turkish state and were therefore strongly opposed to European imperialism. The Kemalist government quickly set about reforming and creating new state institutions, as well as directly funding race science, to support the Republic's territorial claims and policies to socially and culturally homogenize the population. The Iranian nationalism endorsed by Reza Shah had a generally anti-colonial bent, but anthropometric and medical research in Iran was substantially dependent on the initiative and financial resources of Western scholars with access to a semi-colonial infrastructure of archaeological, missionary, and corporate installations.

REINVENTING THE TURK

Edgar Jacob Fisher, a teacher and dean at Istanbul's Robert College between 1917 and 1934, routinely assigned an essay on "a descriptive sample of human society" in the prestigious American-run high school's course "Introduction to the Social Sciences." Students were required to select an ethnic group (e.g., Turks, Greeks, Romanians) and write about its historical, social, economic and physical (i.e., racial) characteristics. By the end of the 1920s, about half of the school's enrollment consisted of "Turks by nationality."[66] Unsurprisingly, then, many of the students chose to describe "the Turks" for their assignment. Nejat Ferit, a student who worked for the student publication *Robert College Herald*, confidently wrote in his October 1929 essay: "The Turks are one of the oldest known nations. The Bible which is one of the oldest historical books speaks of the Turks."[67] Several of his classmates wrote, "Turks are a branch of the white race," or, more defensively, "Although [Turks] are considered from the yellow race as a nation, but [sic] they are white."[68] Zıhni Haldun argued: "Those living in cities can hardly represent the Turkish type, as a result of mingling with many nationalities such as Georgian, Greeks, Circassians. Original Turk belongs to the Turanian Race, but it is not possible to find a pure Turk today."[69]

These school reports offer a glimpse of prevailing ideas circulating about race and Turkish identity in the early years of the Republic. The Turkish students of Robert College, largely the sons of elite urban families, echoed the narratives being formulated by the upper echelons of Turkish academia and disseminated by public lectures and print media. The First Turkish History Congress, held in Ankara in 1932, presented the general outlines of the Turkish History Thesis, which effectively argued that Turks belonged to the Alpine type of the European race; that Central Asia was the cradle of humankind; and that the Turks, its indigenous people, had migrated to populate the world, thus becoming the founders of all ancient civilizations, most importantly the Anatolian Hittites. This narrative allowed the Kemalists to acknowledge the Turks' linguistic-cultural connection to Central Asia while maintaining their territorial claims to be the original inhabitants of

Anatolia. Some established historians dissented against the Thesis, which prompted Atatürk and his supporters to legitimate their vision of an ethnic Turkish nation-state through racial science. The Kemalists marshaled the more "objective" methods of linguistics, archaeology, and physical anthropology to narrate a history of ethnic Turkish autochthony in Anatolia.[70] Research on human genetics in Turkey, using the methods of anthropometry as well as sero-anthropology (see Chapter 2), was initiated specifically to provide supporting evidence for the Turkish History Thesis.

Because these academic efforts were an explicit reaction against existing European representations of Turkish racial identity, the Kemalists also hoped to convince the international scholarly community to accept the Thesis. One of the most important European allies of the Thesis was the Swiss anthropologist Eugène Pittard. Despite claiming to have distanced himself from "constructions of race in the service of nationalist politicians," Pittard became deeply involved with the Kemalist elite and lent his professional and personal support to creating their nationalist narrative of history.[71] In 1928, Pittard met Atatürk in person during his first visit to the Turkish Republic. During this visit, Pittard conducted two major studies: an anthropometric analysis of Turkish soldiers in Ankara and an archaeological investigation of a Paleolithic society in southeastern Anatolia. On the basis of the anthropometric study, Pittard declared that the Turks of Anatolia and the Balkans were racially homogeneous and, moreover, that the Turks were a brachycephalic race—just like the supposed Aryan race that founded European civilization (according to prevailing francophone theories).[72] Upon returning to Europe, Pittard presented his results at international conferences and published them in European and Turkish anthropology journals.[73]

Meanwhile, the first credentialed Turkish anthropologist, Şevket Aziz Kansu, was in the midst of earning a diploma from Pittard's alma mater, the Paris School of Anthropology. Kansu worked as a physician at the University of Istanbul before heading to Paris in 1927, with the support of the Turkish government. He graduated in 1929 after completing a study on the skull morphology of Africans and Neo-Caledonians and returned to the University of Istanbul to take a leading role in the Turkish Institute of Anthropology (*Türk Antropoloji Enstitüsü*), which had been housed within the Faculty of Medicine and staffed exclusively by medical doctors since its founding in 1925.[74] Kansu was not Pittard's student, but the two men enjoyed a close relationship, and together they established the anthropological profession in Turkey. Pittard acted as an esteemed European patron, chairing major conferences (such as the Second Turkish History Congress in 1937) and regularly contributing to the *Turkish Journal of Anthropology*. Kansu taught the first anthropology courses and advised the first cohort of anthropologists within Turkey, beginning with the 1935 relocation of anthropological studies to the newly created Faculty of Language and History-Geography (*Dil ve Tarih-Coğrafya Fakültesi*, hereafter DTCF) at the fledgling Ankara University. Kansu wrote the guidebook for anthropometric measurements, and under his leadership the Anthropology Laboratory at the DTCF measured hundreds of skulls to determine the "racial traits" of the Turks.[75] High school biology textbooks produced by the Turkish Ministry of Education in the 1930s included a chapter on "Turkish racial anthropology" based entirely on Kansu's work. The lesson involved a "practical" activity in which students learned to measure their classmates' heads and calculated the average cephalic index for their class.[76]

Despite, or perhaps because of, his training abroad, Kansu took a strong stance against overreliance on the existing anthropometric data produced by foreign scholars. In the introduction to his 1931 article series in the *Turkish Journal of Anthropology*, Kansu declared: "In my writings, I do not want to make comparisons to the general publications of my Western [*garpli*] colleagues about the anthropology of the Turks. For now, I do not see this procedure as necessary. We should offer our own observations and the results of our own investigations."[77] These nationalist sentiments contrast with his methods—comparative measurements of 200 Turkish men, half from Rumelia (Thrace) and half from Anatolia—which followed the tradition of Western anthropologists who divided "European" and "Asian" Turkey. This tradition emerged from a context of rising Balkan nationalism in the nineteenth century, which emphasized that "European Turks" were simply Slavs or Greeks who had converted to Islam during Ottoman rule and therefore had no ethnic or racial relationship to the true, "Asian" Turks. While Kansu claimed that all the individuals he studied were "pure" (*saf*) Turks, the division of Rumelian and Anatolian sample populations implies that he expected to find biological differences between them.[78] Indeed, this expectation also structured the first Turkish sero-anthropological studies (see Chapter 2).

Afet İnan, one of Atatürk's adopted daughters, also took a leading role in anthropometric research. Deeply committed to the nationalist cause, she had already played an instrumental part in the development of the Turkish History Thesis while employed as a high school teacher. A faculty position was created for her at the DTCF, pending her receipt of a doctoral degree.[79] In 1935, she went to Geneva to earn a doctorate in sociology under Pittard's supervision. She enthusiastically approached her four years in Switzerland as an opportunity to educate Europeans about the true character of the Turks and disseminate information about the research activities of the newly formed Turkish Historical Foundation (*Türk Tarih Kurumu*). In İnan's correspondence with Atatürk, she reported various occasions on which she disabused her classmates or professors of their "misconceptions" about the Turks.[80] According to İnan, Pittard assisted her efforts by "mentioning at every opportunity the civilization of Anatolia and the Turkish race."[81]

Meanwhile, İnan's dissertation (later published in both French and Turkish) analyzed the vast amount of data collected by the Turkish Anthropometric Survey in the last half of 1937.[82] The survey was a major government-funded undertaking to record the measurements of nearly 60,000 Turkish citizens in less than six months.[83] Pittard designed the survey forms, while Kansu spent a week training a team of physicians (civilian and military), nurses, and physical education teachers to collect the data. This team was then dispersed among ten geographically defined regions of the Turkish Republic to record the distribution of cephalic, nasal, and skelic indices; height; and the colors of eyes, hair and skin.[84] The Central Bureau of Statistics processed the data forms. İnan's interpretations of the data, presented in dozens of charts, tables, and maps, obviously aimed to support the claims of the Turkish History Thesis with regard to the Turks' essential racial type and their historical identity as simultaneously migrants from Central Asia, the original inhabitants of Anatolia, and the ancestors of European civilization. However, it would be misleading to suggest that İnan completely conflated this statistically constructed Turkish "race" with the Turkish nation. She was well aware of the varying definitions of race propounded by different European anthropologists, and she explicitly condemned racial determinism. Yet, she argued that race science was a national duty: "It is not that we imagine, in the example of [Arthur de] Gobineau, a racial fatalism [. . .]. But it can be necessary to draw up the racial image of a nation, and to note its degree of biological purity over time."[85]

To define this "racial image," İnan subdivided anthropometric data according only to sex and geographical region. She wrote consistently of two groups, the "Turks of Europe" and the "Turks of Asia," invoking a broader racial whole by comparing her own data to previous studies of "Turks" living outside the Turkish Republic (for example, in Bulgaria). She did not clarify the definition of a "Turk" for the purposes of inclusion in the study, nor did she acknowledge the existence of ethnoreligious minorities within the Republic. Regardless, the emphasis on Turkish race rather than nationality implies the exclusion of any Armenian and Greek communities who survived the ethnic cleansing of genocide and "population exchange" between 1915 and 1923. In contrast, some Kurds were likely included under the Turkish label.[86] İnan specifically credited military doctors for measurements collected in the eastern provinces during precisely the same months that the first military operations moved to suppress the Dersim rebellion of Kurdish tribes against the Turkish state.[87]

In any case, within the designated Anatolian Turks of her study, İnan noted a fair amount of variation that potentially troubled a narrative of Turkish "biological purity." In particular, she conceded that variations in height among brachycephalic individuals meant that not all Turks represented the Alpine type of the white race (which included the French). Some of the population had to be classified as the Dinaric type (the same type assigned to the Greeks and most other Balkan groups). Additionally, she had to explain the geographic distribution of different frequencies of brachycephaly and dolichocephaly across the eastern, western, and central regions of Anatolia, which indicated some amount of historical racial admixture. İnan's narrative choices here are instructive. She interpreted the "much clearer cephalic homogeneity" in Central Anatolia as evidence that this region was a "central ethnic fortress, hardly touched by foreign immigrants," whose local inhabitants represented a direct biological link to the Hittites and even to the most ancient autochthonous "Neolithic-Chalcolithic populations." In contrast, the invasion of "foreign races" was responsible for the dolichocephalic types prevalent in the same regions inhabited by Kurds, Armenians, and other non-Turks.[88]

İnan's reference to dolichocephalic "foreign races" invading Anatolia from the south and east and biologically "interfering" with the cephalic indices of the Turkish population in these regions immediately conjures the specter of the Kurds.[89] The Kurds not only constituted a large and relatively contiguous population in eastern Anatolia, but they also had a national movement recognized by the League of Nations, whose aborted 1920 Treaty of Sèvres provided for the creation of an independent Kurdistan. Although Kemalist forces reached out to the Kurds during the Turkish independence struggle, this reconciliation ended after the establishment of the Turkish Republic and its program of homogenizing Turkish nationalism. The ensuing two decades of intermittent violence between the Kurds and the Republican government led to progressively harsher methods of denying the Kurds' existence as a distinct national or ethnic entity. The first battle was terminological; in the official Republican discourse of the 1920s, generic terms for "nomad" were often used as denationalized euphemisms for the Kurds.[90] Kurdish nationalist publications—and even European books that included the word "Kurdistan"—were banned in 1934, and the term "Mountain Turks" (*Dağ Türkleri*) was coined by the end of 1936, just months before the anthropometric survey began.[91] Accordingly, despite acknowledging the existence of biological variation among Turks in different regions of Anatolia, and even attributing this variation to racial admixture, İnan added a paragraph to the Turkish edition of her thesis arguing that anthropometric data demonstrated the racial unity of the Turkish Republic, calling attention back to the hazy distinctions between race and nation:

> While separate tribal [*kabile*] names are mentioned among the Turks, we wanted to clarify the unity of the Turkish race, even under other names. Studies conducted before this survey, especially in Anatolia, gave figures according to some tribal names (like Laz and Kurd, for example). We have arranged them [according to] geographic region. In this comparison, the existence of a racial unity in Turkey is also seen.[92]

Ultimately, the Turkish preoccupation with racial classification was linked to concerns about political sovereignty. Kansu, İnan, and their colleagues used the language of anthropometry to assert the equal status of Turks and the European powers within a broader scheme of racial hierarchy, rationalized by a belief in innate human differences. But at the same time, as far as the Kemalist elite was concerned, the territorial integrity of the Republic depended on denying such differences within national borders, insisting on the essential homogeneity of the population in order to delegitimize any competing nationalist sentiments.

Race science was a useful instrument of foreign relations because international bodies like the League of Nations accepted the existence of immutable biological races, believed that racial differences were a source of inevitable conflict, and endorsed strategic policies of ethnic cleansing and territorial partition. The League's involvement in negotiating the Greek-Turkish population transfers of 1923 set the stage for further international proposals to resettle refugee populations and allocate territory according to racial identities.[93] Turkish scholars like İnan understood that coupling the discourse on self-determination with arguments about racial demography could resolve territorial disputes in Turkey's favor, driving academic research to prove the Turkish ancestry of key minority groups.

This understanding inspired the "rebranding" of the Nusayri or Arab Alawites (in Turkish, *Nusayriler* or *Arap Alevileri*),[94] a Muslim sect living along the southern Turkish and Levantine coastlines, in the context of the so-called Hatay Crisis of the late 1930s. The former Ottoman Sanjak of Iskenderun (Alexandretta) had been the subject of a heated territorial dispute between Turkey and the French mandate of Syria since 1921. In 1936, a Franco-Syrian treaty promised to recognize an independent Syria that included the Sanjak. The Turkish government quickly moved to derail the treaty, arguing that French High Commission statistics showed that Turks, not Arabs, were the largest single community in the Sanjak. Nonetheless, to achieve a clear demographic majority, Turkish authorities seized upon an obscure suggestion by a German historian, Ewald Banse, that one of the Sanjak's Arabic-speaking religious communities, the Alawites, were not ethnically Arab at all. Rather, like the Turks themselves, they were biological descendants of the ancient Anatolian Hittites. Alawites, who were mostly rural agricultural workers, lived in communities throughout southern Turkey and Syria, particularly the provinces of Mersin, Adana, and Alexandretta. Although they now spoke Arabic, Kemalists claimed, Alawites had merely adopted this language from their neighbors over the centuries, which obscured their Hittite-Turkish ancestry. The Turkish press, supported by the nationalist academic apparatus, began calling the Sanjak "Hatay," after the Hittites' alleged name for the region, and referring to the Alawites as *"Eti-Türkler"* ("Hittite Turks").[95]

In Geneva, İnan attended a December 1936 League of Nations meeting debating the status of the Sanjak. She complained of her impatience with the discussion, wherein the Turkish representative justified Turkey's "legitimate interests" in "Turkey's racial brethren" through recourse to arguments that the Sanjak had been "an inseparable part of the Hittite Empire." This, İnan commented, was the "new politics" that would prevail over the old.[96] The following year, İnan calculated the cephalic indices of 200 Turkish women living near Ankara, comparing her findings to previous anthropometric studies by Kansu, Pittard, and other European anthropologists. Prominent Ankara University historian Hasan Reşit Tankut eagerly incorporated this information into his 1938 polemic *Nusayriler ve Nusayrilik hakkında*. Tankut firmly decreed the racial continuity between ancient Hittite remains and living Alawites and Anatolian Turks according to ethnographic evidence and the Turkish anthropometric measurements compiled by İnan, which closely matched Felix von Luschan's cephalic index calculation for fifteen Alawite men from Antioch (Antakya) within the Sanjak.[97] This evidence was repeated at public conferences sponsored by Atatürk's political party, the Republican People's Party (CHP), such as a 1939 speech by noted linguist Agop Dilaçar called "Alpine Race, Turkish ethnicity, and the People of Hatay."[98]

This "Eti-Türk" campaign strove to convince not only international power brokers at the League of Nations, but also the Alawites themselves, that they were racially distinct from other Arabic speakers who should self-identify as Turks. In May 1937, the League of Nations agreed to grant the Sanjak of Alexandretta a special autonomous status, with its own president, judiciary, and legislative assembly, but sharing a customs regime and diplomatic corps with Syria. The new assembly would have quotas for communal representation, determined according to the proportion of Turks, Alawites, Arabs, Armenians, and Greek Orthodox in the Sanjak's population. All citizens had to register as members of one of these groups to determine the apportionment of representatives. Determined to secure Turkish majority rule, the Turkish government supported a massive propaganda campaign to encourage Alawites to embrace the "pure Turkish blood," "glorious past," and "unity of race" they shared with their Turkish-speaking neighbors.[99] This exacerbated political rifts within the Alawite community along pro-Turkish and pro-Arab lines, which repeatedly devolved into violence.[100] Although some Alawites registered as Turks, it was only through more direct French and Turkish interference—including the suppression of pro-Arab organizations and the movement of Turkish citizens into the Sanjak—that the Turks won majority representation, paving the way for Turkey to eventually annex the territory as the Hatay province in the summer of 1939.[101] The anthropometric transformation of Alawites and Kurds into Turks was thus part of a broader appropriation of European racial worldviews and imperial strategies. Member states of the League of Nations repeatedly manipulated demographic realities to create or deconstruct the degree of ethnic homogeneity necessary to make claims about self-determination, national sovereignty, and territorial integrity under international law. Iran, one of the League's founding members, shared many of Turkey's political concerns and similarly sought to homogenize its population and claim European identity through reference to an Aryan race responsible for the power and prestige of pre-Islamic Iranian empires.

HOMOGENIZING "ARYAN" IRAN

In the 1930s, the Iranian Ministry of Education commissioned a book on the history of the Kurds from Tehran University professor and renowned poet Rashid Yāsamī. Yāsamī, himself of Kurdish background, was a committed Iranian nationalist. His book discussed race science alongside linguistics, religion, geography, and folklore to argue that the Kurds had always been an integral part of the Iranian nation, sharing its cultural, linguistic, and ethnic origins.[102] Of these disciplines, he was particularly skeptical about the value of "empirical race science" (*nizhād shināsī-i āzmāyishī*) for determining Kurdish racial identity. But, he mused, "it is not useless to mention the research of race scientists, and information on some of their mistakes [*ishtibāhāt*] will not be far from interest."[103] Over the next four pages, Yāsamī translated a section about the Kurds from Eugène Pittard's *Race and History*. Pittard had summarized many of the writings on Kurds produced by European travelers and anthropologists between the 1830s and 1920s, noting that accounts differed wildly regarding average cephalic index and frequencies of hair, eye and skin colors. In particular, Pittard enumerated his disagreements with von Luschan. Despite the fact that both scholars had measured Kurds from the same region of eastern Turkey, the former had observed the Kurds to be brachycephalic with predominantly dark hair and eyes, while the latter insisted that they were dolichocephalic with fair features. Based on his translation, Yāsamī argued that "existing research on the

Kurdish people still has no firm judgment other than the one general matter of their being Iranian."[104]

The problem, he explained, was not necessarily flawed anthropometric instruments or calculations. Rather, the categorical assumptions that different race scientists brought to their work meant their methodology supported foregone conclusions: "some of the previous scholars have assumed the Kurd to be an independent race, seeking the attributes and features that differentiate them from the great races, and [the scientists] cannot find such differences. Inevitably they consider a facial feature [ṣafatī-i 'āriẓī] to be a substantial division and on this point arguments occur." Yāsamī perceptively attributed the vitriolic debates of European race scientists to their individual ideological convictions: "If they assume the Kurds are a separate race, they therefore see that it is; [rather, if they assume Kurds to be] a branch of the Iranian race, then their work can reach [that] scientific conclusion."[105] Despite this critique of anthropometry, the overarching consensus of Western anthropologists—that the Kurds belonged to an Iranian racial group—perfectly satisfied Yāsamī's own foregone conclusion.

As in the Turkish Republic, the first two decades of Pahlavi rule in Iran were characterized by the forceful centralization of state power at the expense of de facto regional autonomy and linguistic-cultural diversity. Reza Shah ruthlessly suppressed various separatist and autonomist movements that sprouted along Iran's western provinces, from Gilan and Azerbaijan to Kurdistan and Khuzistan, during the administrative collapse of the Qajar Empire amidst the First World War, the Russian Revolution, and the failed 1919 Anglo-Persian Agreement.[106] Denouncing such movements as the work of foreign agents conspiring against Iran's sovereignty and territorial integrity, he reimposed Tehran's control over the provinces through direct violence, the forced sedentarization of nomadic groups, nationwide military conscription, and a standardized national education system. Meanwhile, Iranian-nationalist historians, including several from "rebellious" provinces, such as Yāsamī and the Azeri scholar Ahmad Kasravi, began to write regional histories with the aim of encouraging their fellow Kurds and Azeris to identify with a united Iranian nation.[107] The first national history textbook commissioned by the Ministry of Education, written by historian and politician Hassan Pirniya, conflated the geographical and ethnic unity of the nation with Aryanism: presenting an etymology of the word "Iran" through earlier varieties of Persian, the book asserted that the country's name meant "Land of the Aryans," and it included a discussion of European racial taxonomy that identified Iranians as belonging to a white-skinned, Indo-European race.[108]

In contrast to the Kemalist elite in Turkey, the nationalist intelligentsia of Pahlavi Iran did not seek to revise European racial taxonomies using anthropometry. Rather, they drew on existing threads of linguistic and archaeological scholarship to reinforce their claims to biological kinship with Europeans.[109] At its extremes, this meant a fetishization of pre-Islamic civilization in Iran and a passionate rejection of all Arabic and Islamic influences on Iranian culture as the impositions of an alien "Semitic" invasion.[110] Meanwhile, anthropological research supported by the Iranian government focused primarily on linguistics, folklore, and the collection of ethnographic artifacts.[111] This is not to say that the Iranian elite had no interest in supporting anthropometric research or failed to perceive anthropometry as epistemically valuable to the legitimacy of the Iranian nation-state. Rather, the scientific methodology was used on an ad hoc basis, interwoven with literary and architectural projects. For example, the Society for National Heritage (*anjuman-i āṣār-i millī*), a group of Iranian politicians, scholars, and intellectuals, undertook the construction of new mausoleums for famous deceased poets, scholars, and statesmen like Firdawsi, Hafiz, Ibn Sina, and Nader Shah, which involved the exhumation of their remains.[112] The Society used the recovered skulls not only to reconstruct portraits and sculptures of these men but also to pronounce them "true Aryans" according to craniometry.[113] Although most Iranian writers did not engage directly in anthropometric research, they selectively referred to the studies of Western anatomists and archaeologists to scientifically validate state-endorsed versions of Iranian history and racial identity. Because a number of such studies, dating back to the nineteenth century, glorified pre-Islamic Iran as the pinnacle of Indo-European civilization, Iranian elites did not need to invest resources in producing a "native" anthropometric tradition to claim European whiteness.

The Iranians who participated most actively in anthropometry, therefore, were not the urban intelligentsia but the community representatives of research subjects who assisted Western anthropologists with their independent projects. Reza Shah's government granted permission to several foreigners, who worked with their own funding and connections to European concessionary installations, to conduct anthropometric research. In the summer of 1931, Beirut-based pathologist Harald Krischner and his wife traveled to Iran and measured hundreds of people identified as "Persians" in eight urban centers. Like their colleagues William M. Shanklin and Cornelius Kappers back in Beirut, they wanted to determine the relationship between their research subjects and the region's "ancient races." They concluded that the Persians living north of Isfahan, whom they designated "Iranians proper," represented "the Medes of the ancient times." Persians from central and southern Iran, particularly the Zoroastrian religious minority based in Yazd and Kerman, belonged to the dolichocephalic "Indo-Persian race" and were "remnants of the ancient Perses."[114] Due to Harald's untimely death in September 1931, his wife entrusted Kappers with publishing their data. The resulting publication offers no information about the Krischners' field experiences or their working relationships with the Iranians.

In contrast, Anglo-American anthropologist Henry Field provides extensive details about the warm welcome he received from Iranian cabinet ministers, governors, and chiefs of police throughout his six-week expedition in 1934 to determine "the physical characters of the modern inhabitants of Iran."[115] The expedition was privately funded by Marshall Field, Henry's great-uncle and the patron of the Field Museum of Natural History in Chicago, and logistics were facilitated by a network of Euro-American diplomats, researchers, and educators, including employees of the Alliance Israélite Universelle (in Isfahan) and the Anglo-Iranian Oil Company (in Shiraz). According to Field, at the end of his research, in September 1934, he met with Ali Asghar Hekmat, the minister of education, to discuss the possibility of "a detailed anthropometric survey of Iran."[116] However, nothing on the scale of the 1937 Turkish Anthropometric Survey ever materialized in the country during Reza Shah's reign.

In Field's account, the friendly cooperation of Iranian government officials is contrasted with the reluctant attitude of the communities subjected to anthropometric measurements. Like many other European and American anthropologists of his time, he attributed this in part to primitive superstitions and a lack of familiarity with the scientific enterprise. However, he also acknowledged that the antagonistic relationship between the government and the populace impeded anthropometric work: "Fear is inevitably aroused that the stranger is working in behalf of the local government to report on the general physique of a group for purposes of military conscription. Also, since the anthropologist is recording numerous physical features as well as scars and [tattoos], it is sometimes suspected that he may be searching for a criminal at the request of the authorities."[117] The close association between foreign researchers like Field and the Iranian government, as well as the resulting mistrust and anxiety surrounding the purpose of anthropometric studies, rendered Field utterly dependent on the cooperation and assistance of local community leaders. Such leaders not only served as interpretive mediators granting the researchers access to their communities, but they also actively participated in the extraction of physical and social data. Field recommended that at least six or seven men be recruited to assist every anthropometric encounter:

> With each group of subjects it is essential to have a competent interpreter, who, whenever feasible, should also be the recorder. [. . .] In addition, two middle-aged and intelligent members of each group are always selected to lend confidence to the other subjects, to assist with the measurement of stature and sitting height, and to check both on the stated age of the individual and on the vital statistics. For example, the subject may reply to the question regarding the number of brothers living or dead, that he has three brothers living and none dead. At this point one of the assistants interrupts with a fierce query regarding his younger brother, who died five years before. The subject will almost invariably reply that he is dead and therefore of no further consequence. Another native assistant is assigned to take hair samples. [. . .] If blood samples are being obtained two local men assist the medical officer with the delicate task of persuading each subject to the minor operation.[118]

Field's differential experiences with Iranians inside and outside the government reflects the framing of his work as a project of salvage anthropology. He chose to measure only marginalized or rural populations who lived in conditions of abject poverty, rather than subjecting government ministers or educated elites to anthropometric study. Evidently, he regarded this elite class as a group that had already been transformed "into a semblance of national unity" by "the rapidly advancing wave of western civilization" and was therefore somehow less representative of an authentic and primordial

Iranian identity.[119] He explicitly described Iran's tribal groups as populations heading toward extinction at the behest of the government, and his description of their history presaged later projects of salvage genetics: "Under the policy of His Imperial Majesty Riza Shah Pahlavi, the tribes are being disbanded, so that within a relatively short span of time tribal divisions will no longer exist and the possibility of tracing the interrelationships of these people will be lost beyond recall."[120]

Nevertheless, during his short time in Iran, Field never managed to study any tribal groups in situ. Ultimately, he studied only 4 groups of men, each under tight time constraints. While waiting in Tehran for travel permits to Isfahan and Fars, he measured 18 Iranian workmen from the Rayy excavation site. Subsequently, he and his local assistants measured 99 Jewish men of the Isfahan ghetto in a single day. Field then traveled south to the village of Yezd-e Khast, a rural settlement in central Iran. Over three days, "despite the general unwillingness of the villagers to submit to anthropometric study, by means of friendly coercion and some bribery," he secured measurements from 48 men.[121] Finally, Field headed for Persepolis, where German archaeologist Ernst Herzfeld (the former director of excavations) sent him to the nearby village of Kinareh. Field spent 2 days there acquiring measurements from 74 men. On the basis of this relatively small (fewer than 300 individuals), heterogeneous, and haphazardly gathered sampling, Field attempted to clarify "the racial position of the modern inhabitants of Iran."[122] As a visiting researcher at Harvard, he and his supervisor, Earnest A. Hooton, used qualitative and quantitative approaches to classify each individual as Mongoloid, Armenoid, Hamitic, or virtually every "type" within the white race (e.g., Mediterranean, Alpine, Nordic, and corresponding subtypes). Amidst this staggering diversity, Field posited that "a new term is desirable to signify the basic population of Iran and in turn to differentiate this basic Irani from allied racial stocks in Iraq and in Afghanistan. I suggest the new term 'Iranian Plateau Race' which [...] refers to the basic Mediterranean type now living in Iran."[123]

Despite the friendly and cooperative relationships Field established with high-ranking officials in the Pahlavi government, it is not apparent that his work was immediately influential on Iranian racial thought.[124] In fact, the racial diversity within Iran highlighted by Field and the Krischners stood at odds with the state's emphasis on unity and homogenization. Mirroring the

Turkish policy toward Kurds and most other Muslim groups, Iranian nationalists generally preferred to "lump" as many human groups as possible under the label "Iranian," hoping to affirm Iranians' racial unity and general membership within a white European or, better yet, Aryan race. The Iranian government's efforts to create national anthropological institutions aspiring to international academic standards reveal a similar dissonance between foreign and local interests in race science.

In March 1935, about six months after Field's departure from Iran, the Ministry of Education held its first meeting to discuss the establishment of an Iranian anthropological museum and research center.[125] A significant presence and speaker at this meeting was Professor Wilhelm Haas, who would later become known in the United States as William S. Haas of Columbia University. Apparently, the French government suspected Haas of being a German spy with a mission to disseminate Aryan racialist ideology and cultivate Iranian sympathy for the Nazis.[126] In contrast, his obituary in the *New York Times* claims he worked for the American Office of Strategic Services after departing Iran in 1939.[127] Haas had left Germany for Paris in 1933 due to the restrictions he faced as an individual of "non-Aryan origin" (*"nicht arischer Abstammung,"* in the Nazi sense, in his case, having Jewish ancestry). In 1934, he obtained an advisory position in the Iranian Ministry of Education through his friendship with noted Iranian nationalist Hassan Taqizadeh, whom Haas met in Berlin during World War I. Archival documents of the German foreign ministry suggest that the Nazi government became aware of Haas's activities in Iran only after he was hired to teach psychology and pedagogy at the newly founded Tehran University.[128]

Although neither a professional anthropologist nor a German agent, Haas encouraged his Iranian colleagues to undertake anthropometric research. In his speech during the March 1935 meeting, he described the duties of an anthropological museum in terms of two disciplinary approaches, social anthropology (*insān-shināsī-i ijtimāʿī*) and physical anthropology (*insān-shināsī-i jismī*). The tasks of physical anthropology in Iran, Haas explained, should be:

> to obtain definitive and comparative data related to the different races of Iran [*nizhādhā-yi mukhtalif-i īrān*] and their influences on each other, define their specific and characteristic types, [and understand] the degree of influence of these races on the physical structure of the Iranian race

[*nizhād-i īrānī*] in a general sense. The different influences that the migration of these different races made on the Iranian race and the degree of resistance to disease acquired as a result of their settlement [should be] a subject of research.¹²⁹

By February 1936, the Ministry of Education charged a committee of literary scholars and historians (including Yāsamī) to oversee the establishment of the Center for Iranian Anthropology and its corresponding museum. In 1937, Reza Shah formally incorporated the center and museum, prompting an article on the definition of anthropology in the ministry's official journal, authored by Fazlallah Haqiqi. Unlike Haas, Haqiqi described the museum as a venue for research *not* on the "different races of Iran" but on only the Aryan race. In his words, the museum would be a site for "gathering works on the ingenuity, life, and civilization of the Aryan people, the civilized race of this country [. . .]. With its genius, this race has been acting for thousands of years to spread knowledge and civilization around the world."¹³⁰ Haqiqi's words reflect the general attitude of his fellow nationalist intellectuals as well as the Pahlavi government. Referring predominantly to francophone scientists and anthropology journals, his article takes for granted an international consensus that Iran's "civilized race" is *the* Aryan race of supreme historical importance.

Accordingly, despite calls by foreigners like Field and Haas for comprehensive anthropometric studies that might expose Iran's racial diversity, the new Iranian anthropological institutions engaged almost exclusively in the collection of folklore and material culture (i.e., what Haas called "social anthropology") as productions of a single, unified Iranian race.¹³¹ The first Iranian anthropology conference, convened in 1938, did not include any presentations of original anthropometric research. Those speakers who did discuss physical anthropology made only vague generalizations about the significance of the Iranian race to human civilizational development, again citing French authors like Ernest Renan and Arthur de Gobineau.¹³² The Center for Iranian Anthropology operated for only five years before it was shut down in the wake of the Anglo-Russian invasion and Reza Shah's abdication in 1941, and it did not reopen until 1958. The outbreak of World War II therefore caused a major interruption of anthropological activities in Iran, as in many other countries.

WORLD WAR II AND RACIAL IDEOLOGY IN THE MIDDLE EAST

The methods and discourses of race science appealed to many among the educated classes in the Middle East, with a legacy that long outlived the years of interwar state-building. Lebanese anthropologist Fuad Khuri recalled an assignment from his introductory anthropology course at the American University of Beirut (AUB) in 1956 that required him to analyze "cephalic variations in the Middle East." He learned to use calipers to measure the heads of his classmates, whose indices he compared to the data collected by Kappers and Shanklin decades earlier. Khuri explained:

> I thought that by matching a cephalic index with Shanklin and Kappers's ethnic scale, I would be able to approximate the "origin" of the person concerned. On this basis, I began, mixing measurements with laughter, to calculate the origin of my fellow students: "You are of Kurdish origin, but you are Armenian," I would tell them. "You're Arab and you Germanic." Soon, I became known in the dorm as an expert on origins.[133]

The fascination with anthropometry was not limited to AUB undergraduates; even Khuri's faculty adviser insisted that Khuri determine his "origin" with the calipers. Like today's consumers of DNA ancestry kits, people in Khuri's social context were intrigued and entertained by the conversion of isolated biological traits into diagnoses of ethnic identity and national belonging. But in the interwar period, the cephalic index and other anthropometric calculations were not simply mechanisms of self-discovery. In an international system long structured by racial hierarchy and newly obsessed with drawing state borders around "unmixed" ethnic groups, anthropometry had immense political significance. The Western search for "living relics" of the Phoenicians and Israelites in the Levant, as well as the differential racialization of Arabic-speaking religious groups, directly fed into local, regional, and global debates about who was entitled to self-determination in the French and British mandates. Likewise, Turkish and Iranian engagement with race science and identification as Hittites and Aryans expanded dramatically in response to the diplomatic norms of the League of Nations, which explicitly rejected the principle of racial equality and resolved territorial disputes in Europe and the Middle East by manipulating the collection of demographic statistics and endorsing procedures of ethnic cleansing.[134]

Many scholars of the Middle East have anachronistically dismissed anthropometric research as pseudoscience, revealing their discomfort with local discourses of race science, particularly that of Aryanism in Iran, that inevitably evoke the racist atrocities of Nazi Germany.[135] However, this chapter has shown that Iranian Aryanism follows a broader pattern of Middle Eastern peoples staking a general claim to European whiteness to assert their national sovereignty, rather than indicating specific enthusiasm for Nazi race science. In fact, German Nordicism stood at odds with Iranians' linguistic understanding of Aryanism, as laid out in the 1937 pamphlet *Nizhād va Zabān* ("Race and Language") by journalist and former parliamentarian Husayn Kay Ustuvan. Kay Ustuvan reviewed the debates of ten European "linguists, anthropologists [*nizhād-shināsān*] and geographers" about the origins of the Aryan race.[136] Anthropology, however, quickly fell by the wayside, as he referred only briefly to the views of Paul Broca and Paul Topinard and made no reference to cephalic indices or other anthropometric data. He focused instead on comparing Persian words to European-language cognates and disparaging the Semitic Arabic language. After describing the scholarly partisans of Central Asian, Scandinavian-Nordic, and Danube hypotheses of Indo-European origins, Kay Ustuvan leveraged philological evidence to argue that "Iran's race and language are Aryan" and that Iran was the "cradle" (*gahvārah*) or "birthplace" (*zānīch*) of the Aryan race, an explicit rebuke to German endorsement of the Nordic hypothesis.[137] Meanwhile, the Kemalist scholars who doggedly insisted upon the Europeanness of the Turkish people consistently identified Turks as belonging to the "Alpine race" rather than the Nordic. In spite of Nazi propaganda broadcasts in Arabic and Persian calculated to reach out to Muslims, most popular sympathy for Germany in the Middle East was driven by resentment of British and French colonialism rather than an ideological commitment to Nazi racial beliefs.[138]

The conceptual adaptation of race science in international diplomacy, as demonstrated in Turkish approaches to the Hatay Crisis, similarly surfaced in Iranian relations with the Nazi regime between 1933 and 1941. One popular textbook story is that the Iranian legation in Berlin suggested the international name change from "Persia" to "Iran," officially implemented on March 22, 1935, with the encouragement of German diplomats, who welcomed the change as a sign of Reza Shah's embrace of Nazi racialism.[139] Both Iranian Foreign Ministry memoranda and contemporary press reports from

Germany and elsewhere identified Reza Shah's primary motivation as a desire to highlight Iran as the homeland of the Aryan race at a time when European countries advocated Aryan racial pride. However, within the Nazi Party, there was no consensus on the "Aryan" status of Iranians or other Middle Eastern groups like Turks and Arabs, and many German diplomats in fact received the news of the name change with much consternation.[140] German diplomatic usage of "Aryan" to describe Iranians was a rhetorical strategy, not a true reflection of how Iranians fit into Nazi racial ideology. The name change represented an Iranian effort to impose their own understanding of Aryanism onto their relations with Germany and convince the Germans to apply this understanding in their racial policies. Individual Iranians, such as diplomat Abdolhossein Sardari, further manipulated the Aryan race concept to protect Jewish Iranian citizens living in occupied France by claiming that they were Jewish only by religion and, as racial Aryans, must be exempted from anti-Jewish policies.[141]

The term "Aryan" proved to be a major ideological liability to the Nazi regime in its diplomatic relations with the neutral nations of the Middle East. The conceptual slippage between different understandings of Aryanism influenced the Nazis' terminological shift from "Aryan" to "German or kindred [*artverwandten*] blood" in the language of the Nuremberg race laws, drafted over the course of 1935 and enacted in September of that year. However, this did not stave off continued confusion within Germany regarding the racial status of Middle Eastern nationalities and how they should be treated under Nazi racial policy. Several Iranians who resided in Germany as secondary-school or university students between 1933 and 1941 recall being treated with respect by Germans as fellow members of the "Aryan race."[142] On the other hand, Johannes Ruppert, the son of a Turkish man and a German woman, was expelled from the Hitler Youth in December 1935 after two years of participation. After he appealed to the Turkish Embassy in Berlin, the German government—that is, its foreign and interior ministries and the Nazi Party's Office of Racial Policy—had no choice but to make an official ruling on Turkish racial identity. Fearful of damaging relations with Turkey, the Foreign Ministry insisted that Ruppert be immediately reinstated to avoid further probing by the Turks into the "Aryan question [*Arierfrage*]."

After months of debate, the official decision to acknowledge Turks as a European people sharing "kindred blood" with Germans was announced in a letter issued by the Foreign Office on April 30, 1936. However, the letter cautioned, this acknowledgement did not extend to other Middle Eastern countries like Iran or Egypt because "these countries have not yet claimed the right to belong to Europe."[143] This line caught the attention of the international press, and by June 1936, several newspapers had refashioned the carefully worded German memo into lines like "Turks are considered to be Aryans [. . .] the citizens of Iraq, Iran and Egypt are not Aryans."[144] Iranian ambassadors to Germany and Turkey were incensed and demanded corrective action—namely, German acknowledgment that Iranians were not only of kindred blood to the Germans but were specifically an Aryan people. The German Foreign Office sidestepped the matter by pointing out that the Nuremberg laws did not contain any references to Aryans and permitted marriages between Iranians and Germans.[145] Nazi politicians were thus already backpedaling from the convoluted academic debates on the precise racial classification of prospective non-European Aryans in favor of a more practicable "kindred blood" concept. The next chapter investigates the development of a new scientific field called sero-anthropology, whose practitioners advocated blood and its discrete biochemical features over imprecise measurements of bones as the default material for genetic studies—and thus for the ultimate "truth" of human identity.

Chapter 2

TRUTH SERUM

IN 1916, POLISH PHYSICIANS LUDWIK AND HANNA HIRSZFELD arrived in Salonika to serve at the Serbian Allied Forces' contagious diseases hospital on the Macedonian Front. The Ottoman Empire had surrendered Salonika to Greece less than four years earlier, in the Balkan Wars. Barely two decades had passed since a teenage Atatürk had left the city, his hometown, for military school. The Hirszfelds were enthralled at first sight with the location of the field hospital: "Once this building had been a Turkish pasha's castle, the tower housed the harem, and nearby was a stone-walled water reservoir that was supplied with springwater and in which *houris* [beautiful women] had bathed not so long ago," Ludwik reminisced in his 1946 autobiography.[1] The Hirszfelds' self-professed thirst for the exotic was indulged not only by the local architecture but also by the human diversity of their working environment. While employed by the Serbian army, they tended to the needs of soldiers and refugees from across the Balkans, in addition to British and French imperial troops, which included Indians, Vietnamese, Arabs, and sub-Saharan Africans (see Figure 3).

In this milieu, Ludwik perceived the chance to pursue a thread of his serological research that he had previously dismissed as logistically impossible. Before the war, he studied the ABO antigens that appear in blood serum, which cause red blood cells to clump together after an incompatible transfusion. He had played a leading role in discovering the pattern by

FIGURE 3. Christmas postcard featuring "the Salonica Army, Historic Group of Allied Comrades in Arms," including colonial troops from Africa and Asia alongside European officers, 1916. Courtesy of the Imperial War Museum.

which ABO blood groups are inherited in humans, and he became interested in applying this discovery to questions of racial anthropology. Specifically, he thought that a global survey of blood groups might reveal "hitherto unknown and anatomically invisible relationships between different races." As he and his wife explained to the Salonika Medical Society in June 1918, "Through the accident of the war we happened to come to a part of the globe where more than elsewhere various races and peoples are brought together," enabling them to test this idea without many years of international travel.[2] Over the course of 1917, with the permission of military officers, the Hirszfelds took finger-prick blood samples from a diverse array of Allied soldiers, Balkan refugees, and prisoners of war from the Central Powers, ultimately tabulating the results of more than 8,000 individuals.[3]

The Hirszfelds mixed the blood samples with two different kinds of prepared blood sera, one containing A antigens, the other B antigens. Depending on whether the blood coagulated with one, both, or neither of these antisera, it was classified as type A, B, O, or AB. The Hirszfelds then calculated the frequency of types and found that the frequencies varied considerably "for each people and race."[4] Indeed, the Hirszfelds treated their fourteen population categories interchangeably as representing "people," "nationality," and "race" without much concern in their research reports. Their primary interest was the detection of inherited difference, which they assumed to have significant anthropological and historical meaning. Their colleagues in Salonika agreed, and in 1919, after some effort, the Hirszfelds' work was published in the British medical journal the *Lancet* and the French journal *L'anthropologie*. These publications are now acknowledged as foundational to a field variously called sero-anthropology or racial serology and, later, anthropological genetics.[5]

The Hirszfelds took most of the population categories they used completely for granted. They did not provide descriptive explanations for their European national categories (e.g., English, French, Italian, Serb, Bulgarian, and Russian) and the more broadly construed ethnoracial categories of the colonial troops (labeled "Arabs," "Malagasies," "Negroes (Senegal)," "Annamese" [i.e., Vietnamese] and "Indians"). The Hirszfelds offered notes of clarification only for the groups they named "Jews," "Greeks" and "Turks":

> For the Jews we used the refugees from Monastir belonging to a people which came from Spain about 400 years ago. For the Greeks we examined 300 soldiers from Old Greece and the Islands and 200 refugees from Asia Minor and Thrace. For the Turks we used Macedonian Mahommedans. These last must certainly contain a large admixture of Slav blood, and the statistics should be confirmed in Turkey.[6]

Science studies scholars have critiqued the categories by which the Hirszfelds calculated and interpreted their data.[7] In particular, Lisa Gannett and James Griesemer addressed the post-Ottoman context that made Jews, Greeks, and Turks "quite contested identities" in Salonika, emphasizing the inconsistency of using these three "religious" categories as equivalent to the "nationalities" of the other groupings. They suggested that the Hirszfelds' use of these designations was an appropriation of the local history of the Ottoman *millet* system, which classified certain religious minorities as legally autonomous communities.[8] However, the millet system itself, and accordingly the religious associations of the national labels "Greek" and "Turk," was not solely a product of local history; rather, it had been and would continue to be shaped by a long transregional history of interactions between the

Ottoman Empire and the rest of Europe. A similarly complex history underlies the slippery usage of "Jew" as simultaneously a religious, national, and racial category, and the Hirszfelds' own Jewish roots surely influenced their concept of Jews as a national-racial group. The Hirszfelds recognized that Greek, Turkish, and Jewish populations were not completely satisfactory in biological terms; but they did align perfectly with an international process of racialization that enshrined these categories in political and legal terms.[9]

At the San Remo Conference in April 1920, and in the Treaty of Sèvres, signed in August of that year, the victorious Allied powers attempted to formalize their plans for an aggressive partition and occupation of nearly all former Ottoman territories, including the Turkish "heartland" in Anatolia. This treaty galvanized the Turkish nationalist movement under Atatürk's leadership. The ensuing three years of war in Anatolia, now commonly known as the National Campaign (*milli mücadele*) in Turkey, precipitated the massive displacement of Muslims residing in Greece (who, as the Hirszfelds noted, were identified as Turks, regardless of ancestry) and Christians residing within the newly defined borders of the Turkish Republic. The displacement of refugees during the conflict was rendered into a formal "population exchange" by the Turkish and Greek governments, whose representatives signed the Convention Concerning the Exchange of Greek and Turkish Populations on January 30, 1923, at Lausanne. By May 1923, the "Macedonian Mohammedans" sampled by the Hirszfelds in Salonika would have been compelled to leave for Turkey, if they had not already done so. The Jewish refugees from Monastir tested by the Hirszfelds did not find permanent respite in Salonika either. Along with the rest of Salonika's Jewish population, they suffered economically and politically at the expense of the newly arrived "Greeks" from Anatolia and endured increasing expressions of antisemitism in the city. By the early 1930s, these conditions bolstered support for Zionism among Salonika Jews, who gradually began to emigrate to Mandatory Palestine, where the British had promised the development of a Jewish national home.[10]

The field of sero-anthropology was thus born at precisely the same time, and out of the same human material, as the new paradigm of nation-states in the Middle East. In the first decades after the Hirszfelds' pioneering study, sero-anthropology on Middle Eastern populations was carried out concurrently with the anthropometric research described in Chapter 1.

Accordingly, the physicians who collected and analyzed blood across the region were preoccupied with the same kinds of anthropological questions: Which communities descended from the region's ancient civilizations? What was a given nation's characteristic blend of ABO blood groups? What was the racial position of Middle Eastern nations vis-à-vis Europe? A number of doctors from Egypt, Syria, Turkey, and the New Yishuv in Palestine turned to blood serum for answers, comparing the ABO frequencies of hospital patients with their own beliefs about local social history.

Meanwhile, although the overarching conceptual framework of racial classification went unchallenged, the emergence of sero-anthropology prompted global methodological debates about how best to evaluate historical claims of relative purity and admixture. Hematologists pitted the relatively simple testing procedures and calculations applied to discrete phenotypes like blood antigens against physical anthropologists' tedious skull measurements and the statistical inconsistencies produced by continuous morphological traits. These methodological disputes had high stakes for the relative disciplinary authority of anthropology versus medicine. Medical practitioners claimed to be able to supersede anthropologists with more accurate and objective data for racial classification. Anthropologists countered that because all human populations contain some proportion of the four blood groups, the ABO antigens could not offer any clear diagnostic feature on which to divide populations. During the late 1920s and 1930s, American supporters of sero-anthropology trekked across Egypt, the Syrian Mandate, and Iraq, hoping to substantiate their methodology using socially or geographically isolated groups like Copts, Samaritans, and Bedouin tribes. Accordingly, Middle Eastern blood became significant not only for local nationalist projects of racial classification but also for the resolution of an international scientific debate.

SERO-ANTHROPOLOGY VS. ANTHROPOMETRY IN RACIAL CLASSIFICATION

Serological race research caught on quickly in continental Europe, particularly in France and Germany, with hundreds of studies conducted on Europeans published during the 1920s.[11] Initially, physicians and medical scientists like the Hirszfelds, as well as anthropologists like Henry Field, treated ABO blood group frequencies as simply another layer of anthropometric data. Many interpreted their findings within existing European racial classification schemes that had been developed according to other techniques. Furthermore, in analogy to craniometry's cephalic index, the Hirszfelds had formulated a "biochemical index" calculation (also known as the "racial index"): a simple ratio of the frequency of group A to group B present in a given population. On the basis of this calculation, they argued that their sampled populations could be divided into three broad "biochemical races." Europeans had index values above 2, the "Asio-African" category had values below 1, and "Intermediate" types (which included Arabs, Turks, Jews and Russians) had values between 1 and 2. Although the Hirszfelds' racial categories were far from universally accepted, the concept and calculation of a "biochemical race" was readily adopted in much of the subsequent literature on blood group distribution.[12] In general, serologists began to associate higher frequencies of group A with Europeans, group B with South and East Asians, and group O with genetic isolates or "pure races," such as the indigenous populations of North and South America. To explain the very high frequency of group O in Native Americans, serologists hypothesized that O was the original blood type of all humans before the later evolution of genes for A and B antigens within Europe and Asia, respectively. In other words, "human biohistory had involved ancient invasions by 'pure' A and B races, imposing themselves on an indigenous O race,"[13] and the proportions of each blood type in a population represented proportions of ancestry from these imagined original races.

Within this conceptual framework, the American physician Reuben Ottenberg argued that the Hirszfelds' calculation was inadequate for racial classification given its omission of group O, and he suggested a new classification system based on direct comparison of frequencies of the four blood types, subdividing the Hirszfelds' "Asian" racial type into three distinct races and adding a "Pacific American" category.[14] A fellow American biologist, Laurence Snyder, also denounced the biochemical index calculation as "worthless" and insisted that the most accurate racial classification system required comparing the frequencies of the three ABO alleles, rather than of the four blood groups they produce. Snyder's system added a seventh racial type ("Australian") to Ottenberg's, but Snyder took pains to clarify that "the grouping of peoples into 'types' is purely arbitrary, merely for the sake of convenience in dealing with the data. [...] Because two peoples occur in the same type, it is not implied that they have the same racial history."[15] Due to this arbitrariness, Snyder eventually discarded the use of racial "types" altogether, simply discussing the data in terms of "peoples" who had undergone different degrees of "racial crossing" with one another throughout history.[16]

Despite these challenges from Ottenberg and Snyder, the Hirszfelds' biochemical index system remained in common use for sero-anthropological work in the Middle East, whose populations remained drifting in the "Intermediate" category of the Hirszfelds' European and Asian-African typology. For example, in 1927 the bacteriologist Ali Tawfiq Shusha Pasha, the deputy director of the Public Health Laboratories in Cairo, calculated the race index for 417 Egyptian specimens and claimed that it lay "between the Intermediate and Asia-African type" (the number actually came closest to the Hirszfelds' Malagasies). Evidently unsure how to interpret these results, Shusha chose not to draw any anthropological conclusions from the Egyptians' biochemical race index, instead deferring to the existing historical narrative: "The Egyptian people is beyond all questions a mixed race [...]. In the ancient days they were mixed with Ethiopians from the South, Libyans from the West and Semites from the North-East, and, in more recent dates, either by foreign occupation or immigration they were mixed again with the Arabs and Turks."[17]

Employees of medical institutions in the mandate-governed Levant similarly took an interest in how ABO frequencies might be related to racial mixture. These institutions included the Altounyan Hospital in Aleppo, the School of Medicine at the American University of Beirut, and the Hadassah hospitals in Mandate Palestine. The responses to the early reports of these institutions, published in British and American medical journals, illuminated the significant dissent among Western scholars over whether the racial distribution of blood groups superseded anthropometric techniques. Middle Eastern data was especially contested when variations in blood group frequencies did not correlate well with the received racial histories Europeans had constructed on the basis of anatomical differences. Furthermore, the academic methodological debate highlighted the extent to which Middle Eastern communities had very different ideas from foreign scholars regarding their own history and level of admixture with neighboring populations.

In 1928, physician Ernest Altounyan wrote to the *British Medical Journal* reporting the blood group frequencies of 1,758 Arabs, Armenians, and

Jews who had been treated at his family's renowned Altounyan Hospital in Aleppo in the preceding five years. Alongside a simple table of the numbers and percentages, he noted briefly,

> Armenians and Jews represent well-defined unmixed racial groups; under the heading of Arabs a certain number of Turks, Turkomans, and Kurds have been unavoidably included. Their number is relatively small, and the analysis of a small series of "pure" Bedouin Arabs has given practically identical figures for the distinguishing Groups [A] and [O]. It is hoped in the course of the next few years to compile a series for the Kurds, about whom, ethnologically, little is known.[18]

Altounyan considered himself British and had trained in medicine at the University of Cambridge, but he was a native of and spent most of his life in Aleppo. He was the son of the Armenian physician Aram Assadour Altounyan (the founder of the hospital) and a mother from Northern Ireland. Aram had cultivated the patronage of Ottoman leaders, including the infamous Cemal Pasha, which enabled him to rescue and protect refugees of the Armenian Genocide. The Altounyans treated and subsequently employed many such refugees in their hospital, as well as in their households, which were located in Aleppo's Jewish quarter.[19] Altounyan became a close friend of T. E. Lawrence before serving as a medical officer, as well as an "expert adviser on Middle Eastern affairs," for the British military during both world wars.[20] The Altounyans, along with Lawrence, attended the 1919 Paris Peace Conference to support the Armenian and Arab national causes.[21] Altounyan's beliefs about the relative degree of "racial mixture" in different Middle Eastern groups reflected his own lived experiences as well as those of his neighbors.

However, Altounyan's note was not well received by his British audience, who regarded sero-anthropology with particular skepticism relative to physicians and anthropologists in continental Europe and the United States.[22] Within a month, the *Journal* printed a letter in which British physician and eugenicist James Stewart Mackintosh contested all of Altounyan's descriptive claims:

> Dr. Altounyan of Aleppo contributes an analysis of blood group percentages for Arabs, Armenians, and Jews, and remarks that they "represent

well-defined unmixed racial groups." Here we have again the familiar confusion between "race" and nation or people. The Jews, by their own accounts, have proselytized freely both before and after the dispersion, and the Armenians, by reason of their geographical position and vicissitudes, can hardly have kept their Hittite germ-plasm intact, if that in its origin were racially pure. The Arabs are notoriously a racial mixture. Dr. Altounyan also mentions the Kurds as a folk "about whom, ethnologically, little is known." On the contrary, it is well known that the Western Kurds are a blonde, dolichocephalic folk, speaking an Aryan dialect, and are therefore probably an outlier of the Nordic race—inhabiting, I may add, a plateau with climatic conditions of low rainfall, a dry summer, and a bright frosty winter. These conditions... closely resemble those in which the Nordic type was evolved, and therefore may account for these people maintaining their racial traits, despite their isolation.[23]

Mackintosh's sociopolitical commitments certainly differed from Altounyan's; he had a documented preoccupation with the "Nordic blonde" elements of the English nation, and I have found no evidence that he had ever visited the Middle East, much less seen a Kurdish person.[24] What Mackintosh calls "well known" facts about Kurds are the particular views of Austrian anthropologist Felix von Luschan, based on anthropometric work on Kurdish groups residing in Eastern Anatolia. Mackintosh's critique of Altounyan exposes the incommensurability of analytical time scales produced by anthropometric and serological methods. Anthropometric racial classification compared cephalic indices of living populations to cranial indices of ancient human remains to track ancient migrations, whereas seroanthropology, which by definition could work only on living populations, allowed comparison within just a few generations.[25] Mackintosh speaks in terms of evolutionary time when he suggests that the Kurds have maintained their ancient, prehistoric Nordic "racial traits" due to the long-term effects of climate. Meanwhile, Jews, Armenians, and Arabs are not "racially pure" (if they ever were) due to their intermarriage with other races during the past several millennia. Altounyan, on the other hand, operates on a much narrower time scale when he describes the Jews and Armenians as racially unmixed. Having faced discriminatory segregation, the Jewish and Armenian communities in Syria served by his hospital were indeed relatively well-defined by religion, communal language, and urban location, as opposed to the "Arab" category. The term "Arab" could be broadly applied to anyone speaking Arabic (that is, most of the Syrian population), but it was also associated to a certain degree with nomadic groups. Residents of post-Ottoman Syria could be roughly divided into urban dwellers (which included people descended from a Turkish-speaking administrative class), rural peasants, and nomadic populations (which included the tribes Altounyan called "Turkoman" and Kurdish in addition to "pure" Bedouin Arabs). Altounyan's categories, and his perception of their relative purity or admixture, therefore reflect more or less fluid patterns of group recognition in urban Mandate Syria, rather than any academically established European system of racial classification.

Despite the temporal limitations inherent in sero-anthropology, the projection of the "biochemical race" concept onto ancient human history captured the racial imagination of both scientists and the literate public. For example, American University of Beirut (AUB) anatomist William M. Shanklin used the ABO blood groups to proclaim the racial purity and ancient origins of certain Bedouin tribes in Syria. Specifically, he argued, the blood group frequencies of the Ruwalla Bedouins showed that "they are not related to any of the settled populations of the Near East. [. . .] They, however, like the American Indians and some island aborigines, are very high in group O and should, probably, be classified with [Ottenberg's] Pacific American type. Although surrounded for at least 2,000 years by peoples high in groups A and B, the Rwala have maintained a high degree of purity. They clearly represent a race of considerable antiquity."[26] These unexpected results spurred ongoing anthropometric expeditions in the Middle East, such as Henry Field's adventures in Iraq and Iran, to begin collecting blood samples in the course of their measurements.[27] Meanwhile, the *Palestine Post* reported Shanklin's findings in a way that further overemphasized the relevance of blood group frequencies to human prehistory. *Post* readers learned that the high frequency of group O in the Ruwalla, living in "the deepest parts of the [Syrian] desert," could be interpreted as evidence that "pure blooded Arabs are closely related to the purest blood American Indians" because "the blood types of the two races are the same."[28] The alluring simplicity of such assertions of racial affinity led foreign and local researchers alike to conduct their own sero-anthropological studies across the interwar Middle East.

IMAGINING ENDOGAMOUS COMMUNITIES
IN EGYPT AND THE LEVANTINE MANDATES

William M. Shanklin was able to incorporate blood-typing into his routine anthropometric fieldwork in the depths of the Syrian desert thanks to the serological laboratory developed by his colleague Leland W. Parr, a bacteriologist at AUB's School of Medicine. Concurrently with Altounyan's work in Aleppo, Parr worked throughout the 1920s to assemble a comprehensive data set of Middle Eastern blood group frequencies (see Figure 4). In 1931, the *American Journal of Physical Anthropology* published his findings on nine Egyptian and Levantine populations, with special attention to Egyptian Copts, Samaritans, Armenians, and Syrian Christians.[29] Parr considered the first two groups to be of interest a priori, due to their claims of direct and undiluted descent from ancient Egyptians and Israelites, respectively.[30] Parr argued that Egyptians were distinct from "the peoples called Arabs" in the rest of the Middle East and North Africa, based on the blood he collected from 130 Egyptian patients and students at the American University Hospital in Cairo. This sample included both Muslims and Copts, and Parr noted that the latter "lay claim to being the purest descendants of the ancient Egyptians."[31] Although he decided that his blood-typing results were insufficient to pass judgment on the Copts' historical antiquity, he encouraged future investigators to search for serological differences between Copts and Muslims. His writing also indicates his acceptance of Coptic endogamy: "Whatever may have happened a thousand years ago, certain it is that in recent decades there have been practically no marriages between Copts and Moslems."[32]

He seems to have accepted an even longer time frame of reproductive isolation for the Samaritans, in spite of evidence to the contrary he witnessed with his own eyes: "The Samaritans are alleged not to marry outside their own group, although at the time we visited them [May 1928] we found two young men with English-speaking Jewish wives."[33] Parr did not clarify whether these Jewish women were excluded from his Samaritan sample, but he nevertheless explained his results as representing "the effects of inbreeding," implying a permanent and almost hermetic seal on the community. Parr lamented that because of this extensive inbreeding, sero-anthropology could not answer the question of Samaritan origins.

His conclusions on Armenians and Arabs were decidedly less cautious.

FIGURE 4. Tamir Nassar collecting a finger-prick blood sample from a villager in the Akkar mountains of Lebanon, July 1935. Courtesy of American University of Beirut Libraries.

The group of Armenians Parr typed in Beirut had significantly different ABO frequencies than either Altounyan's Aleppo Armenians or Armenian émigrés in Paris. Despite acknowledging these variations, Parr designated the Armenians "the largest pure group" of his study, effectively dismissing the blood data to insist upon their "high degree of unity" based on "social and economic traits."[34] Yet the variation in ABO frequencies he calculated between groups of Armenians in several Syrian cities was quite similar to that between Syrian Christians and Muslims—a difference that Parr trumpeted as his "most significant finding" with deep historical implications.

Parr argued that strict religious divisions among the Arabs actually reflected a racial divide: while the ABO frequencies of Muslim and Druze Syrians matched the "Intermediate" Arabs studied by the Hirszfelds, Syria's Christian communities (including Maronites, Greek Orthodox, Greek and Roman Catholics, and Protestants) constituted a distinct racial type with a "European" index. Most impressive is Parr's confidence in the antiquity of the genetic differences he observed: "Intermarriage with the Crusaders might account for some of the observed blood picture, but it is far more reasonable to suppose that the Christians of the area are descended from an earlier Mediterranean race and have maintained their earlier characteristics."[35] Parr suggested that his findings provided serological support for the anthropometric arguments of AUB's visiting Dutch professor, Cornelius Kappers. As described in Chapter 1, based on cephalic indices, Kappers claimed that Arab Christians, Armenians, and Assyrians all originated from the same ancestral population inhabiting the Fertile Crescent.[36] The Druze biochemical index, however, did not match the origin narrative Kappers attached to their cephalic indices—a "curiosity" that Parr blamed, as with the Samaritans, on extensive inbreeding.[37]

In contrast to his eager racialization of Syrian Christians, Parr wrote that on the basis of serological evidence, Jews were "a religion rather than a race."[38] For this, he cited data on Ashkenazi and Sephardi Jews he had obtained by personal communication with physician Rina Younovitch in 1929. Younovitch (née Goldberg), an employee of the Hadassah Hospital in Tel Aviv, undertook multiple ABO typing surveys of Jewish immigrants to Mandate Palestine and was perhaps the first in her organization to publish serological data using an anthropological perspective. In 1932 and 1933, she published data on Yemenite Jews, "Asiatic Jews" (*juifs asiatiques*), and Samaritans, introducing standard categorical practices of Jewish anthropometry into sero-anthropology by treating Asiatic Jews (including Moroccans) as a population unit in contradistinction to Ashkenazim.

Although Younovitch's work was performed in the Zionist context of the Hadassah Hospital, she did not share the same essential assumptions as contemporary Jewish anthropologists who sought out an Israelite *Urtypus*. Rather, she routinely cited the French-Serbian physician Nicolas Kossovitch's 1932 work on Jews in Morocco, in which he argued that Jews shared only a religious unity, not a racial one.[39] Younovitch herself framed her own work as a challenge to the idea of some scholars (specifically naming German geographer Richard Andree) that Jews represented "a pure, unmixed people."[40] Furthermore, she discredited attempts to identify any particular Jewish community as untainted representatives of the ancient Israelites. For example, while she designated the Yemenite Jews as an ethnic isolate, she argued: "Weissenberg, who studied these Jews from an anthropological perspective, considers them to be the true descendants of the Hebrews. But the history of this people leaves no doubt of their mixture with other peoples. [...] The Jews of today are a very mixed type of people."[41]

She approached her study of other groups of Asiatic Jews with the same skepticism toward claims of maintaining unbroken lines of descent from the Hebrews of the biblical period. Noting that Ashkenazim and Sephardim in Europe demonstrated considerable admixture with neighboring peoples, she expected the same of other Jewish groups, despite the fact that "the Asiatic Jews inhabit the ghettos of Muslim towns, remain isolated without having any occasion to mingle with their coreligionists of Europe, and have a custom of saying that they represent islands [*îlots*] of the primitive Jewish race."[42] Younovitch sorted her calculations on eight distinct communities into three broad categories—autochthonous Arabic- or Ladino-speaking Jews (including communities from Syria, Palestine, and Morocco); Persian Jews (from Iran, Bukhara, and Dagestan, in the Caucasus); and Mesopotamian Jews (from Baghdad and Kurdistan). She argued that the last group had a definite history of local admixture, while Persian Jews did not: "The inferiority of [Persian Jews'] social status renders their mixture with the Persian population very improbable. . . . [Babylonian] Jews, exiled by the Babylonian kings, maintain after hundreds of generations the tradition of their ancient origin, although they have experienced in certain epochs an influx of proselytes."[43] Ultimately, Younovitch concluded three major points based on the ABO frequency data, which generally conformed to her historical impressions: (1) all five groups of Asiatic Jews were serologically similar and represented a mixture of the Asiatic and European races; (2) the serological characteristics of Asiatic Jews were very different from those of the Ashkenazim; and (3) the Jews of Kurdistan and Baghdad had experienced "an intense infiltration by the European race."[44]

Younovitch emphasized the evidence of admixture between European and Asian "races" in various Jewish communities, especially among the Jews of Mesopotamia, as proof against their traditional claims of historical endogamy. Very unlike Parr's conclusions on Arab Christians, she contradicted the projection of existing socioreligious divides between Middle Eastern Jewish communities and their neighbors into an eternal history of racial separation. Of more long-term significance is her second, seemingly straightforward conclusion that her data showed a major difference between the different Asiatic Jewish communities and Ashkenazim, whose frequencies she presented in a single group even while acknowledging that the Ashkenazi patients originated from different countries, denoted as "Russians, Polish, German, etc." Curiously, she did not amalgamate the eight distinct communities included in her Asiatic data in a similar fashion for the purpose of the comparison, which would indeed have yielded very different frequencies from the Ashkenazi data—as well as very different frequencies from any of the Asiatic communities, separately considered. This is because combining the Asiatic community data effectively averages out the proportions of blood groups that otherwise distinguish the separate communities, which display a high range of variation. A similar range of variation was observed between the various national communities of Ashkenazi Jews in Europe (so much so that Reuben Ottenberg and Laurence Snyder had classified German, Romanian, and Polish Jews as completely separate racial types!).[45] Why, then, did Younovitch provide only combined data for her Ashkenazi patients in Palestine?

The answer lies in the nature of Jewish settlement in the New Yishuv. By the time of Younovitch's research, a social rift had already formed between ideologically driven Zionist settlers (predominantly Russian, Polish, and German Jews) and the "Asiatic" communities. The former, through their commitment to Zionist economic, social, and cultural institutions, were perceived to assimilate into a generic Ashkenazi melting pot with relative ease, subscribing to the ideals of a new national culture based on secularized European norms. The latter had either lived in Palestine for centuries or had immigrated more recently, sometimes self-motivated by religion and sometimes (especially in the case of the Yemenites) recruited by Ashkenazi Zionists to serve as cheap Jewish labor in place of non-Jewish Palestinians. These recruits were employed as agricultural laborers and domestic servants for the Ashkenazi landholders in Zionist settlements, lived in small family groups in squalid conditions, and were kept at arm's length from participating in the very same Zionist institutions that had convinced them to immigrate, leaving them unable to buy land or claim benefits.[46]

This socioeconomic segregation ultimately amounted to biological segregation because "white Jews" married only among one another, refusing to do so with Yemenites or other Middle Eastern Jews. The ensuing waves of Ashkenazi *aliyah* during the 1920s and early 1930s only reinforced this process of segregation. Meanwhile, the smaller groups of non-Ashkenazi immigrants, whose social interactions were restricted mostly to others from their individual countries of origin, found themselves unable to effectively combat this discrimination collectively across the Yishuv.[47] Ashkenazi academics, such as those working at Hadassah's Department of School Hygiene in the 1920s, incorporated the discriminatory features of this social structure into scientific research. Ashkenazim were consistently treated as a single racial or biological category, while non-Ashkenazi data were divided into anywhere from five to twelve separate communities.[48] Younovitch's work, like that of other Ashkenazi medical practitioners in Palestine, therefore reflected the ethnic stratification manufactured by the Orientalist colonial logic of Zionism in the Yishuv.[49]

Younovitch's report on the blood group data of the Samaritans, collected in cooperation with the Italian Corrado Gini's anthropometric survey (described in Chapter 1), reveals a significant ambivalence toward this community so enthusiastically studied by Jewish anthropologists. Younovitch began collecting her Samaritan data independently, in advance of the rest of the Italian-Yishuv joint team.[50] A major concern was the inclusion of only "pure" Samaritans, whom she regarded as indistinguishable from their non-Samaritan neighbors. Younovitch noted that the cooperation of Amram Cohen, the high priest of the Samaritan community, proved crucial to the integrity of the research:

> We were able to perform the serological study of the Samaritans thanks to the friendly aid of the High Priest of this people [. . .]. It was easy to confuse the Samaritans with Arabs or Spanish Jews, who speak the same language (Syrian Arabic) and who present themselves to travelers as Samaritans. We have examined only individuals whose origin was certified by the High Priest.[51]

Despite remarking on the "anthropological" resemblance of Samaritans and Sephardic Jews, and even titling her work a "Serological Study of Samaritan Jews," Younovitch did not compare the Samaritan blood group data to that of the Jewish immigrant populations she worked with. In a significant break from anthropometric studies of Samaritans, she compared them only to Syrian Arabs of the same province. On the basis of this comparison, she emphasized a racial distinction between the Arabs and the Samaritans: "The percentage of group A is quite high in the Arabs and reaches that of European or intermediate people, while that of the Samaritans [. . .] is close to Asiatic peoples." However, the racial indices she calculated were very similar (1.5 for the Samaritans and 1.59 for Arabs), and the results diverged sharply from Parr's Samaritan index of 2.35, a solidly "European" number, calculated just a few years earlier. The latter phenomenon could result from the different number (and likely different combination) of community members studied: Parr obtained blood from 83 Samaritans; Younovitch, 108. To represent Samaritans as "Asiatic" in contrast to "European" Arabs was an interpretive choice to disregard the Hirszfeld index in favor of the raw statistics: compared with Parr's results, Younovitch had found a much higher proportion of type O individuals, and a much lower proportion of type A.

Like most scholars studying the Samaritans, Younovitch perceived this "very ancient people" to be "in a state of complete disappearance." Both the novelty appeal of the Samaritan community to anthropologists and the apparent imminence of Samaritan extinction were predicated upon "the tradition of the Samaritans and several testimonies of foreign observers [according to whom] the Samaritans never marry out [*ne se mésallient jamais*]."[52] Younovitch made no mention of the three Jewish wives described in Gini's report of the Italian study, nor did she discuss the possible absorption of Samaritans into the Zionist "body politic," as reported by Gini. Instead, she emphasized the Samaritan commitment to endogamy and the community's inevitable extinction. The significance of her insistent grouping of the Samaritans with "Asiatic peoples" is unclear. Perhaps she meant to emphasize the genetic differences between "Samaritan Jews" and Syrian Arabs in a way that supported the former's indigeneity and shared ancestry with the "Asiatic Jews" she had studied. Yet she had already vigorously contested the other communities' claims to ancestral purity, and the Samaritan ABO frequencies and biochemical index came nowhere close to resembling those of the various Asiatic Jews. So perhaps her conclusions simply reflect a more pessimistic perspective than that of Yitzhak Ben-Zvi, one of her colleagues on the Gini expedition. Ben-Zvi was determined to rehabilitate the Samaritans' reputation as authentic Israelites and ensure their protection through the "progressive" activities of Zionist settler-colonialism. In contrast, Younovitch's writing suggests that she perceived the Samaritans as beyond salvation. Regardless of their kinship with Jews, they represented a backward and primitive human landscape that could not survive in the modern, Western-facing national culture envisioned by Ashkenazi Zionists.

BRINGING TYPE A TO EUROPE: SERO-ANTHROPOLOGY IN THE TURKISH REPUBLIC

As described in the previous chapter, the first generation of anthropologists in the Turkish Republic carried out their research to provide an "objective" body of evidence supporting the racial claims of the Turkish History Thesis. Like the early anthropometric studies, blood group studies on the Turkish population were conducted by medical doctors in Istanbul. Although Şevket Aziz Kansu declared the Turkish Institute of Anthropology's interest in this research as early as 1931, until the end of that decade, most work in sero-anthropology was carried out under the auspices of the Istanbul University Medical Faculty.[53] While Turkish expertise in anthropometry came via Eugène Pittard and other francophone anthropologists, expertise in serology derived instead from German Jewish refugee scientists working in Istanbul, especially the microbiologist Hugo Braun and the hematologist Erich Frank, who arrived in 1933. The first sero-anthropological works in Turkey aimed to correct the 1919 Hirszfeld results by shifting Turks from the "Intermediate" to the "European" biochemical classification. However, due to the inconsistencies generated by the index calculation, Turkish researchers ultimately settled on a dynamic historical narrative equally aligned with the Thesis: that the Turks had introduced the A blood type, along with civilization, to Europe.

In September 1931, physician Ahmed Şükrü Dimen presented the results of Turkey's first sero-anthropological investigation at the Turkish National Medical Congress. On the basis of data from 1,200 individuals sampled at the Bakırköy psychiatric hospital near Istanbul, he calculated the Turks' biochemical index at 2.4 (as opposed to the Hirszfelds' 1.8), nudging Turks into the European category. However, these results were later deemed unsatisfactory because the population of the asylum included "a mixture of Turks, Greeks, Armenians and Jews, with the result that it does not therefore characterize the Turkish population."[54] Subsequent studies strove to include only "pure" Turks, especially those from Anatolia. Ethem Babacan's 1936 study on 2,000 Turkish medical students and soldiers posted in Istanbul, conducted under the guidance of Braun, clarified that his procedure included questioning subjects to ensure that they were "pure Turks and not mixed or members of another people."[55] Babacan calculated an overall biochemical index of 2.5 for the Turkish Republic, but he found that subdividing the population according to region (Istanbul 2.2, versus Anatolia 2.94), or to provinces of birth within Anatolia, resulted in a fairly broad range of values (from 2.3 to 4.2), a phenomenon for which he had no clear explanation.

This phenomenon, in fact, often arose in studies worldwide during the first decades of sero-anthropology because of the inherent problems of the Hirszfelds' biochemical index calculation. As alluded to above in the discussion of Ottenberg's and Snyder's critiques of the index, the simple ratio of A to B blood group frequencies was genetically meaningless. The ratio relied on inaccurate models of, firstly, the underlying genetic mechanisms (e.g., we now know that multiple A alleles exist to generate the same phenotype) and, secondly, human evolution (e.g., the ABO antigens also exist in apes, therefore predating the existence of all humans). Finally, and most significantly, the classificatory function of the ratio hinged on the assumption that a given race's proportion of blood groups would remain the same over time and for any sample size. In other words, the populations defined by the researchers must be closed to immigration and emigration, must marry only among themselves, and must maintain a consistent number of individuals. The fact that Babacan took all his samples of "Anatolian" blood from individuals living (at least temporarily) in Istanbul obviously belies this fundamental assumption. Furthermore, these migrants were all soldiers or university students, highlighting the role of state institutions in confounding the very "natural" boundaries of human populations that they sought to classify.

Nevertheless, Turkish physicians selectively deployed the Hirszfelds' racial index alongside other statistics. Nureddin Onur, a physician working at the Hygiene Laboratory at the University of Istanbul, synthesized information from a total of 3,729 blood samples taken from Turkish soldiers and sent to the Hygiene Laboratory and the Institute of Legal Medicine.[56] Onur's overall index calculation was 2.11. Unlike Babacan, he did not subdivide the sample population into provinces; he used only the broad regions of "Europe" and "Asia Minor." Among his samples, he identified "1,473 persons whose origins were incontestably established" and calculated a major index difference between the Turks of Europe (2.44) and Asia Minor (1.80).

Although the difference between the values is similar to Babacan's calculations for Istanbul versus Anatolia, the values themselves point to an opposite classification, with Onur's Asia Minor having a lower (i.e., less "European") index than Babacan's. Glossing over this difference, he ignored the index values in favor of Ottenberg's technique of phenotype frequency, where the variations were less pronounced: "The [frequency of] group A is not sensibly different among the Turks of European Turkey and of Asia Minor, while group B is a little more elevated among the Turks of Asia Minor," but these results were insufficient for any "ethno-anthropological conclusion."[57] Many more studies were needed, he explained, taking into account other blood factors and testing a completely pure Turkish sample, before the Turkish race could be accurately characterized in serological terms.[58] Onur at first demurred from offering a nationalistic interpretation of his results, but he presented the same data to the Second Turkish History Congress (discussed below) as validating the racial migration hypotheses of the Turkish History Thesis.

The search for "pure Turks" led physicians to study the blood of the Yürüks, semi-nomadic tribes living in western and southern Anatolia. Since the late Ottoman period, the Yürüks had been romanticized as representatives of the original physiology and culture of the Turkish tribes who had emigrated from Central Asia to conquer Anatolia from the Byzantines.[59] In 1937, Sadi Irmak analyzed blood samples from two southern Anatolian Yürük tribes, the Boahmatlı and Eseli. Like Onur, Irmak was a physician and eugenicist based in Istanbul. He was also a Nazi sympathizer intermittently active in electoral politics.[60] Irmak explained his interest in the Yürüks as "tribes [aşiretler] who have preserved the same biology and culture" since their migration from Central Asia. He believed that because "almost no intermixing" had occurred among the Yürüks, they were an ideal population for studying "the special characteristics of pure races and, as such, of the biology of the Turkish race."[61] Indeed, Yürük blood fulfilled the desire for scientific evidence to classify Turks as a European race. Irmak's biochemical index value for the 400 Yürüks he studied was 5.2, staggeringly more "European" than the averages calculated by Babacan and Onur for settled Anatolian and Istanbul Turks, due to the Yürüks' much higher frequency of group O and very low group B. Ignoring this complication, Irmak considered only his own value from the Yürüks to conclude that the entire Turkish race belonged to the Alpine type, in congruence with the claims of contemporary

Turkish anthropometric studies, rather than the Mediterranean (Dinaric) type as suggested by the Hirszfelds in 1919.[62]

Both Irmak and Onur presented their blood group data to the Second Turkish History Congress in September 1937. Irmak began with an explanation of the inheritance of blood groups (as understood at the time) before briefly summarizing his original research. Irmak emphasized that the very high frequency of O in Yürüks relative to Turks living in Istanbul or Anatolian cities indicated their ethnic purity, like that of Mayas and Eskimos in North America. The rest of his remarks stressed the biological similarities (in terms of blood group and fingerprint pattern frequencies) of Turks with Northern and Western Europeans in contrast to Arabs, Iranians, and Indians: "According to the general state of the blood groups [. . .] our nation departs from the nations of south and east Asia and shows proximity to the nations of northern Europe."[63]

The message of Onur's presentation was essentially the same, but unlike Irmak, he articulated a specific historical explanation for the data using the language of racial sources or stocks (*kök*). Citing various national percentages of group A and B frequencies, he reiterated the common argument that since B was highest in Asia and A highest in Europe, "the source of the white race [*beyaz ırk*] must be group A and the source of the colored race [*renkli ırk*] must be group B."[64] He extended this logic to explain how "pure, incontestably established" Turks could present such geographical diversity in their ABO frequencies across Anatolia and Thrace: "I mentioned the slight increase of the A group when passing from Asia to Europe, while in contrast the B group decreased. We also see this in our country when passing from Anatolia to Rumelia. . . . These results, [which] if compared with historical findings would not be far off, push us to this conclusion: *the Turkish race is the primary stock that brought type A to Europe.*"[65] In other words, the gradient of A and B blood groups across the Turkish Republic supposedly reflected the prehistoric advance of mostly type-A Turkish ancestors toward Europe, introducing the A allele to that continent while migrating away from Asian races made up mostly of type-B individuals.

Even for its time, the biological logic of this conclusion is thin, and it appears that Onur was purposely stretching his interpretations for the benefit of the conference's goals to confirm the Turkish History Thesis with scientific data. In a later general compilation of information on blood groups, including a review of their use in sero-anthropology, Onur dropped these explicit references to historical Turkish migrations from Asia to Europe. However, his preface to the book confirmed his nationalist ideals: "The world is obligated to rewrite its history, [which] is partly based on doubtful narratives, in the light of science. [. . .] But still the most important and sacred duty of the Turkish son is to search, find, and introduce to the world this merit and honor of the Turk *in his own noble blood*, which we believe to be the oldest and most adept architect of the civilized world."[66]

MIDDLE EASTERN BLOOD WITHIN AN INTERNATIONAL DEBATE

By the mid-1930s, as physicians and anthropologists across the globe calculated ABO blood group frequencies at an ever-increasing rate, the accumulation of sero-anthropological data only cast further doubt on the early claims that blood group frequencies could clarify problems in racial classification. On the contrary, the more that data on far-flung populations became available, the more the waters were muddied by frequency similarities between peoples from Africa and China, or Australia and Greenland. William C. Boyd, an immunologist based at Boston University and an enthusiastic advocate of sero-anthropology, teamed with anthropologist Leland C. Wyman to explain this "failure" in 1935. Writing in *American Anthropologist*, they argued that the discovery of a blood antigen system analogous to ABO in apes suggested that the current distribution of ABO frequencies in humans was the result of dispersions so ancient that they predated the divergence of the existing racial types classified by anthropometry. Far from being useless, then, sero-anthropology promised to detect historical human migrations that preceded race itself! The authors acknowledged the perennial scholarly caveat that much more research was necessary to prove this hypothesis, "especially on all groups of people giving evidence of long isolation and all of the rapidly vanishing primitive tribes."[67]

In pursuit of such research, at the end of 1935, Boyd and his wife, Lyle, an experienced biochemist, set off for Europe and the Middle East with funding from a Guggenheim fellowship. Over the next two years, they traveled to Ireland, Wales, the USSR, Spain, Egypt, the Syrian Mandate, and Iraq. In each locale, they tested for ABO frequencies in addition to collecting data on two newly discovered hereditary traits, the MN blood group system and the ability to taste a chemical called phenylthiocarbamide (PTC).

In Egypt and Syria, the Boyds retraced the steps of AUB's Leland W. Parr and William M. Shanklin, whose published results they found particularly

FIGURE 5. The field laboratory equipment used by William and Lyle Boyd at the American Mission schools in Asyut, Egypt, January 1936. Courtesy of Schlesinger Library, Radcliffe Institute, Harvard University.

compelling. They began their work at the Qasr al-Aini Medical School in Cairo, where physiology professor Gleb von Anrep provided them laboratory space and Arab staff who served as interpreters and test tube washers. In letters to her parents, Lyle gushed about the hospital's diverse and exotic patients, who were "every shade of color," but ultimately the Boyds were in pursuit of the Copts, "who are presumably a more pure racial group than the others here."[68] After consulting unpublished data shared by Egyptian medical professor Dawood Matta and working some days in a Coptic hospital, they concluded that Copts differed from Muslims only in a lighter average skin color. The Boyds relocated to the Nile town of Asyut, where the majority Coptic population was "said to represent as pure a strain as can now be found" of the alleged descendants of ancient Egyptians. Yet despite sampling hundreds of Copts at the American Mission Hospital and mission-affiliated schools in Asyut, the Boyds, like Parr, could not find any significant difference between Coptic and Muslim blood group frequencies (see Figure 5).[69]

Syria presented the Boyds with logistical and ethical challenges. Keen to follow up on Shanklin's work with the Ruwalla Bedouins, they abandoned their usual modus operandi of working in local hospitals to mount an expensive desert expedition. They hired a driver for a week, loading his car with food, antisera stashed in an iced thermos bottle, and gallons of water to wash their racks of test tubes. However, keeping their supplies clean and cool turned out to be the least of their problems. Their chauffeur and the young tribal leaders who escorted them to remote campsites proved less interested in the scientific merits of the excursion than in coaxing the Boyds to pay per head for blood samples from Bedouins who turned out to belong to the wrong tribe.

Frustrated, the Boyds enlisted the help of Adib Tayyar, an AUB pharmacy graduate involved with the rural Village Welfare Service.[70] When Tayyar accompanied them to other camps, Lyle marveled that "he could persuade those old Bedouins to part with their blood in a miraculous way" through cajolery and outright misrepresentation of the Boyds as doctors on a government medical mission. However, the Boyds were incensed when they discovered that Tayyar had schemes of his own to help the chauffeur and tribal leaders secure more fees. Without any sense of irony, Lyle mourned Tayyar's "oriental" ethical standards, calling him "a small tragedy" from an affluent Syrian family whose Americanized education had left him a "permanent misfit" unsuited to both Western and Arab society.[71] In other words, if Tayyar lied to research subjects to obtain their blood, that was acceptable scientific practice, but his deceiving American researchers was evidence of a serious moral failing.

The Boyds were further alarmed when, despite extensive testing and retesting of their hard-won Ruwalla samples, they could not reproduce Shanklin's dramatic results (on the basis of which he had argued for the Ruwalla to be classified alongside Native Americans). Instead, they found that the Ruwalla did not have an inordinately high O frequency and closely resembled neighboring Arab tribes after all.[72] The Boyds interpreted their new data on the Ruwalla as evidence for their own hypothesis of pre-racial ABO evolution.[73] After working with the rural villagers of Boarij and Meshghara, they could not find consistent serological differences between the religious groups. Accordingly, they also disputed Parr's claim that Syrian Arab Christians might be racially distinct from their Muslim neighbors.[74]

Following their difficulties in Syria, the Boyds were relieved to move on to Baghdad, where they enjoyed the facilities of a laboratory at the Royal College of Medicine. Ahmed Kayssi, the director of the forensic medicine department, escorted them to a number of Bedouin villages on the outskirts of the city. This ease of access to Iraqi Bedouin communities seemed to foreshadow the blood results, which showed that the tribesmen were not particularly different from the "city Arabs."[75] With regard to the Jewish community of Baghdad, the Boyds announced that their data matched well with that of Rina Younovitch's Mesopotamian immigrants to the Yishuv. More significantly, they concurred with her emphasis on the divide between Middle Eastern Jews and Ashkenazim.[76] Despite noting that Baghdadi Jews and Christians shared similar frequencies, the Boyds excluded only the Jews when calculating an average ABO distribution for the city of Baghdad to be compared against surrounding localities. This implies that they considered the Baghdadi Jews to be "a people apart" from their neighbors, possibly reflecting their field experiences working at the Jewish Hospital and Polyclinic. In contrast to the cooperative patients they found at Baghdad's Royal Hospital, Lyle complained that the Jewish polyclinic patients were "mostly women, and very emotional, and they shouted and yelled and the babies cried—pure pandemonium. Many of them refused to submit, so the doctor would refuse them medicine, and they'd come back to us and stick out a finger, but looking daggers at us. Some went away without treatment, rather than give a drop of blood."[77] The Boyds regarded the Jewish patients' level of resistance as resembling the primitiveness and ignorance of the Syrian Bedouins and rural villagers. Perhaps this inclined them toward accepting the traditional claims that their patients were relics of the Babylonian exile.

Upon their return to Boston, William Boyd first organized an updated compilation of worldwide blood group data. He undertook this thankless task, necessitated by the constant deluge of serological publications that rendered all such compilations obsolete within a few years, to establish the geographical basis for evolutionary and historical claims he planned to make in his next book.[78] In addition to his own data from the Middle East, Boyd incorporated data from the publications of Younovitch, Ethem Babacan, and Sadi Irmak into his 1939 compilation for *Tabulae Biologicae*.[79] Boyd's work caught the attention of J. B. S. Haldane, the eminent British population geneticist, who sought to defend sero-anthropology against skeptical Anglo-American physical anthropologists. Haldane, an expert at evaluating genetic hypotheses through mathematical modeling, complained that the anthropologists' favored traits for classification, such as skull shape and skin pigmentation, were not reliable for reconstructing evolutionary history. The former was subject to a great degree of environmental influence, and the latter to rapid natural and possibly sexual selection. Blood groups, on the other hand, were exclusively inherited, and at that time there was no evidence that they offered any kind of selective advantage. Therefore, he reasoned, sero-anthropology was more likely to yield accurate information about prehistoric human migrations. Like Boyd, Haldane argued that the blood of living populations was more useful than ancient skeletal remains to determine the movement and mixture of humans in the distant past.

Boyd's work provided Haldane with the necessary serological data to develop a genetic hypothesis on the peopling of Europe, prompting a correspondence between the two men during the summer of 1939, concerning mainly the detection of possible calculation errors and population duplications in Boyd's compiled frequency tables.[80] Upon hearing the news of Germany's invasion of Poland, Haldane predicted imminent German air raids over the United Kingdom and sent Boyd an incomplete manuscript, requesting assistance to proofread the material and submit it for publication in the United States.[81] Boyd obliged, and by the end of March 1940, Haldane's paper (as revised by Boyd) had been accepted for publication in the journal *Human Biology*.[82] Meanwhile, the correspondence with Haldane motivated Boyd to dust off and revise drafts written in Cairo of his similar arguments regarding the value of sero-anthropology, which he then submitted to the *American Journal of Physical Anthropology* in April 1940.[83] Boyd's paper accepted Haldane's migration hypotheses to explain the distribution of ABO blood group frequencies in Europe and expanded on it to propose an evolutionary narrative of periodic population isolation and migration that could account for the ABO distribution of the entire world. The Haldane and Boyd papers initially drew favorable responses from anthropologists in the United States, and by the beginning of 1941, there was even talk of establishing a new national society for the study of human genetics and its application to anthropology. Although the proposal quickly fell through—due to a lack of interest among American geneticists and the prevailing mood of wartime uncertainty, according to Boyd—he believed that the papers had attracted the desired support of American anthropologists.[84]

Because the Middle East, broadly defined, delineates Europe's southeastern borders, Middle Eastern serological data figured prominently in Haldane's paper on the racial origins of Europeans. As Haldane noted, he had "arbitrarily drawn the frontier of Europe so as to exclude the peoples of the Caucasus," relegating them to a separate table and discussion section on "neighboring non-European populations" alongside Central Asians, Anatolian Turks, and North Africans from Morocco, Tunisia, and Egypt.[85] Meanwhile, Haldane's European table included data on the Jewish population of Odessa, as well as on 1,000 "Turks." In a lengthy discussion explaining his selection of population data and describing possible sources of error, Haldane singled out the work of Turkish physician Babacan as an example of obviously faulty calculations among the data, such as making "the truly remarkable assertion that of 100 Turks examined at Ankara, 9.3 percent belong to group B."[86] However, Haldane's own calculations of Turkish frequencies reflect an uneven imposition of nationalist assumptions of ethnic uniformity on certain populations (primarily those of the Middle East) and not on others (especially Germans). His "European" Turkish frequencies are an amalgamation of the data from 500 individuals Babacan had tested in Istanbul with the 500 "Macedonian Mohammedans" sampled by the Hirszfelds in Salonika—two populations that, considered separately, had quite different ABO frequencies. Similarly, Haldane (re-)created a combined figure for "Anatolian Turks" from Babacan's provincial breakdown (Boyd had excluded Babacan's original Anatolian calculation from the *Tabulae Biologicae* compilation).

Boyd felt that Haldane's creation of such amalgamated populations was methodologically suspect. In their correspondence, Boyd commented disapprovingly, "I am very doubtful if one ought to make an overall figure for [an ethnic group], unless they all live in the same locality. [...] The only way I see of knowing that it is justifiable to combine figures relating to a given ethnic group is by seeing that the results are none of them significantly different. And then you do not need a combined figure!"[87] Despite this criticism, Boyd used his editorial prerogative over Haldane's manuscript to add a similarly problematic set of data to Haldane's table of "neighboring non-European peoples": that of Younovitch's Ashkenazim in Palestine, a modification that Haldane was not able to explicitly approve before publication.[88] The logic of this addition is not clear. Haldane certainly did not consider Jews, especially

Ashkenazim, as non-European; he had included Odessa Jews on the European table. Boyd seems to have relegated Ashkenazi settlers in Palestine to the "non-European" table only because of their geographic location in the Middle East, ignoring Younovitch's own description that these were recent immigrants originating from multiple parts of Europe.

These statistical and conceptual inconsistencies buried within well-received work by prominent Anglo-American geneticists demonstrate how identity concepts fostered by Middle Eastern nationalisms and incorporated into local practices of sero-anthropology became normalized within the international sphere of human genetics research. Haldane replicated Babacan's and Nureddin Onur's presentations of Turks as simultaneously a singular biological population whose frequency data could "legitimately be pooled" yet could also be neatly divided into European and non-European components, along the geographical boundary of the Bosphorus.[89] Boyd's insertion of Younovitch's data implies an acceptance that Ashkenazi settlers in Palestine were indeed creating a new "non-European" population, as well as an expectation that the ABO frequencies of this group of "New Jews" in Palestine would no longer match those of their ancestors segregated into distinct German, Russian, and Polish communities. Yet, as discussed above, these notions reflect the conceptual ideology and physical population movements provoked by Republican Turkish and Zionist nationalism. Haldane and Boyd used Middle Eastern ABO frequency data to argue that blood groups "give information of a more fundamental character on racial structure" than skull shape or skin pigmentation. But in fact, these frequencies were just as much artifacts of the "immediate past" as anthropometric measurements.[90]

THE LEGACY OF RACIAL SERO-ANTHROPOLOGY

The methods and racial conclusions of sero-anthropology had never gone uncontested by physical anthropologists. Although its most enthusiastic supporters, such as William C. Boyd and J. B. S. Haldane, consistently and confidently predicted that serological data would replace anthropometric methods, they were never able to convince most anthropologists of the epistemological supremacy of blood-typing techniques. The persistent marginality of sero-anthropology to the field was cemented by the accumulation of serological data that overwhelmingly contradicted existing systems of racial classification.⁹¹ The global pattern of ABO frequencies could not be reconciled with the other phenotypic data of classical anthropometry, nor did it match prevailing hypotheses about human evolution and racial differentiation. Furthermore, the gradual discovery of other blood antigen systems alongside ABO (such as the MN, Rh, Lewis, and Kell systems) from the late 1920s onward complicated earlier attempts to narrate population histories on the basis of ABO frequencies alone.

Middle Eastern anthropologists, too, challenged various aspects of sero-anthropological research. In Egypt, social scientists like Abbas Mustafa Ammar rejected the Western preoccupation with locating biological differences between Muslims and Copts that might substantiate "pharaonic" discourses of Egyptian identity. In contrast, Ammar's anthropometric and serological studies of the northeastern Sharqiya province during the late 1930s and early 1940s emphasized the historical migration of Arabs into Egypt. He engaged the help of Dawood Matta, Boyd's contact at Cairo University, to determine the ABO types of more than 1,000 Sharqiya residents. Based on Shanklin's work on the Syrian Bedouins, Ammar interpreted the east-to-west gradient of decreasing O frequencies as representing the Arab "invasion" of Egypt.⁹² His anthropological interest in Egypt's Arabness reflected a general shift in Egyptian nationalist thought, which in this period turned increasingly toward pan-Arabist and pan-Islamist ideas. It also showed the political interest of Ammar and his circle of Egyptian intellectuals in justifying Egyptian colonial rule over the Sudan. British claims to the territory highlighted racial differences between Sudan and Egypt. Instead, Ammar argued that the whole Nile Valley shared Hamitic racial origins and a historical experience of Arabization.⁹³

In Turkey, sero-anthropology sustained some popularity after the Second Turkish History Congress because it was perceived to support a Kemalist nationalist narrative.⁹⁴ However, in 1943, Şevket Aziz Kansu's student Nermin Aygen singlehandedly unraveled the enterprise of Turkish sero-anthropology. Her dissertation research on 500 soldiers of the Presidential Guard in Ankara demonstrated that ABO blood groups had no correlation with anthropometric traits (specifically, she examined cephalic, facial, and nasal indices; height; and color of hair, eyes, and skin) or the racial classifications supposedly determined by these traits (e.g., Alpine or Dinaric). Therefore, Aygen argued, the previous work by her Turkish colleagues

Sadi Irmak and Nureddin Onur—in fact, any work using the classification schemes of the Hirszfelds, Reuben Ottenberg, and Laurence Snyder—had to be rejected.[95] Although her dissertation was published only in Turkish with an English summary, it attracted some favorable attention abroad.[96] After the publication of her thesis, Aygen retrained as a social anthropologist (publishing later work under her married name, Erdentuğ), following a common shift in postwar anthropology away from studies of racial classification in favor of social and cultural approaches to human difference. Subsequently, Turkish physical anthropologists abandoned the study of blood groups for nearly three decades in favor of anthropometric methods. Worldwide, postwar human blood group research was initially dominated by physicians within an emergent paradigm of medical genetics research, which primarily aimed to trace the inheritance of disease rather than define racial characteristics.

Meanwhile, the end of the Second World War heralded the decline of eugenics and racial classification as dominant ideological paradigms in both politics and science. During the interwar period, a vocal minority of anthropologists and biologists had already begun to deconstruct prevalent race theories in their individual research.[97] With the rise of the Nazis in Germany, many American and British scientists and several of their professional organizations signed manifestos that publicly refuted and condemned Nazi Aryanism.[98] Regardless, only after the war and the formation of the United Nations did the reformation of race science become a concerted effort under the auspices of the United Nations Educational, Scientific and Cultural Organization (UNESCO, founded in 1945). UNESCO's first president, the evolutionary biologist Julian S. Huxley, had advocated since the 1930s for scientists to abandon the term "race" in favor of allegedly more accurate terms like "ethnic group." At last, in his new position, he had the power to realize his dream of organizing an "international inquiry ... which would result in an impartial scientific pronouncement on the subject [of race]."[99] However, the UNESCO Statements on Race, issued in 1950 and 1951—although now often interpreted as scientists' declaration of the nonexistence of biological races—in fact merely condemned the use of race as a basis for social policy and ideology, rather than its use within scientific circles.[100] According to the committees of sociologists, anthropologists, and geneticists convened by UNESCO, scientific studies of race could not be held accountable for the great human tragedies of the early twentieth century, ranging from eugenic sterilization programs to the Nazi Holocaust. Rather, they claimed, these events were caused by the ideological misuse or abuse of legitimate, objective science.

Regardless, "race" as a scientific term had become so fraught that many biologists did increasingly employ terms like "ethnic group" or the particularly neutral-sounding "population." Additionally, the term "eugenics" was purged from the names of professional journals and societies, as in the 1954 name change of the journal *Annals of Eugenics* to *Annals of Human Genetics*. Such nominal modifications seemed to signal the end of the ideologically biased approaches of racial sero-anthropology and the beginning of a purely objective human population genetics, even though there was no major transformation in many of the field's operative assumptions. While blood-based genetics was supposedly divorced from its initial race science applications through this transition, it never actually lost its relevance to nationalist politics. Middle Eastern geneticists had already internalized nationalist origin narratives and accepted the sociopolitical need to subsume ethnic, linguistic, and religious differences into homogeneous national citizenries. In fact, the debates about the proper meaning of "race" as a genetic category, as well as the emergence of international scientific organizations like UNESCO, soon made it possible to transform Middle Eastern political and social identities into standard biological categories accepted by geneticists around the world.

Part II
MEDICINE AS ANTHROPOLOGY

Chapter 3

THE TRAFFIC IN BLOOD

IN THE SUMMER OF 1956, AMBITIOUS ENGLISH UNDERGRADUATE Douglas Botting mustered all the prestige of the Oxford University Exploration Club to mount an expedition to Socotra, an island in the Arabian Sea and a remote frontier of Britain's Aden Protectorate. According to Botting, "one of the most important aspects" of their expedition was the collection of blood samples "from any bedu the doctors could lay their hands on," for the purpose of determining the "racial origins and affinities" of the little-known indigenous people living in Socotra's isolated, mountainous hinterlands.¹ Neil Orr and Richard Lister, two newly minted doctors from London, took charge of the "blood-sucking" work, which was beset at every stage with difficulties. Firstly, Botting recounted, the Socotrans had to be persuaded to allow "the odd and fanatic Englishmen" who traversed the countryside or pulled them into the small clinic at Hadibo town to prod them with venule needles (see Figure 6). The students regularly resorted to bribes of money and biscuits, as well as endorsing the "medieval idea" that bloodletting improved health; "the arguments were unethical," Botting admitted, "but time was short."²

    The blood had to be stored immediately in a portable refrigerator that the expedition team had brought from England to the Hadibo clinic. In order to get the blood off the island, the team assembled the sample tubes in boxes full of ice. They rowed the boxes in canoes to Socotra's lone airstrip,

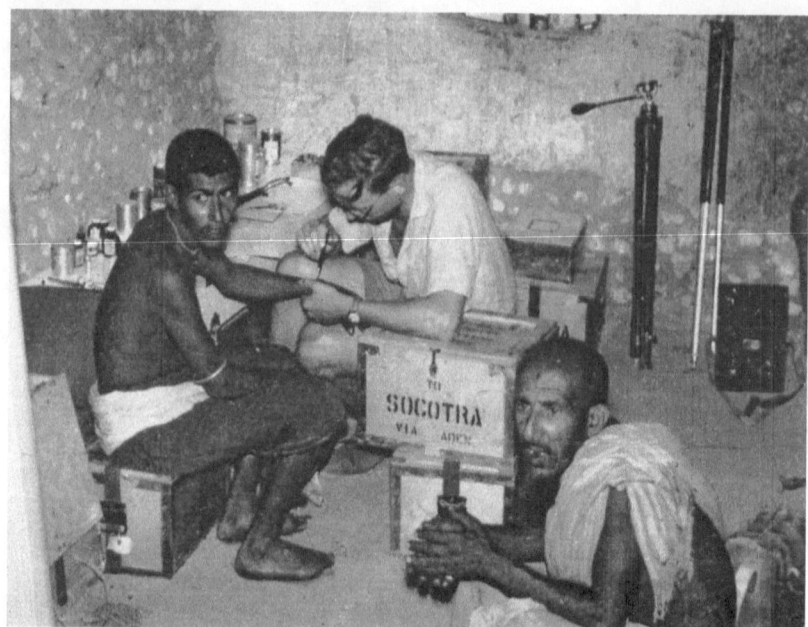

FIGURE 6. Richard Lister treating indigenous Socotrans at the Oxford expedition's clinic in Hadibo, 1956. Photograph by Douglas Botting. Reproduced with permission from the Botting family.

where they had arranged with the Royal Air Force (RAF) to fly them back to Aden. Upon reaching Aden (via the RAF base at Riyan, near Mukalla), Orr, Lister and Botting packed the chilled blood as an air-freight shipment to London. Here, at the end of a journey of more than 4,000 miles, 100 Socotran blood samples reached the hands of Elizabeth W. Ikin and Arthur E. Mourant of the Blood Group Reference Laboratory (BGRL). Ikin, an expert technician, tested the Socotran blood not only for the ABO antigens but also for those of five other blood group systems, mostly discovered after World War II. Mourant, the laboratory's director, had expected to see an "African blood picture" among the Socotrans, because the island was relatively closer to Somalia than Arabia.[3] However, based on Ikin's results, he determined that not only were the Socotrans more "Asiatic" than he anticipated, they also had "a considerably lower degree of African ancestry than their continental Arabian neighbours"[4] and likely represented "an ancient Arab isolate, somewhat modified by genetic drift."[5]

The tortuous path through which blood left the veins of indigenous

Socotrans and came to rest in a London laboratory, where its genetic characteristics were then interpreted for anglophone scientific audiences, represents a typical mode of postwar human genetic research that both departed from and continued certain interwar practices of sero-anthropology. Firstly, the stated purposes of this study, with its explicit invocations of racial identity and group origins, are indistinguishable from those of research in prior decades. The United Nations Educational, Scientific, and Cultural Organization (UNESCO) statements on race, particularly the 1951 version (which Mourant was involved in drafting), did not actually advocate for the total erasure of race from scientific discourse. In fact, these statements enshrined three "major races" of white/European/Caucasoid, black/African/Negroid, and yellow/Asian/Mongoloid as legitimate classificatory labels.[6] Furthermore, although the Hirszfelds' biochemical/racial index had now been thoroughly discarded, dedicated sero-anthropologists like William C. Boyd continued to publish new racial classification schemes based on serology until the early 1960s.[7] These conceptual continuities underlie Mourant's depiction of Socotran ancestry in terms of relative proportions of "African" and "Asiatic" blood.

Secondly, the basic modus operandi of Botting's expedition relied on familiar colonial institutions and networks, like British military bases and medical facilities and the social cachet of the Oxford club. Blood-collecting adventures conducted by British, American, French, and other Western researchers in the Middle East continued to make use of the space, equipment, and staff of local hospitals and clinics run by Christian missionaries and oil companies. Extracting blood from exotic research subjects like the indigenous Socotrans, while done under the auspices and guise of healthcare provision, frequently gave way to bald market transactions of food or cash. The cajolery involved in getting such research subjects to part with their bodily fluids further reinforced researchers' perceptions of these peoples as having primitive, superstitious, and irrational mindsets. Accordingly, scientists tended to express skepticism about the oral histories of the communities they studied, portraying such histories as unreliable self-identifications that concealed the underlying truth about the ancestral origins of tribal, religious, or national groups.

There were, of course, much more convenient sources of Middle Eastern blood. In less remote areas, government hospital patients, blood bank donors, and soldiers conscripted into national military services offered suitably large, random samples of populations at local and statewide levels. These remained the preferred research subjects of the Middle Eastern physicians who contributed to sero-anthropological studies and whose development of local laboratory facilities had made possible the interwar work of roving Westerners like William and Lyle Boyd. However, World War II radically transformed the landscape of hematology by spurring the use of new methods for blood storage and transport, the discovery of new hereditary traits in blood and serum, and the emergence of an international health infrastructure to regulate and coordinate blood research. These changes enabled and even forced blood collected from remote and isolated areas like Socotra to bypass nearby laboratories in Aden, Egypt, and the Levant in favor of distant metropole "centers of calculation" like Mourant's BGRL.[8]

It was these technological and infrastructural changes, more than any shift in the conceptualization of human biological variation, that differentiated interwar racial sero-anthropology from postwar human population genetics. Wartime medical emergencies meant the need for blood transfusions skyrocketed, prompting the growth of local blood banks to collect, test, and regulate blood supplies on the home front.[9] To store and transport donated blood, existing refrigeration technologies were appropriated for medical purposes, creating "cold chains" to supply chilled blood to soldiers stationed in remote theaters of war like the Pacific Islands. After the war, Botting's expedition, as well as dozens of others, relied on similar military and medical channels of iced air freight to carry blood in the reverse direction—not for transfusion but for sero-anthropology.[10]

Increased investment in blood research during the war also drove a rapid succession of discoveries of heritable blood group and serum proteins, which greatly expanded the number of genetic markers applied to anthropological questions. For example, the 1940 discovery of the Rhesus (Rh) antigen system, which, like the ABO system, is clinically significant for blood transfusion, was followed within a few years by studies of the racial variations of its genotype frequencies. Although all populations worldwide have a majority of Rh-positive individuals, one particular chromosome type conferring the Rh-positive phenotype—known as Rh_o or cDe, depending on one's preferred nomenclature—was found to be ten to twenty times more common in African Americans than white Americans. One of the discoverers of the Rh system, Alexander S. Wiener, accordingly suggested that "tests for the Rh blood types might be used as an index of the purity of a Negro population."[11] Mourant, who had worked on the Rh system as a member of the Galton Serum Laboratory in Cambridge during the war, quickly adopted this suggestion for his anthropological work at the BGRL. Mourant concluded that the indigenous Socotrans had many fewer African ancestors than mainland Arabs, based primarily on the former group's much lower proportion of *cDe*-carrying individuals.

Before long, the ABO and Rh systems alone were considered insufficient to accurately determine the ancestral origins of any given population. Serious researchers also had to mine their blood samples for the P, Lewis, Duffy, Lutheran, Kell, and Diego antigens as well as for rare varieties of hemoglobin, haptoglobin, transferrin, and immunoglobulins.[12] Interwar seroanthropologists, like the Boyds, had considerable success working in remote locations because ABO antigens could be detected with simple agglutination tests amenable to field conditions, using antisera readily available from local hospitals or blood banks. However, because many newly discovered antigens were not relevant to blood transfusion or other routine clinical procedures, the necessary antisera and other reagents were much harder to come by. The detection of variant blood proteins, like sickle-cell hemoglobin, required specialized and sometimes expensive laboratory equipment, namely apparatuses for chromatography or electrophoresis. In many cases, these factors forced field workers to preserve and ship their samples to large, centralized laboratories and serum banks that specialized in such analyses and could store the samples for anticipated future testing.[13]

The World Health Organization (WHO), founded in 1948, was responsible for authorizing the status of such labs and serum banks, designating them as "international reference centers" with a broad set of tasks. The directors of these centers drafted the protocols for how human blood should be collected, transported, stored, and analyzed, and center staff produced the standardized reagents to be used in field and laboratory testing, trained researchers in protocol techniques, and performed analyses for researchers who did not have the equipment or facilities to do so themselves.[14] These centers, located overwhelmingly in Europe and North America, were sometimes generated anew, but more often they were reorganized out of existing "national reference" labs developed for "utilitarian wartime purposes," like Mourant's BGRL, which was recognized by the WHO in 1952.[15]

The WHO and other international agencies with missions to monitor, regulate, and coordinate global health and scientific research (e.g., UNESCO and ICSU, the International Council of Scientific Unions) promoted visions of global cooperation rooted in a form of universalist humanism embodied in medical and scientific progress. However, this veneer of universalism obscured these organizations' reinforcement of a Cold War geopolitical order that configured scientists from regions on the receiving ends of international development schemes as "lacking expertise" in laboratory methods and therefore in need of tutelage from Western scientists.[16] Mourant was thus a major beneficiary of the international authority conferred on his laboratory by the WHO. Upon being appointed director of the laboratory in 1946, he sought to establish himself as an expert in "the anthropological application of blood group surveys," feeling that this would finally distinguish his research activities from being a "pale reflection" of those of his mentor, Robert R. Race.[17] With this motivation, Mourant spent five years updating Boyd's 1939 compilation of worldwide blood group data. The resulting book, *The Distribution of the Human Blood Groups*, debuted in 1954 to much acclaim. Reviews in leading anthropology journals pronounced the tome "indispensable" for the study of human races and national populations.[18] The text was not only, to date, the largest collection of blood group data compiled, synthesizing 1,716 reports on nearly half a million people, but also an agenda-setting document that placed Mourant at the center of an enormous global web of scientific correspondents. Over the next two decades, these correspondents supplied his laboratory with thousands of blood samples from peoples he had never heard of living in countries he had never visited.

This chapter examines the analytical effects of this international "traffic in blood"—my term for a new regime of exchange and knowledge-control among scientists that reflected the increasing division of labor between the collection of samples from individual bodies in the field or clinic and the work of testing these disembodied samples in the laboratory. Just as in the interwar period, field-workers engaged in transactional relationships with research subjects, trading blood and other body parts for money or material goods. But now that these hard-won samples could be preserved, transported, and shared with other researchers, they took on an additional value for cultivating professional relationships in the sciences. Clinical and blood bank workers, too, had things of value to share with metropole reference laboratories: blood samples from isolated populations or patients with rare blood types, as well as massive data sets of blood group frequencies. In exchange for these valued gifts, prominent scientists who managed reference laboratories offered reagents, equipment, medication, advice, coauthorship or published acknowledgment, and access to other scientific contacts and funding.[19]

I explore the logistical, institutional, and social infrastructures that made this scientific trafficking possible, as well as the relationships between the scientific actors at either end of the "cold chains" that funneled Middle Eastern blood across the region and the world. Hematologists and other medical practitioners in Egypt, Iraq, Saudi Arabia, Israel, and Iran all became embedded in a transnational project to share data and specimens with European and American scientists and laboratories. This project took the form of numerous informal and ad hoc collaborative relationships: a knowledge-control regime in which Western-based figures like Mourant took the lion's share of credit for interpreting and publishing genetic accounts of the racial ancestry and historical origins of national groups. Yet while Mourant tended to substantially marginalize the role of field-workers and other Middle Eastern medical institutions in his publications, he was utterly dependent on their collaboration not only to obtain blood samples from far-flung communities but also to imagine anonymized vials of fluid as recognizable and cohesive human groups.[20] The invaluable social and historical knowledge conveyed by Middle Eastern scientific collaborators, of course, was itself constrained by the concurrent reimagining of ethnic, religious, and national identities intrinsic to the process of building new nation-states.[21] I trace how the postwar traffic in blood brought together semi-colonial medical infrastructures and professional networks with a range of nationalist institutions and ideologies and show how the logistics of these interactions manifested in the production of genetic information.

CREATING BLOOD SUPPLY CHAINS
Egypt and Iraq: Colonial Legacies

The earliest anthropological ventures of Mourant's Blood Group Reference Laboratory emphasized the determination of Rh blood type frequencies to complement the existing data compilations on ABO and MN groups produced by interwar sero-anthropology. Outside Europe, Mourant's first samples trickled in from populations accessible through British colonial institutions and networks, including those based in nominally independent Middle Eastern states like Egypt and Iraq, which had been occupied by the British military during the war. Mourant, like most sero-anthropologists before him, had a particular interest in such "antique" populations as Egyptian Copts, and he seized upon any rare opportunity to have blood from those communities directly shipped to his laboratory. At the same time, the BGRL provided antisera and serological training to the staff of Egypt's State Serum Institute (in Abbassiya, Cairo) and Iraq's Medico-Legal Institute (in Baghdad), who in turn provided Mourant with data on the typical "mixed" Arab majorities in their countries. For example, in the late 1940s, Ahmed I. Kayssi, the director of the Medico-Legal Institute and a friend of the Boyds', traveled to London to learn the technique for Rh-typing. After returning to Baghdad with four varieties of BGRL antisera in tow, he tested 300 residents of the city and dutifully reported the resulting frequencies to Mourant.[22]

Around the same time, the BGRL hosted Karima A. Ibrahim, who worked for the Egyptian State Serum Institute. Ibrahim collected blood samples from more than eighty Egyptians resident in London, presumably as a training exercise that also suited Mourant's interests. The results she obtained were statistically combined with those of "20 Moslems and 20 Coptic Christians of Egypt, from specimens of blood kindly sent through the courtesy of Brigadier H. T. Findlay" to the laboratory of Robert Race in 1947.[23] Finally, Mourant received shipments of blood collected by one Professor J. H. Fisher from 62 Copts in Upper Egypt, who had been "specially selected from the anthropological point of view"—that is, rigorously questioned to ensure that none of their ancestors had non-Egyptian origins.[24] In the analysis of the ABO, MN, and Rh frequencies of this agglomeration of samples, Mourant and his collaborators found no significant differences between Coptic and Muslim Egyptians. Taken together, however, Egyptians' Rh group distribution differed greatly from all populations studied to that date, except for a "moderate degree of resemblance" to the Baghdadis examined by Kayssi.[25] In June 1950, Mourant traveled to the annual Symposium on Quantitative Biology at the Cold Spring Harbor Laboratory, in New York, to present his research at a panel on the "genetic analysis of racial traits" alongside his mentor Race and colleague Boyd. Here, in his discussion of the "blood groups of the peoples of the Mediterranean area," he concluded that the primary genetic difference of Egyptians and Iraqis from Southern Europeans (namely, the former's higher proportion of *cDe*) could be attributed to "negro admixture," which was "easily explained by the progressive assimilation of negro slaves which is known to have taken place" in those countries.[26]

Mourant's blood supply chain from Egypt operated parallel to, not in collaboration with, local researchers who had ready access to large numbers of specimens: Egyptian military physicians. Indeed, even as Mourant painstakingly cobbled together a data set amounting to fewer than 200 specimens, Egyptian army laboratories had amassed ABO frequencies from thousands of Egyptian soldiers since 1941, when the routine typing of recruits for transfusion purposes had begun.[27] By the end of the decade, Salah El-Dewi of the Koubba Blood Transfusion Centre began producing some of his own Rh antisera (importing other varieties from the United States) and started typing hundreds more soldiers for the Rh system.[28] The Egyptian hematologists at these labs were well aware of the potential anthropological import of their work. Like Mourant and previous sero-anthropologists, they invoked the question of genetic difference between Copts and Muslims, taking for granted that Copts "descend mainly from ancient Egyptians," while Muslims "descended from Asiatic Arabs and Copts."[29]

Yet whereas Mourant enthusiastically linked Rh frequencies in a small, haphazard assemblage of specimens to Egyptian history and anthropology, his Egyptian counterparts, with their thousands of samples, consistently challenged the validity of such claims. They justified their skepticism through statistical tests showing the enormous effect of sampling errors on ABO frequencies, contending that any sample size below 3,000 individuals rendered "quite useless" results.[30] In turn, they quickly discounted their own finding that Copts and Muslims might have significantly different ABO frequencies on the basis that they had not tested a sufficient number of Copts. As for the Rh system, they (like Mourant) had not found a significant difference between Copts and Muslims. However, as for the anthropological relevance of this, or the elevated incidence of the "negroid" *cDe* chromosome, they insisted that nothing could be confirmed before thousands more Egyptians were tested.[31]

Mourant responded negatively to the Egyptians' research in his 1954 *Distribution of the Human Blood Groups*, accusing their results of "inconsistency" and suggesting that their antisera may have been "unreliable" and "giving some false positives."[32] In this way, Mourant asserted the epistemological superiority of the results found by his own laboratory, reinforced by its WHO recognition, and downplayed the importance of statistical errors necessarily introduced by the much smaller sample sizes of Egyptian blood he could import to London. Therefore, the traffic in blood from former colonies in the Middle East and other "third world" locales to metropole labs was never caused solely by a lack of adequate facilities to carry out genetic tests. Rather, it represented the desire of Western scientists to control the creation and production of genetic knowledge in a manner that legitimized their professional standing in the international sphere.

Saudi Arabia: Corporate Genetics

Not all of the Middle Eastern blood samples that arrived in Mourant's laboratory came from the lingering threads of British colonial networks. In fact, at the time that the BGRL received the hard-won specimens from Douglas Botting's Socotra expedition, its largest single supplier of Arab blood was the Arabian American Oil Company (Aramco), based in Dhahran, near the eastern coastline of Saudi Arabia. Late in 1955, George Maranjian, a Harvard-trained Armenian-American physical anthropologist working for Aramco, began collecting blood samples from the company's Saudi workforce and shipping them to London. Although it is unclear who initiated the collaboration, most of the scientific recognition accrued to the British end. By the end of 1957, Elizabeth Ikin was processing 1,384 specimens that Maranjian had collected from Aramco's health clinics, which became the backbone of her doctoral thesis on the blood groups of Mediterranean and Middle Eastern populations. Mourant took charge of publishing the results in *Nature* and *Human Biology*. At some point during the early 1960s, Mourant sent Maranjian a draft of the latter paper for his approval but received no response. As he later recounted to an Aramco physician, Armand Phillip Gelpi, "After several unsuccessful attempts to get into touch with [Maranjian], I published the paper with his name on it, as he had done much of the work, both collecting and transmitting the blood samples, supplying very detailed information on birthplaces and carrying out much of the statistical work."[33]

Other than these papers published on his behalf, Maranjian left little trace of his nearly twenty-five years of work for Aramco in any academic journals. Regardless, as Mourant acknowledged, the BGRL staff relied heavily on Maranjian's guidance, as well as Aramco's corporate infrastructure, to obtain and interpret Arab blood. This infrastructure was distinct from—sometimes even markedly at odds with—the social policies of the Saudi government, reflecting Aramco's status as a colonialesque enclave that operated its own public health initiatives and anthropological research department like a state within a state. For example, after massive outbreaks of malaria in the oases of al-Hasa and Qatif, from where Aramco drew the majority of its native employees, Aramco began an intense malaria eradication campaign in 1948. On paper, the Saudi government contracted Aramco to provide the necessary services of DDT spraying and parasite infection surveys, and Aramco staff merely performed contractual duties (to the point of halting DDT spraying whenever the government fell behind in paying its bills). On the ground, Aramco's early malaria surveys acclimatized the oasis villagers to the presence of foreign researchers who seemed to have "no particular therapeutic role" yet collected drops of their blood (initially in exchange for Fig Newton cookies). Over the next decade, the villagers became generally cooperative with regard to blood extractions in clinical settings, enabling Maranjian to collect a large number of samples to ship to London.³⁴

However, Aramco's anthropological "intelligence-gathering" rather than its medical services demonstrated Aramco's autonomy from the Saudi state. In the early 1950s, Federico Vidal, a social anthropologist employed by Aramco's Arabian Research Division, conducted ethnographic fieldwork to study the large Shi'i population in the al-Hasa oasis. For a number of reasons ultimately related to the ruling family's legitimacy, official Saudi policy ignored the Shi'is' existence, refusing to acknowledge their presence within the country in any government records. In contrast, Vidal wrote a comprehensive report that detailed economic and social relations between Shi'is and Sunnis in al-Hasa's towns and agricultural production. He noted that Shi'is faced discrimination in land tenure and everyday relations, working primarily as agricultural labor for Sunni landowners and often choosing to live in their own sectarian villages rather than in mixed towns. Meanwhile, Vidal pointed out, due to the Saudi government's settlement of Sunni Bedouin tribes in al-Hasa, the proportion of the Sunni population in the oasis was growing at the expense of Shi'i land and wages.³⁵

Although Aramco's ability to collect such "politically sensitive" sociological information had little effect on its technocratic approaches to environmental and labor issues, it unwittingly provided the analytical framework for interpreting the biological relationships of its Arab employees. The tubes of blood Maranjian had shipped to London were labeled only with anonymous numbers, whereas Mourant was eager for him to supply "a breakdown into tribes," which would allow the BGRL to compare the Saudi data to other Arab groups tested in the facility, like Yemenis and Socotrans.[36] Because Maranjian had not recorded "tribal" identities from patients at the time he collected the blood, he turned to Aramco's personnel records, from which he "extracted" details of each sampled employee's background—namely, his birthplace and religious affiliation as a Sunni or Shi'i Muslim.[37] Accordingly, the "breakdown" Maranjian provided to Mourant and Ikin divided the samples into 11 population groups, separating Shi'i, Sunni, and Bedouin "classes" into province of birth, with the Eastern Province further subdivided into the oases of Qatif and al-Hasa. However, as the sample sizes of these categories ranged from 22 to 394 individuals, they were not suitable for the routine statistical analyses the BGRL used to interpret genetic data. In her thesis, Ikin explains how she collapsed the 11 categories into 5 larger groups, in "consultation with Dr. Maranjian," who confirmed that the new arrangement still "agree[d] with the known anthropology of the country."[38] Under the combined groupings, the Shi'is and Bedouins were each treated as single groups, with Sunnis divided into 3 broad regions of Eastern, central (Najd), and western (Hijaz, Asir, Najran) provinces.

The categorical breakdown and related information that Maranjian provided to the BGRL clearly incorporated Vidal's insights. To justify the separation of data along religious lines, Ikin explained, "Where both Shia and Sunni live side by side, intermarriage does not usually occur, so that each sect retains its individuality."[39] The ultimate effect of this sectarian categorization was to highlight the biological difference of the Shi'is. On the basis of her blood tests, which involved twenty antigens from eleven blood group systems, Ikin suggested that the Shi'i communities had experienced more African admixture than the Sunnis, with "the Bedouin least affected by such an admixture."[40] The most remarkable genetic distinction the BGRL found among the Shi'i, Sunni, and Bedouin communities was their respective frequencies of an inherited blood condition called sickle cell disease, which had recently been discovered to provide adaptive protection against malaria (see Chapter 4). Although nearly a quarter of Shi'is carried the gene for the disease, it was much less common among Sunnis and completely absent among the Bedouin. Mourant and his colleague Hermann Lehmann, a specialist in sickle cell disease, argued that these frequency differences confirmed the basic demographic patterns observed by Vidal: the Shiʿis had more of the disease genes because many more generations of their ancestors had lived in highly malarious oases. This effect of natural selection had had less time to manifest in the Sunnis or Bedouins because they had more recently arrived at these oases from malaria-free parts of the country.[41] The role of the Saudi state in driving this migration, however, is not discussed. This omission highlights the outsize role of the oil company in providing the source of research subjects, the means and equipment to collect their blood, and the information about individual and community identities. At every stage of the research, representatives of Aramco, and not the state, guided the production of genetic knowledge about Arabs in the Saudi kingdom.

BLOOD BANKS AS LOCAL CENTERS OF CALCULATION

The traffic in blood involved more than the literal circulation of blood samples between the Middle East and the West. For example, although Arthur Mourant corresponded with the directors of blood banks in Israel and Iran, they rarely shipped samples to London. Instead, Mourant provided rare antisera to the Hadassah blood bank director in Israel, who then shared his results from Mizrahi Jewish immigrants on paper to benefit from the skills of Mourant's resident statistician, Ada Kopeć. Meanwhile, the case of Iran shows the limitations of Mourant's efforts to monopolize the Middle Eastern blood traffic. Iranian hematologists shared their blood group data with Mourant, but the blood and serum samples that left the country long evaded his laboratory, traveling instead to Paris and Amsterdam. In this section, I examine how Israeli and Iranian blood banks operated as local centers of calculation with distinctly nationalist agendas for genetic research. The ethnic narratives outlined in their scientific publications are explicit products of volatile social transformations and territorial disputes. Yet the international discourse on human genetics, as funneled through Mourant and other Western geneticists, stripped away the national politics of these institutional contexts. The exchange of blood through paper inscriptions, just as much as through chilled liquid in vials, normalized the use of local population categories on a global scale, thereby transforming politicized processes of identity formation into allegedly objective biological realities.

Israel and the Management of Mizrahi Immigrants

The first local research group to investigate the blood types and hereditary diseases of the immigrants who flooded into Israel after 1948 formed under the direction of physician Joseph Gurevitch at the Hadassah Hospital in Jerusalem. Born in Minsk, Gurevitch first joined the Hadassah staff in 1930 and became the head of the hospital's department of bacteriology and serology in 1943. When the Hadassah blood bank was established in 1944, Gurevitch was appointed its director.[42] Like Rina Younovitch, Gurevitch made a professional name for himself through work on "non-Ashkenazi" Jews. But while at the time of Younovitch's publications the "Asiatic" Jews numbered only a few hundred in mandate Palestine, when Gurevitch conducted his work, immigrants from the Middle East and North Africa had suddenly become the majority of Israel's Jewish population. Between 1949 and 1951, the new state's population effectively doubled. The rapid arrival of more than 600,000 Jewish immigrants, mostly from the Balkans and the Middle East (including essentially the entire Jewish populations of Bulgaria, Yemen, Libya, and Iraq), "required an absorption effort practically unequaled in recent world demographic history."[43] All immigrants were quarantined upon arrival and subjected to medical examinations, involving X-rays and blood tests, before being released to transit camps (*ma'abarot*) managed by a combination of Israeli government and Jewish Agency personnel.[44]

The Yishuv-era prejudice against Mizrahi immigrants rapidly intensified. Whereas European and Balkan immigrants were prioritized for quick transfers into permanent housing, Mizrahi families frequently wallowed for years in the crowded, poorly equipped camps. Many were eventually settled into "development towns" (*'ayarot pituaḥ*) along the remote borders of the state, assigned to factory work or other manual labor, due to the belief of Ashkenazi immigration officials that the "Orientals" as a whole were backward, uneducated, and incapable of holding white-collar jobs.[45] According to the ideological plan of "blending the exiles" (*mizug galuyot*), one of the primary mechanisms through which Mizrahim were to be culturally and biologically absorbed as Israelis was through intermarriage with Ashkenazim. Yet, in practice, Ashkenazim rejected the notion that their children should marry "those black Jews."[46] Mizrahi communities differed vastly in their socioeconomic, cultural, and religious characteristics according to their countries and cities of origin. However, the discriminatory Israeli settlement process elided such differences such that Mizrahim came to form a "sociological block" contrasted against Ashkenazim.[47] Ashkenazi journalists repeatedly claimed that they were unable to tell the difference either between different Mizrahi communities or between Mizrahi Jews and Palestinian Arabs.[48] Throughout the early 1950s, articles in the daily *Haaretz* encouraged the adoption and enforcement of medical selection policies tailored to reduce and delay the immigration of Mizrahi communities, especially those from North Africa, India, and Iran.[49]

Within this context of medical scrutiny, Gurevitch and his Ashkenazi colleagues published on the blood group frequencies and incidence of hereditary blood diseases among the immigrant groups who required their care. Before the post-statehood mass immigrations, Gurevitch had already taken an interest in the different frequencies of ABO and Rh blood groups within different Jewish communities. In 1947, he divided a group of nearly 2,000 Hadassah patients into five categories: Ashkenazim, Sephardim (groups from North Africa and the Balkans), Kurds and Persians, Yemenites, and "combined minor groups of oriental origin (Georgians, Babylonians, Assyrians, and others)."[50] His first post-statehood publication maintained the categorical divisions of his 1947 study, but he soon turned his attention to the blood groups of individual "Oriental" communities, regularly citing Younovitch's work and comparing her 1933 data to his own.[51] In fact, the overall formula of his publications between 1951 and 1960 tended to match hers. He described each study population with a historical narrative, often drawn from articles of the Jewish Encyclopedia, the *Jerusalem Post*, and even the Old Testament, that traced the group's original dispersion from the rest of the Jewish people. Unlike Younovitch, who explicitly interpreted her data as evidence against many of the Mizrahi communities' own narratives of endogamy, Gurevitch and his colleagues simply presented the data without any interpretation as to which versions of history might be (in)validated by the results. Generally, however, Gurevitch's assessments of the relative "purity of stock" or "anthropological purity" of each community were similar to those of Younovitch; he disagreed with her only regarding the level of admixture among the Yemenites, whom he considered less admixed than she did.

Both the historian Nurit Kirsh and the anthropologist Nadia Abu El-Haj have closely analyzed Gurevitch's publications and the assumptions underlying his division of population categories, selection of community narratives, and presentation of comparative results. In particular, they note how his work constituted the "Ashkenazim" and "Sephardim" as singular units representing all of Northern and Southern Europe, as opposed to the "Oriental communities" analyzed by individual country of origin.[52] They attribute this inconsistency to Gurevitch's "internalization" of Zionist ideology as well as to broader Israeli social politics in the first decade of mass immigration. But as Chapter 2 shows, there is nothing novel about his representation of Jewish origins or population boundaries; the roots of these patterns lie in the Yishuv period. Instead, Gurevitch's work is remarkable for mobilizing local and transnational medical infrastructures capable of reinforcing and exporting these Zionist notions of Jewish biology.

Locally, Gurevitch's sampling procedures drew on his authority at Hadassah to access the network of local health clinics hastily assembled to serve the flood of new immigrants. Unlike other Israeli physicians who occasionally published similar blood group studies,[53] he and his colleagues did not simply compile samples from hospital patients or blood donors, but also selectively gathered data from the transit camps and development towns where the communities of his interest had been collectively settled (see Figure 7). For example, in 1950, he requested blood samples from the Hadassah children's hospital branch serving Yemenites living at the Rosh Ha'ayin *ma'abarah* (which he and his coauthors euphemistically termed "a village," although the inhabitants were still dwelling in tents).[54] With the assistance of fellow Hadassah physicians Emmanuel Margolis and David Hermoni, Gurevitch collected blood belonging to Jewish immigrants from Cochin, Iran, Morocco, and Tunisia, transporting the samples on ice back to Jerusalem, where they were stored in the blood bank's refrigerators. One exotic community, the former residents of Gebbel Gefren (Libya), who had lived a "primitive life in caves," motivated the doctors to drive "several hours" into the Negev Desert to visit the "villages" where they had been settled.[55]

This type of fieldwork, enabled by the discriminatory practices of Israeli immigrant absorption officials, made Gurevitch indispensable to Western-based researchers, including Arthur Mourant, who had direct access only to Ashkenazi or smaller Sephardi communities. Although Gurevitch sourced most of his Rh antisera from commercial producers in the United States (the home base of the Hadassah organization), he requested the rarer varieties, relevant less for medicine and more for anthropology, from Mourant's

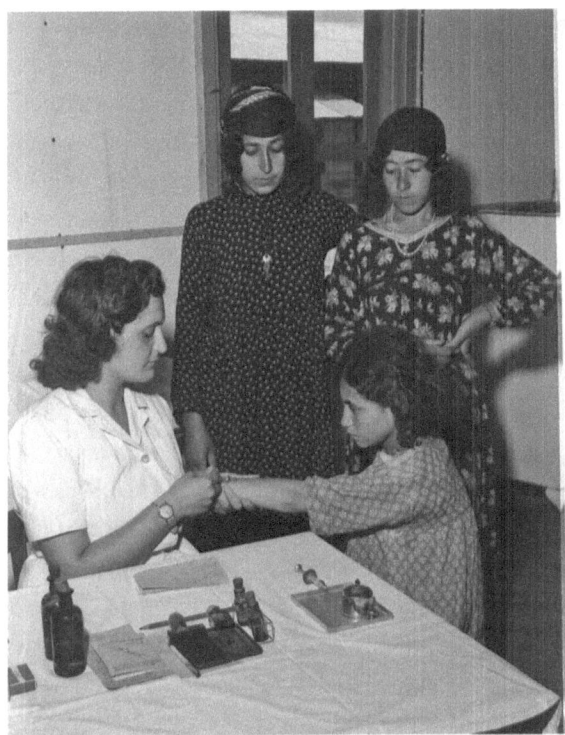

FIGURE 7. A nurse at the Atlit *ma'abarah* administering a tuberculosis vaccine to Nadia Bet Salim, a thirteen-year-old immigrant (likely from Iraq), while her aunts look on, June 1951. Photograph by Teddy Brauner. Courtesy of the Government Press Office, Israel National Photo Collection.

BGRL in London. In certain cases—for example, when Hadassah researchers believed they had found sickle cell disease among the Yemenites at Rosh Ha'ayin (see Chapter 4)—they shipped blood samples directly to London for further investigation. More importantly, Gurevitch began to share unpublished data with Mourant, enabling the latter to incorporate new information on Kurdistani and Baghdadi Jews in his 1954 book.[56] The scientific exchange between Hadassah and the BGRL is a crucial reminder that Gurevitch wrote for a transnational audience of geneticists interested in exotic, socially isolated groups, including a broad range of Jewish communities.[57] His research practices must thus be contextualized in these terms, while still acknowledging that he took advantage of his position as an Ashkenazi physician in Israel for professional gain.

Both Kirsh and Abu El-Haj pose Mourant as a foil to Gurevitch and other contemporary Israeli geneticists, showing how the latter downplayed the potential genetic evidence of Jewish proselytization and Jewish-Gentile intermarriages throughout history and portrayed immigrants to Israel as representative samples of all Jewish communities. They contrast these underlying Zionist assumptions with Mourant's explicit consideration of these factors in his reviews of Jewish genetic data.[58] However, Mourant was not more objective or neutral in his interpretations of Jewish genetics. As seen above from his attitude toward the Egyptian blood banks, his concerns about representative sampling were not consistently applied to his own work. Rather, the institutional frameworks that drove Israeli data into British hands affected publication trends in both countries and ultimately reinforced Zionist notions of Jewish common origins in the Eastern Mediterranean. Gurevitch and his colleagues began routinely citing Jewish data from other countries compiled by Mourant, using his interpretations about the geographical origins of Rh chromosomes and the typical frequencies of a "Mediterranean" population. In 1958, they shifted away from investigating individual "Oriental" communities in favor of two studies on composite groups, involving immigrants from multiple countries who had arrived between 1949 and 1957: the Sephardim of the Balkans (mainly Bulgaria and Turkey) and the Ashkenazim of Europe, "the descendants of the German and French Jews."[59] These aggregate studies had greater comparative value for the data in Mourant's compilation, and the Israelis sent their blood-typing results to be analyzed by Mourant's statistician, Ada Kopeć, at the Nuffield Blood Centre. It was only *after* this engagement with Mourant and his staff that Gurevitch and his Israeli colleagues published their most explicit assertions about the "common Mediterranean origin of the Jewish people."[60]

Mourant tended to be more forthcoming in his publications about caveats surrounding the interpretation of Jewish genetics, but he ultimately promoted the Israeli data as authoritative. In 1959, the managing editor of the *Jewish Journal of Sociology* requested that Mourant write a review of all Jewish genetic research to that date. In that review, Mourant diagrammed a detailed breakdown of European Ashkenazi populations, revealing a broad range of ABO frequencies—certainly as broad as the range he calculated for the "Jews of Asia"—which he attributed to different levels of intermarriage with European non-Jews in certain communities.[61] Yet in his final conclusions, Mourant gave a greater weight to the Rh system, where the available data (all from Gurevitch and a Canadian research team) treated Ashkenazim and Sephardim as aggregate populations. Although he acknowledged that the Canadian Jews tested were "scarcely distinguishable from their former European non-Jewish neighbours by their ABO groups," he claimed that their Rh distribution showed that "physically they are more nearly related to their Mediterranean ancestors than to these European neighbours."[62] Mourant's heavy reliance on Gurevitch's data—and implicit endorsement of Gurevitch's population categories—foreshadowed the increasing adoption by non-Israeli geneticists of Zionist assumptions about a shared Jewish "gene pool" originating in the Levant.

Iran: The Elusive Majority

Iran is a useful case to understand how the contingencies of institutional, logistical, and social infrastructures of blood transfusion within a single country could produce remarkably different kinds of genetic populations. Until the 1970s, transfusion services in Iran were highly decentralized and uncoordinated. Reports on Iranian blood group frequencies were produced by regional hospitals and medical schools; the Red Lion and Sun Society (Iran's Red Crescent); military medical facilities; and national institutions that regularly hosted foreign physicians, like the Tehran branch of the Pasteur Institute and (after 1954) the Iranian Oil Refining Company in Abadan.

Among these, the Tehran-based institutions lay at the top of a hierarchy of prestige, resources, and international recognition. This meant that genetic data calculated from the population of Tehran blood donors often came to stand in for the population of Iran as a whole. For example, when Mourant sought out Iranian data for the first edition of *The Distribution of the Human Blood Groups*, he relied on figures provided through correspondence with Manuchehr Motamed, a laboratory director at the Tehran Faculty of Medicine, and Ahmad Azhir, the founder of the Red Lion and Sun Blood Transfusion Service. Mourant published their "Tehran series" of frequencies, calculated from more than 10,000 people, as representative of Iranians generally.[63] Subsequently, personnel at regional medical centers, such as the cities of Tabriz and Mashhad, occasionally published reports on the ABO, Rh, and MN blood group frequencies of their patients. Rather than identifying their blood donors as members of a national "Iranian population," they specified their numbers as representing the population of the province (i.e., Azerbaijan or Khorasan). This practice suggests that local physicians perceived or anticipated regional variations across Iran. However, they did not explicitly attribute this variation to the differential distribution of ethnolinguistic or religious groups across the country. In fact, they avoided mentioning ethnicity, language, religion, and other identity labels for their patients; the "populations" they describe are defined solely by Iranian administrative geography.[64]

This avoidance reflected both pragmatic and political issues. Firstly, the blood banks did not routinely collect from their donors any demographic information deemed not medically relevant. Secondly, as a continuity from the interwar period, the Iranian government's ongoing fears about ethnic and tribal separatism meant that research into these identity categories was discouraged and closely monitored by the authorities. In the immediate aftermath of the Second World War, the Soviet Union supported the establishment of two breakaway states: the People's Republic of Azerbaijan and the Kurdish Republic of Mahabad. Around the same time, Qashqa'i and Bakhtiyari tribal leaders in Fars province revolted against the central government, who blamed the British for arming the tribes and instigating unrest. These political movements rejected the exclusionary Persian nationalism of the Pahlavi state, demanding provincial autonomy, recognition of languages other than Persian, and tribal self-rule. The 1946 labor strikes at the Abadan oil refinery inflamed local Arab-Persian ethnic tensions, which Persian strikers and the Iranian government similarly blamed on the incitement of British management at the Anglo-Iranian Oil Company. These incidents set the tone for postwar Iranian politicians who tarred any expression of legitimate grievances by ethnic minorities as a foreign conspiracy to divide and rule Iran.[65]

Because Tehran was the center of gravity for the international traffic in blood, Tehran-based researchers could largely ignore the question of provincial or ethnic variation. At the end of the 1950s, the Iranian army physician Mirza Eftekhari shipped serum samples taken from 296 Iranian soldiers to the Pasteur Institute in Paris. There, hematologists Luba Podliachouk and André Eyquem tested the sera for the presence of the Gm^a immunoglobulin. The samples, collected by military doctors at the Iranian Army Blood Transfusion Center in Tehran, did not include any data on ethnic, linguistic, or religious categories, only the individuals' province of birth. Podliachouk and Eyquem accordingly calculated their results by province but did not find the regional breakdown worthy of discussion, concluding only that "the distribution of Gm^a factors among the Iranians is nearly the same as that previously found for persons of the white race."[66] In a further study of the same samples, the acceptance of a unitary Iranian population was even more pronounced. Podliachouk and Eyquem not only reiterated the samples' racial provenance and abandoned any regional analysis, but they also used the Iranian population to test a three-allele inheritance model for the Gm serum factors according to the Hardy-Weinberg equilibrium,[67] indicating their assumption of an Iranian society sealed from emigration and immigration and marked by a high degree of social homogeneity. Such an imagined Iranian population could take shape only where blood samples were collected from Tehran-based institutions, particularly from individuals subjected to compulsory national integration through military conscription.

In contrast, Western medical researchers who spent time working in Iran's provinces routinely marveled at the country's human diversity. Rather than accepting the notion of a singular, homogeneous Iranian population, they defined their Iranian research subjects according to two consistent practices: firstly, emphasizing ethnolinguistic and religious categories to identify minority groups, and secondly, describing the "dominant" Iranian majority as a highly admixed population. For example, French researchers affiliated with Iran's national branch of the Pasteur Institute secured permission to collect blood from Kurdish and Turkmen tribes in the early 1950s. Like interwar anthropometrists who searched for representatives of the pure ancient Iranians, they also demonstrated a preoccupation with the Zoroastrian religious minority.[68] Similarly, Philip H. T. Beckett, an Oxford undergraduate like Botting, targeted Zoroastrians' blood on a 1950 university expedition to Kerman. At the local Morsalin Hospital (established by the British Church Missionary Society), he began a project to compare the ABO frequencies of Muslim and Zoroastrian patients. He reasoned that Zoroastrian endogamy and rejection of converts rendered their blood "fairly representative of that of Sassanian Persia and its composition relatively unaffected by the incursions of Arabs, Mongols, Turkomans, Afghans, etc., that have increased the complications of the pattern of racial origins in Persia during historical times." However, he was able to collect only six Zoroastrian samples before "local [anti-British] political feeling made it advisable to discontinue the work."[69] The notion that the Zoroastrians represented "uncontaminated" biological descendants of the pre-Islamic Persian empire became increasingly pervasive in Iranian genetics during the late 1950s and 1960s. The alleged purity of Zoroastrian communities was routinely contrasted with the majoritarian population's promiscuous admixture, even when genetic markers did not show clear differences along religious lines.

The lack of diagnostic genetic variations between religious and ethnic categories in these early results intensified the problem of how to define the "elusive majority" of Iranian research subjects. As a population category, "Iranian" reliably referred to native Persian speakers, most often Shi'i Muslims. But European researchers frequently achieved these criteria through a process of elimination: "Iranian" blood samples came from any individual who could not be assigned to a recognized religious or ethnolinguistic minority. In other words, these "default" Iranians were so labeled to differentiate them from the ostensibly more "pure" minority groups, who served as genetic representatives of ancestral populations who had originally settled in Iran. This phenomenon replicates what Rasmus Elling has identified as the Iranian "ethnic commonsense." "Iranian" was simply an umbrella term for everyone—including most of the local medical staff—who did not belong to the (allegedly) clearly bounded minority communities that fascinated many foreign researchers.[70] In biological terms, this practice was no more precise than the administrative definition used by Iranian blood transfusion staff in Tehran: "Iranians" were all those with Iranian citizenship whose blood passed through national medical services, which aggregated geographical data at the level of regions, provinces, and cities, but never collected information on donors' language, religion, and ethnic identification. In both circumstances, "Iranians" were thus a "supra-ethnicity"—that is, "a reified Iranian national identity grounded in Persian culture" whose supporters alternately emphasized a shared national language and literature, a shared ancient history, a shared Shi'i religious identity, and a shared Aryan racial ancestry.[71] The competing genetic data produced by the different institutions who collected Iranian blood, however, regularly underlined an inconvenient truth: Iranians, no matter how they were (un)defined, did not constitute a uniform, homogeneous category.

The predominant role of Iranian scientists in constituting themselves as a genetically legible population unit with a distinct national ancestry is best exposed through an example of collaborative breakdown, wherein a set of Iranian blood samples was effectively orphaned in a European lab.

Between 1956 and 1958, Dr. W. de Graaf, a medical officer employed by the Dutch-managed Iranian Oil Refining Company in Abadan, shipped about 950 blood samples to the Central Laboratory of the Netherlands Red Cross Blood Transfusion Service in Amsterdam. De Graaf, who worked mostly with Arabic- and Persian-speaking Iranians at the company facilities, had managed to sample seven populations: Kurds from Kermanshah; Qashqa'is from the environs of Shiraz; Bakhtiyari nomads (from the present-day province of Chahārmaḥāl va Bakhtiyārī); and Armenians, Assyrians, Arabs, and Iranians from the "strongly mixed" city of Abadan. When de Graaf suddenly passed away before completing the study, Lourens E. Nijenhuis, a serologist, was left to interpret the lab results alongside the "anthropological notes" de Graaf had included in their prior correspondence. Nijenhuis assumed that "probably only the group 'Iranians' is somewhat heterogeneous," as opposed to the other populations, whose religious or tribal affiliations were clearly defined and thus hindered intermarriage.[72] However, he struggled to piece together the categorical logic by which de Graaf had labeled 348 samples "Iranian." He cited the various "impressions" de Graaf reported based on his interactions with the local company employees:

> The people indicated as Iranian originate from various parts of the country. For part of them the origin could be determined precisely, for another part this was not possible because of the high frequency of migration from one part of the country to another. Dr. de Graaf got the impression, however, that intermixing between Northern and Southern Iranians did not occur as a rule, this being a result of the differences of climate between the Northern and Southern halves of the country. The habitus of the Iranians living in the areas around the Persian Gulf is strongly like that of Pakistanis; Northern Iranians are more like Turkish People.[73]

In terms of ABO and Rh frequencies, Nijenhuis observed that "the 'Iranians', which must be a group with mixed origin, are intermediate between the Bachtiiari [sic] and the Kurds." However, he cautioned, this did not mean that contemporary "Iranians" were simply the product of a fusion between two ancestral Bakhtiyari-like and Kurd-like populations.[74] He believed that de Graaf's small sample size, of individuals who originated mostly from southeastern Iran, precluded any valid interpretations of the average genetic composition of the Iranian population. De Graaf's "Iranians," he explained, were representative not of the country as whole, but only of the assorted individuals he had ready access to in Abadan: his own hospital and laboratory staff, hospital patients, oil company administrative staff, Army employees, and members of the city's fire brigade. Divorced from this local social infrastructure, which had helped de Graaf make sense of Iranian ethnic and regional identities, Nijenhuis chose not to speculate on what the "original components" of the Iranian "ethnic group" might be.[75] In cases where Western medical researchers remained embedded among their Iranian colleagues, they displayed no such reticence. As shown in Chapter 5, several American and European scientists readily adopted Iranian nationalists' anthropological self-definition and even promoted their historical narratives about Iran's pre-Islamic past. Such collaborative developments hinged on the study of specific inherited diseases, and not simply blood groups, as markers of racial ancestry.

GENETICS AS AN INTERFACE BETWEEN MEDICINE AND ANTHROPOLOGY

The true extent of Middle Eastern participation in genetic research has long been underestimated because the roles of particular actors—local physicians and medical technicians, military officers and soldiers, oil company employees, research subjects and their community leaders—are not evenly or thoroughly documented in the scientific publications and archival sources left behind by their Western collaborators. When characters like Botting and Mourant explicitly acknowledge Middle Eastern actors, they frequently describe how the locals hindered their work. For adventurers like Botting, the problems mainly concerned actual blood collection from suspicious research subjects and red tape from local politicians or community leaders. For laboratory directors like Mourant, local scientists, too, could present challenges. Scientific actors in Egypt, Saudi Arabia, Israel, and Iran resisted the collaborative expectations of his knowledge-control regime by collecting samples and producing competing data according to their own interests, often using population categories incommensurable with Mourant's existing data sets or anthropological preoccupations. In the case of the Egyptians, this competing data (and its larger sample sizes) even threatened the credibility of Mourant's work. Yet much of his research agenda relied on an exchange regime made possible through the many Middle Eastern individuals enmeshed within his collaborative networks. These involved not only the direct correspondents with whom he developed transactional relationships but also members of broader social infrastructures beyond the medical profession.[76] Physicians invoked social obligations and personal favors from all kinds of family members, childhood friends, former schoolmates, and military comrades, just to set in motion the lengthy processes of coordinating blood collection and analyzing gene frequencies.

These intricate social infrastructures provided the means for Mourant and other well-placed Western scientists to enlist far-flung institutions into the international traffic in blood. In 1957, Mourant was finally able to acquire coveted Assyrian Christian blood samples through a chain of personal contacts linking the hospitals of London and Baghdad via the Iraqi royal family. Mourant's colleague, the hematologist Hermann Lehmann, had a student at St. Bartholomew's Hospital (recorded as "G. R. Hobday") who had attended Harrow School with the young King Faisal II. Through Hobday's intercession, the king arranged for his minister of health, Dr. A. A. Allawi, to collect Assyrian blood, with ninety-nine samples ultimately finding their way to the BGRL in London.[77] Mourant solicited access to this religious community because the Assyrians (like Copts and Zoroastrians) claimed descent from an ancient civilization, the Assyrian Empire of Mesopotamia. Although "only ten per cent are said to be of a physical type resembling the ancient Assyrians," Mourant argued that their distinct physical differences from "neighbouring Arabs" suggested they derived not only "religious and cultural traditions" but also "probably to a large extent their physical ancestry from the original Christian Church of Mesopotamia, founded in A.D. 70."[78] These precious specimens underwent a thorough investigation: Elizabeth Ikin tested the blood for nine types of antigens, while Lehmann searched for rare hemoglobins. However, the BGRL concluded that although the Iraqi Assyrians had recently suffered from serious ethnic violence and discrimination, their blood group frequencies did not differ significantly from surrounding populations, suggesting that "they have probably not been long subjected to any extreme degree of reproductive isolation."[79]

The fascination with the Assyrians, and the social network leveraged to access their blood, were both products of the British mandate that had established the Iraqi monarchy in 1932.[80] As shown in the previous chapters, the desire for exotic and socially isolated groups whose hereditary traits could be correlated with known history was certainly not new. The distinguishing features between interwar sero-anthropology and the BGRL's postwar research were predominantly technological and infrastructural, enabling Assyrian blood to be shipped by air and tested with fully twenty different varieties of antisera and paper electrophoresis. New international health organizations helped the Iraqi king and his ministers to perceive this trafficking in blood as a medical endeavor worthy of their support. Naturally, the publication of the Assyrians' genetic data largely erased the "middlemen," the Iraqi field-workers and technical staff who directly extracted blood from research subjects. Yet these neglected figures had the most influential role in interpreting ethnic belonging and group origins: it was they who arbitrated the identity of research subjects at both individual and communal levels and who therefore circumscribed the possible meanings of the blood group frequencies found in the laboratory.

This chapter highlights the variety of people and institutions involved in the extraction, storage, and transport of Middle Eastern blood. Blood banks, forensic laboratories, and oil companies all contributed to the local and transregional circulation of liquid blood samples as well as data on paper. The work of many Middle Eastern physicians appeared in the updated and massively revised second edition of Mourant's *Distribution of the Human Blood Groups*.[81] But as shown in Mourant's many tables and charts, the collection practices of each particular institutional and national setting generated incommensurable sets of data, sorting individual humans into groups based variously on birthplace, religion, language, ethnic or tribal affiliation, and even the region served by a given clinic. These inconsistencies emerged from each institution's differing functions, facilities, and material resources. They were also products of the social backgrounds and political commitments of the local medical staff who mediated the interactions between Western scientists and Middle Eastern research subjects. The next two chapters explore how the institutional and social infrastructures examined here enabled the emergence of medical genetics, a field that has historically made grandiose anthropological claims without much ethnographic reflexivity. ABO and Rh blood groups and their relevance to blood transfusion had made it logistically and intellectually possible for medical practitioners to take a leading role in debates about racial classification. Specialist hematologists soon joined forces with biochemists and population geneticists as they began to uncover the molecular basis of inherited blood disorders like sickle cell disease and favism. Unlike ABO and Rh blood groups, these disorders did not occur in every population worldwide but clustered in particular geographical regions and social groups. Budding medical geneticists therefore embraced these disorders as more precise diagnostic tools with which to stake claims about the origins of Middle Eastern peoples.

Chapter 4
SICKLING SOCIOLOGIES

IN 1993, A HEBREW UNIVERSITY ORAL HISTORY PROJECT INTERviewed Dr. George Mendel about his work as the former director of the Hadassah Children's Hospital in Rosh Ha'ayin, a major transit camp for Yemenite Jewish immigrants, between 1949 and 1951. His revelations about the hospital's activities eventually brought him to the attention of the State Commission of Inquiry into the Disappearance of Yemenite Children, 1948–1954, who called him into the Knesset to repeat his testimony under oath in 1996. The Cohen-Kedmi Commission, as it became known, was the Israeli government's third investigation into the so-called Yemenite Children Affair, wherein Ashkenazi immigration authorities, especially medical practitioners, separated hundreds of Yemenite infants and young children from their newly arrived families. Grieving Yemenite parents, refusing to believe official reports that their children died from illnesses in the camp hospitals, have alleged that racially prejudiced authorities kidnapped their children and placed them for adoption with Ashkenazi families. To support their claims, Yemenite parents cite a range of mysterious circumstances, such as being denied permission to view their children's bodies, graves, and death certificates, and receiving army draft notices for children who had been pronounced dead eighteen years earlier. The parents' quest to uncover the fate of their lost children has frequently been characterized by the mainstream Israeli media as a conspiracy theory.[1] However, Mendel's reflections lent credibility to the charge that medical treatment in Rosh Ha'ayin was marked by ethnic discrimination.

In the 1993 interview, Mendel recalled the visit of prominent American-Jewish hematologist William Dameshek to the camp at the end of 1949. Dameshek, himself of Russian Ashkenazi background, "looked at the Yemenites and thought that they must have Negro blood [*dam kushi*]" and that they might have a disease known as sickle cell anemia. Dameshek recruited interns to search for evidence of the disease in blood samples from Rosh Ha'ayin patients being tested for malaria. After months of searching, the interns discovered what they believed to be sickle cells in a few Yemenite children. They shared their findings with Fritz Dreyfuss, the director of internal medicine at Hadassah Hospital in Jerusalem, who "became excited and immediately published an article in a medical journal with a grand theory about how the Yemenites arrived [in Yemen] and where they migrated in the world and such." However, the discovery was soon revealed to be erroneous; the apparently sickle-shaped cells had been produced by laboratory conditions, not by an actual disease. Dreyfuss had to write to the journals acknowledging that his previous reports were inaccurate. But in the meantime, Mendel remarked, the Israeli researchers "had already told the Yemenites that they had Negro blood [*kevar amru le-temanim she-yesh bahem dam kushim*]!"[2]

In December 2016, Mendel's testimony was one of hundreds of thousands of documents related to the Yemenite Children Affair that was released by the Israeli government as part of a fourth official investigation. Over the following months, the Israeli media reported a renewed sense of public outrage surrounding this specific segment of Mendel's testimony as well as other evidence of medical experimentation conducted in Rosh Ha'ayin without the consent of the Yemenite patients. The phrase "Negro blood" made its way into the headlines, reinforcing the racialized nature of the injustices experienced by Yemenite and other Mizrahi Jews.[3] The media took Mendel's use of this phrase, which has derogatory valences in Hebrew as it does in English, to mean that the Ashkenazi medical staff in the transit camps had exploited the new immigrants to conduct research that would "prove" the innate inferiority of Mizrahim. This perception is bolstered by Mendel's testimony, which implies that the search for Yemenite sickle cells was motivated solely by racial beliefs, not by concern for patients' welfare. Yet, paradoxically, the intersection of the Yemenite Children Affair with sickle cell research highlights the temporal moment that scientists began to doubt that sickle cell disease was a sure marker of "Negro blood." Far from demonstrating that Yemenite Jews had African ancestry, the interns at Rosh Ha'ayin unwittingly changed the global course of studies on this hereditary disease such that the Middle East, and particularly Yemen itself, became the focal point of heated scientific debates about human evolution and racial mixture.

In today's clinical usage, "sickle cell disease" actually refers to several inherited conditions affecting red blood cells caused by mutations in the gene that codes for hemoglobin, the protein responsible for transporting oxygen. When these cells are depleted of oxygen, the mutated form of hemoglobin, called HbS, deforms the cells from their normal round shape into elongated crescents, hence the name "sickle cell." If too many cells take on this shape, they can block blood vessels, causing strokes, organ failure, skin ulcers, and other problems. The individuals most at risk for these deadly complications are those who inherit two copies of the gene for HbS; they are considered to have sickle cell anemia and may not survive infancy without medical intervention. On the other hand, an individual who inherits only one copy of this gene—known as a "carrier" of the sickle cell trait—often presents no clinical symptoms at all. In such cases, one's status as a carrier can be determined only through blood screenings that find sickle cells mixed in with the normal red cells or detect the presence of HbS using biochemical tests.

The history of research on sickle cell disease is important for understanding the emergence of medical genetics as a discipline. In the first half of the twentieth century, knowledge of genetics was not considered relevant for the clinical work of medical practitioners, and medical schools did not include genetics in their curricula. In the postwar period, sickle cell disease was the instrumental case upon which American biochemists like Linus Pauling and Harvey Itano built the concept of "molecular disease"—that is, a disease caused by an abnormality in a single molecule whose patterns of inheritance could then be traced and anticipated.[4] This basic concept demonstrated that genetic research could have enormous implications not only for human evolution and anthropology but also for medicine and public health.[5] However, as Mendel's testimony to the Cohen-Kedmi Commission reveals, there has never been a fixed line distinguishing between medical and anthropological genetics, and the history of research on sickle cell disease is tightly linked to racial discourses on Africans.

Existing accounts of this history focus on the distinctly North American context of sickle cell discovery and racialization.[6] The disease was first described by a Chicago physician in a 1910 publication, which identified "peculiar elongated and sickle-shaped red blood corpuscles" in a medical student from the Caribbean island of Grenada.[7] Other physicians soon found further cases of sickle cells, predominantly in African American patients. By the 1920s, clinicians in the United States had developed a medical-anthropological discourse that fixed sickle cells to African ancestry. In fact, their belief that sickling was a racial condition became so strong that when they discovered sickle cells in "white" patients, they challenged not their own assumptions about the disease but rather the patients' ancestry. In other words, physicians argued that individuals of European background who had sickle cells could not be purely white but must have unknown "Negro" admixture in their family trees. Eventually, as evidence accumulated that most white sicklers had Italian or Greek origins, this discourse stretched beyond individual family histories into the distant past, with medical publications invoking Carthage and the African slave trade in the days of the Roman Empire. Based on these convictions about the racial source of sickle cells, American physicians urged their counterparts working in the British and French colonial medical services to search for the origins of the disease in African tribespeople.[8] In the late 1940s, a number of doctors working in the British colonies of Uganda and Rhodesia heeded this call, initiating sickle cell screening procedures in their hospitals. The research of one of these men, Hermann Lehmann, played a key role in reorienting narratives of sickle cell disease from a problem concerning solely black Africans to a grander vision of human evolution linking Africa to Asia through the Middle East.

This chapter explores Lehmann's journey to the British colony of Aden, his subsequent collaboration with a Turkish hematologist named Muzaffer Aksoy, and the historical and anthropological narratives they used to interpret their discovery of sickle cells in marginalized communities living in Yemen and Turkey. During the 1950s and 1960s, sickle cell research in the Middle East unfolded along the fractured social politics of race, both at the local scale of health-care provision and the international scale of scientific collaboration. Within a politically charged atmosphere of rising pan-Arabism, the vexed question of Arab identity became entangled with international debates about sickle cell disease and its implications for human evolutionary history. The work of Lehmann and Aksoy prompted sickle cell investigations in Lebanon, Egypt, and Saudi Arabia, which all struggled to reconcile racial, historical, and sociological boundaries between white and African, Arab and Turk. As the parameters of Turkish and Arab nationalism shifted in the Cold War–era Middle East, so did the explanatory narratives for the variable rates of sickle cell disease in different communities, according different degrees of importance to African ancestry, socially enforced endogamy, and evolutionary adaptations to malaria. These narratives reflected varying levels of engagement with, or resistance to, regional Middle Eastern collaborative networks versus the powerful transregional networks centered in Britain and the United States.

LEHMANN OF ARABIA:
REWRITING THE EVOLUTION OF SICKLE CELL DISEASE

Hermann Lehmann was born to a Jewish family in Halle, Germany, in 1910. He studied clinical medicine in Frankfurt and Heidelberg, but after the Nazis took power, he finished his medical degree at the University of Basel in 1934. Unable to work as a physician in either Switzerland or Germany, he left the continent permanently in 1936 to work in a biochemistry laboratory at the University of Cambridge. In 1943, he obtained an officer's commission in the Royal Army Medical Corps, which sent him to India. It was there that Lehmann first studied blood disorders, specifically anemia caused by iron deficiency in Indian soldiers. After the war, he was awarded a three-year research fellowship with the Colonial Medical Service to study malnutrition and anemia in Uganda.[9] Based at a Kampala hospital, Lehmann applied himself to anthropological as well as medical problems. He sought out tribal groups across the countryside and tested their members for sickle cells, using the anthropological convention of sorting these tribes into categories of "Hamitic," "Nilotic," or "Bantu," based on their spoken language. In general, Lehmann found, Hamitic groups had the lowest rates of sickle cell disease and Bantu groups the highest. However, different tribes within each of these linguistic categories had significant differences in their sickle cell frequency, which he correlated to particular histories of migration and language replacement. Accordingly, he argued, sickle cell genetics could answer anthropologists' long-standing questions about African tribal origins.[10]

When Lehmann returned to the UK, he still believed—as did most other physicians and hematologists aware of sickle cell disease—that the condition had originated in Africa and therefore indicated African descent. This belief was profoundly shaken in the spring of 1951. First, a team of Greek researchers announced that they had discovered sickle cell disease among Greek villagers at a higher rate than had been discovered in African Americans.[11] This was such a dramatic finding that Lehmann wrote a letter to the editor of the *Lancet* expressing his skepticism, suggesting that the Greek researchers might instead be observing symptoms of malaria.[12] But just one month after the Greeks, Fritz Dreyfuss—one of Lehmann's former classmates from the University of Basel—published his first report that Yemenite Jews also carried the sickle cell trait.[13] The investigations in Rosh Ha'ayin had been driven by a search for "Negro blood," and the sickle cell discovery might have been interpreted as further evidence for Rina Younovitch's mandate-era sero-anthropology, which insisted the Yemenites were a highly admixed community. Instead, in his scientific publications, Dreyfuss turned to the work of German-Israeli historian Shelomo Dov Goitein and American physical anthropologist Carleton S. Coon. Coon was well known for his expeditions to North Africa and the Middle East, which included anthropometric research in Yemen in the 1930s.[14] Goitein and Coon identified the Yemenite Jews as descendants of migrants from Judaea and representatives of a "purely" Mediterranean race. Based on their accounts, Dreyfuss argued that because geography and religion had insulated the Yemenite Jews from intermarriage with Africans, their sickle cell disease must come from a separate and independent genetic "reservoir" than that of "Negroes."[15]

The close succession of the Greek and Israeli findings, as well as Dreyfuss's proposed hypothesis, prompted Lehmann to rethink his assumptions about the origins of sickle cell disease. At this point in his career, Lehmann was a clinical pathologist at the London teaching hospital of Saint Bartholomew's and had limited means to carry out medically relevant biochemical research. Anthropological studies, particularly anything requiring overseas fieldwork, were well beyond his purview. Motivated by personal enthusiasm alone, he appealed to the Nuffield Foundation to fund a trip to southern India. The voyage was inspired by an idea Lehmann evidently heard in the 1930s from the Cambridge anthropologist Alfred Cort Haddon, who suggested that prehistoric South Asian peoples had migrated to East Africa before the Neolithic period.[16] Lehmann proposed to test "the idea of a racial link between Africa and India" by searching for allegedly African genetic traits, including sickle cells, among indigenous tribes living in southwestern India's Nilgiri Hills.[17] This search, in turn, could illuminate where—and in what race of people—the sickle cell mutation first originated.

Early in 1952, Lehmann traveled to the Nilgiri Hills and, true to his expectations, found sickle cells among the native tribespeople. In his excitement, he immediately telegraphed colleagues across the globe and prepared publications for both medical and anthropological journals in which he pronounced that sickle cells were "not an essentially Negroid feature" after all and thus could no longer be perceived as a marker of African ancestry.[18] Meanwhile, on his way back to England from India, Lehmann stopped in Israel to corroborate the Dreyfuss reports about Yemenite Jews. Using Lehmann's more sophisticated chemical techniques, they discovered that the strangely shaped blood cells observed by the interns in Rosh Ha'ayin were not true sickle cells after all; the Yemenites' other blood traits could neither confirm nor deny potential African ancestry.[19] The evidence for Dreyfuss's "grand theory" about the multiple origins of sickle cells thus evaporated.

Regardless, the case of the Yemenite Jews sparked Lehmann's interest in southern Arabia as a possible cradle of sickle cell evolution, especially given its location as a migratory pathway connecting Africa and Asia. In 1953, he received funding from the Wenner-Gren Foundation to travel to the British colony of Aden with Elizabeth Ikin of the Blood Group Reference Laboratory (BGRL). Together, they set up shop in the pathology laboratory of Aden's Civil Hospital. While Lehmann searched for sickle cells, Ikin tested for a broad range of other inherited blood traits as part of her doctoral research on Middle Eastern and Mediterranean populations. Taking 111 samples from people they identified as "Yemenite Arabs," Lehmann found only two to possess sickle cells. However, in a socially marginalized ethnic group that other Yemenis call *akhdām* (Arabic for "servants"), almost one quarter (23 percent) of the 104 tested individuals carried sickle cells. Based on this discovery, Lehmann developed a new hypothesis that sickle cells had first evolved in southern Arabia among the ancestors of the *akhdām*, whose migrations carried the gene to India, Africa, and perhaps the rest of the Mediterranean.[20] This new explanation for the geographical distribution of sickle cell disease was the pinnacle of Lehmann's contribution to the subject, and he propagated this hypothesis for the rest of his award-winning career.

Perhaps ironically, Lehmann developed a new racial theory about sickle cells at the same time that an Oxford researcher, Anthony C. Allison, produced groundbreaking evidence that the evolution of the disease was driven primarily by environmental conditions, not by racial admixture. In 1954, Allison published his discovery that carriers of the sickle cell trait have a natural immunity to malaria.[21] For many geneticists, this was a transformative development showing that humans, like other organisms, are subject to the forces of natural selection. Some scientists extrapolated this finding to argue that the sickle cell gene had evolved separately and independently in different parts of the world with endemic malaria. This "multicentric" hypothesis meant that sickle cell carriers in Africa, Europe, and Asia need not share an ancestral or racial connection—essentially the same hypothesis advanced by Dreyfuss in relation to the Yemenite Jews. In contrast, Lehmann argued, although malaria clearly regulated the success of the sickle cell gene within a population, it could not cause the original mutation of the gene and therefore could not resolve the question of its origins. Well into the 1970s, Lehmann continued to defend his hypothesis that individuals with sickle cell disease carried a gene with a single geographical and racial origin in southern Arabia.

Effectively, Lehmann staked his vision of human evolution on a group of people whose racial and ethnic identity had been contested in Yemen for several hundred years. Since the nineteenth century, British and French visitors had described the class known as *akhdām* as occupying the very bottom rung of Yemen's highly stratified urban society, even below that of former slaves (in Arabic, *'abīd*). This class of people filled occupations perceived to be ritually polluting, such as musicians, blacksmiths, rubbish collectors, and street sweepers. The ensuing social ostracism both produces, and is justified by, a racializing narrative that ascribes non-Arab, African origins to the *akhdām*.[22] According to ethnographic research conducted in the 1980s and 1990s, Yemeni Arabs refer to the *akhdām* and other lower-class groups as "black" on account of their darker skin color, but they legitimate the pariah status of the *akhdām* in genealogical and religious terms. Specifically, they claim that the *akhdām* are the descendants of the Abyssinian/Ethiopian soldiers serving in the army of Abraha, the viceroy of Aksum, who ruled Yemen in the mid-sixth century C.E. In Islamic tradition, Abraha's army is associated with the antagonists of the Qur'an's Verse of the Elephant, which describes God destroying the army before it can attack Mecca.

In this view, the social and economic marginalization of the *akhdām* befits their status as a religiously accursed and foreign minority. Meanwhile, the *akhdām* reject this narrative, self-identifying as Muslims and ethnic Arabs with origins in the city of Zabid, where their largest community historically resided. Some of them also explicitly reject the derogatory term "*akhdām*," preferring other demonyms such as "Zabidi."[23]

British colonial administrators were responsible for establishing the particular *akhdām* community tested by Lehmann, having brought *akhdām* from Zabid to Aden to serve as street sweepers beginning in the 1850s. Between this period and the 1940s, British observers in Aden noted that most other social groups refused to intermarry with the *akhdām* and accordingly continued to identify them as a separate race with distinct physical features. Such writers commented that the ancestral origins of the *akhdām* were "obscure" or disputed" but favored the belief that their roots lay in Africa.[24] At this time, ethnic discourses throughout Yemen contrasted Arabs with Ethiopians, Sudanese, and Somalis, construing the region as an interface between Mediterranean and African, white and black. For example, when the Zaydi Imam of Yemen permitted Coon to measure his subjects in 1936, his son Abdullah bin Yahya repeatedly interrupted Coon's work to complain that he was "measuring too many Negroes" and should only use "white" Yemenis, so as to improve Yemen's image abroad.[25] Lehmann's sickle cell research therefore departed radically from local narratives of African/Arab dichotomy when he identified the *akhdām* as members of a third racial group: the Veddoids.

In December 1953, not long after he and Ikin had returned from Aden, Lehmann presented a paper to the Eugenics Society. Its title promised not only to review the global distribution of the sickle cell gene but also to shed "a new light on the origin of the East Africans." This new light was the role of the Veddoids, allegedly one of the "most primitive races of mankind," whose typology had been developed in the 1920s to classify the indigenous tribal peoples of South Asia. Lehmann, like Dreyfuss, turned to Coon, who had argued that the Bedouin tribes of the Yemeni region of Hadhramaut should be classified as Veddoids. When Lehmann found sickle cells in the *akhdām*, he wrote directly to Coon, who "confirmed that [the Achdam] very likely represent a Veddoid survival."[26] Lehmann's enthusiasm for Coon's racial typology is perhaps ironic, given that the former is sometimes billed as a "founder of molecular anthropology," while the latter called blood typing a "fad" and a politically correct distraction from anthropometric studies.[27] Nevertheless, with Coon's endorsement, Lehmann argued that sickle cells had evolved within a Neolithic Veddoid population in Southern Arabia, which had migrated to both India and East Africa, bringing the sickle cell trait with them. In his model, the Veddoids might have even migrated to southern Europe, which could explain how Greeks without African ancestry had the sickle cell gene.

Lehmann's portrayal of his research subjects is key to his racial hypothesis. Lehmann described the "Achdam" as "a dark-skinned people with Veddoid features living near Zabid in the southern half of Yemen." Lehmann argued, despite the prejudices of what he called "the local Arabs," that the *akhdām* had not been "imported" from Africa. He drew on other genetic traits, especially a reduced frequency of the Rh system chromosome Rh_0 or *cDe*, to support this contention. However, neither did Lehmann support the claims of the *akhdām* to be Arabs. Although they spoke Arabic, practiced Islam, and indeed were indigenous to their country, Lehmann pronounced the *akhdām* to be a race apart. Because intermarriage with Arabs was "out of the question," Lehmann portrayed the *akhdām* as the biologically isolated living remnants of a prehistoric gene pool.[28]

Following his anthropological observations, Lehmann concluded his paper with a sociological one. He noted that in East Africa, India, Arabia, and even Greece, "the highest sickling rates were always found in the socially lowest stratum of a population. [. . .] These people are forced to live where they are, they are unable to travel and they do not intermarry with their more fortunate neighbours. Inbreeding which usually goes hand in hand with a slave-like condition in life seems to be an important factor in producing a high sickling rate."[29] Lehmann mentioned Allison's recent research breakthrough that sickle cell hemoglobin conferred resistance to malaria, acknowledging that "some such mechanism must play a part in explaining the presence of the gene." However, he insisted, natural selection could account for only one of three factors determining the rate of sickle cell disease in any given population. In his mind, the other two factors—namely, a population's "anthropological derivation" and a "social situation which produces inbreeding"—were far more significant to understanding which population groups were disproportionately afflicted with the disease. This dismissal of environmental determinism dovetailed with the next major discovery of sickle cell disease outside Africa in 1954. In the city of Mersin, in southern Turkey, physician Muzaffer Aksoy found that a particular group of his patients "possess[ed] the highest incidence of sickling in the white race."[30] Like the *akhdām* in Aden, these people faced discrimination from the Turkish majority and were derisively called *Arap uşakları* ("Arab servants").

MUZAFFER AKSOY AND THE REVIVAL OF THE "ETI-TURKS"

Muzaffer Aksoy was born in 1916 in Antalya into a family that became devoted to Kemalism. He was sent to the prestigious Istanbul Boys' High School in 1931, the same year his father was elected to the Turkish parliament as a member of the Republican People's Party (CHP), Atatürk's political party. In an interview conducted late in his life, Aksoy recalled learning as early as middle school about Atatürk's theories concerning the Turkish race (*ırk*).[31] He subsequently entered Istanbul University's medical school, where he studied under the noted German-Jewish hematologist Erich Frank. After graduating with honors, Aksoy fulfilled his compulsory military service as a ship's doctor in the Navy, then returned to Istanbul for a residency in hematology. In 1947, Aksoy was dispatched to Mersin State Hospital, along the Mediterranean coast, to work in the internal medicine department. His transformation from provincial clinician to international researcher began with a sabbatical in Boston in 1952 and 1953. There, he worked with William Dameshek, the same famous hematologist who had recently prompted the search for sickle cell anemia among Yemenite Jews. Aksoy's connection to Dameshek catalyzed his learning of English and helped him secure research funding from the Blood Research Foundation in Washington, D.C., to buy specialized equipment for a pathology laboratory back in Mersin (Figure 8). After his Mersin colleagues discovered a case of sickle cell anemia in a single patient, Aksoy spent 1954 searching for the disease in the city and its neighboring villages. He unearthed fifteen cases of anemia, all belonging to a single Arabic-speaking religious community: the Nusayri Alawites.

As described in Chapter 1, the Hatay Crisis of the 1930s brought the Alawites under close scrutiny by the Turkish Republic. In order to convince the League of Nations that the former Sanjak of Alexandretta should not be absorbed into an independent Syria, the Turkish government contended that most of the territory's inhabitants were ethnic Turks, not Arabs. Turkish

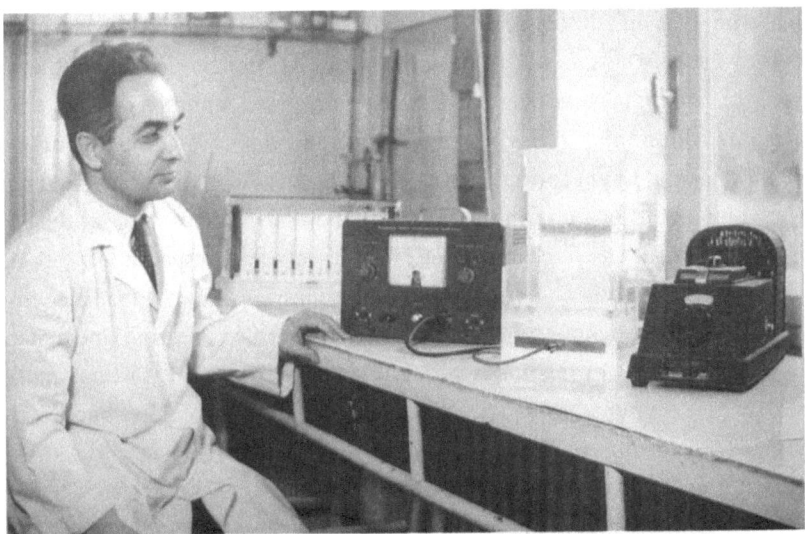

FIGURE 8. Muzaffer Aksoy in his laboratory at Mersin State Hospital, 1956. Photograph published in Nursel Duruel, Çiğdem Altay, and Orhan N. Ulutin, eds., *Bilime Adanmış Bir Ömür: Muzaffer Aksoy* (Ankara: Türkiye Bilimler Akademisi, 2005). Reprinted with permission from Türkiye Bilimler Akademisi (Turkish Academy of Sciences).

political organs in the territory targeted the Sanjak's Alawites, rebranding them from Arabic-speaking peasants into *Eti-Türkleri*, "Hittite Turks," drawing on historical, archaeological, and anthropometric discourses. Meanwhile, Alawites in the Turkish provinces of Mersin and Adana faced intense state programming to "assimilate" them into the Sunni Turkish population, with the government funding not only special Turkish language schools but even the "trousseaus of mixed marriages" arranged between Alawites and Turks.[32] The Alawites responded to these efforts, particularly the Eti-Turk discourse, with ambivalence. Recent historical and ethnographic work shows that some Alawites have and continue to embrace these narratives of Hittite or Turkish ancestry, but "Eti-Turk" was never widely adopted as a self-identification.[33] Meanwhile, local Turkish communities have largely continued to discriminate against them as both ethnic Arabs and heterodox Muslims, using the derogatory terms *fellah* and *Arap uşakları*.[34]

Aksoy, on the other hand, was reluctant to use these contemptuous labels for his patients. His first English publication in 1955 introduced his study population without any communal name at all, offering only a vague description of the group's historical origins: "At Mersin, on the southern coast of Turkey, 15 cases of sickle-cell anaemia were found in a distinct and separate community [...] whose ancestors had been immigrants to Turkey from Syria and Egypt several centuries ago and, so far as is known, had no Negro blood but had intermarried among themselves, becoming inbred."[35] Unlike for the *akhdām* in Yemen, the racial status of the research subjects as whites or Africans was not fundamentally in question. Aksoy introduced his conclusion with the query "How can we explain this fairly high incidence of sickle-cell trait in the white community?" He presented "the two most widely held opinions" as equally plausible: perhaps the Alawites had experienced "an admixture of Negro blood during either the Roman or the Turkish empires, or else in the era of medieval Mohammedan expansion." Or, citing Hermann Lehmann, Aksoy mentioned that "tribes from India" may have spread the sickle cell trait to the Mediterranean and East Africa. Further research, Aksoy concluded, would be necessary to determine the more likely pathway.

Lehmann immediately took note of Aksoy's report. Within a week, the *Lancet* printed Lehmann's letter to the editor, which expressed his interest in Aksoy's findings and a correction that it was not "Indian tribes" who had carried the sickle cell gene to Africa but rather "the Veddoids of South Arabia."[36] Both Lehmann and Anthony Allison wrote to Aksoy, expressing interest in a research collaboration; Aksoy chose to work with Lehmann.[37] By the end of 1955, Aksoy had shared with Lehmann Turkish and Alawite blood samples containing rare hemoglobin variants (namely, hemoglobin D and E). Furthermore, in an extended version of his sickle cell article submitted to the American journal *Blood*, Aksoy corrected his account of Lehmann's hypothesis about the Arabian Veddoids. But this was not his only addition to the material. He now gave a collective name to his research subjects: "[The patients] belonged to a distinct and separate community which emigrated some centuries ago from Turkey's southern neighbors, Syria and Egypt, and settled exclusively in the southern part of Turkey bordering the Mediterranean Sea. [...] This community is officially known as Eti-Turk."[38]

Aksoy thus introduced the term "Eti-Turk" to anglophone science, and the label was immediately adopted by Aksoy's non-Turkish colleagues, including Lehmann. But did Aksoy's use of this term merely reflect his Kemalist bona fides or his actual beliefs about the biological ancestry of this population? His writing leaves this unclear; he neither clarifies the "Hittite" ancestry narrative associated with the term, nor does he mention the religious and linguistic features that made this community "distinct and separate" within Turkey and the Levant. However, within three subsequent publications coauthored with Lehmann in 1956 and 1957, the research subjects are consistently described as "Eti-Turks, a small Arabic-speaking group" living in southern Turkey.³⁹ Aksoy's involvement in transnational scientific collaboration, a marker of career achievement, forced him to clarify features of Alawite identity that the Turkish government wished to ignore and undermine. Collaborators like Lehmann were interested in large-scale regional distributions of genes according to patterns of social differentiation, like ethnicity, religion, language, and tribal affiliation. Aksoy had to navigate his collaborators' interests in categorizing the Alawites as Arabs, Turks, or a race apart, like the *akhdām*. At the same time, Aksoy's transnational connections dramatically magnified the long-term impact of his decision to revive the terminology of "Eti-Turk" alongside an acknowledgment of the Alawites' Arabic language. His new recognition in the anglophone scientific community offered him a platform to disseminate Turkish nationalist views that rejected any ethnic relationship between Arabs and Alawites.

In February 1957, Aksoy left Mersin for Istanbul, where he worked as a physician and faculty member at Istanbul University's medical school. He also served on the local organizing committee for a major international symposium on "Abnormal Haemoglobins," held in Istanbul in September 1957. Funded by the United Nations Educational, Scientific, and Cultural Organization (UNESCO), the World Health Organization (WHO), and the Rockefeller Foundation, the choice of the symposium's location was explained both as an opportunity to "provide a stimulus for research" in the region and because "the Near East appears to be a sort of haemoglobin crossroads."⁴⁰ Prominent American biochemists and geneticists like Harvey Itano and James Neel attended, along with Lehmann and other hematologists from across Europe and Asia. At the symposium, Lehmann presented his hypothesis on the origins of sickle cell disease. Notably, he described the sickle cell carriers of Aden not as *akhdām* but, more obliquely, as a "certain group" that "differ[ed] from the present-day representative [. . .] Arabs." When elaborating his model, he referred only to the "Veddoids of Arabia" living in the late Neolithic period, when the region was "green and fertile." According to Lehmann, due to the arrival of new populations and to climate change, these Veddoids with their sickle cell gene dispersed to India and East Africa, and perhaps even to Egypt and Europe, where "prehistoric remains ... have been claimed to show Veddoid features." Such Veddoid features, he commented, "still persist in coastal Arabia where the sickling rate is high [. . .] some of these sickling genes may have come from an admixture by intermarriage with African slaves," but in his view, this explanation could not account for the high incidence of sickle cell disease in Indian Veddoids.[41] Again, Lehmann's model is predicated upon racial typologies external to the Yemeni social structure: he minimizes both the self-identification of the *akhdām* as Arabs and their neighbors' identification of them as Africans.

Immediately following Lehmann, Aksoy presented a summary of his work on abnormal hemoglobins in Turkey, in which he clarified the earlier ambiguous description of his research subjects. He explained that the patients described in his presentation belonged to an "Arabic-speaking population, known officially as Eti-Turks," which had emigrated centuries ago from Syria and Egypt to southern Turkey. Despite these geographic origins, Aksoy insisted, the Eti-Turks were not directly related to the majority population of the contemporary Levant: "The racial background of this community is obscure. *Although they speak Arabic, they are not Arab.* Some people of Syria and Lebanon, called Fellahs, might have the same racial background."[42] This is the first recorded instance wherein Aksoy explicitly denies that the Alawites in Turkey are Arabs. Although he does not delve into the Kemalist narratives of the 1930s claiming the Alawites are actually Turkish or Hittite, he announces that they are not racially Arab, whatever that might mean. Summarily, Lehmann and Aksoy reported high frequencies of sickle cell disease in two socially marginalized Arabic-speaking groups, yet they claimed that from a racial or genetic perspective, neither of these groups was Arab. These claims were driven by shifting and competing notions of Arab identity operating at multiple scales: within Lehmann and Aksoy's British collaborative network; within the social climate in Mersin; and within the political conflict emerging between Turkey and its neighbors in response to Nasserism and Ba'athist pan-Arabism.

Firstly, Lehmann and Aksoy's portrayals of their research subjects contrasted notably with those of their colleague Elizabeth Ikin, who worked with Lehmann in Aden. As described in Chapter 3, Ikin, a senior technician at the Blood Group Reference Laboratory (BGRL) in London, used the data she produced there to write a doctoral thesis on the blood group systems of Near Eastern populations. Unlike Lehmann and Aksoy, Ikin had no special interest in sickle cell disease; rather, she examined a broad range of inherited blood traits to interpret anthropological relationships between different populations. During the same years in which she was blood-typing hundreds of Arab samples from Aramco, she tested Aksoy's Eti-Turks along with a so-called "control group" of Turkish hospital patients from Mersin. In a paper she coauthored with Aksoy and Lehmann, Ikin argued that the two populations were substantially similar in all aspects except for the Eti-Turks' high frequency of sickle cell anemia, an extremely rare condition in the Turks.[43]

Ikin's presentation to the Seventh Congress of the International Society of Blood Transfusion in September 1958 revealed how she contextualized her work on the Eti-Turks alongside data from the Arabian Peninsula. Firstly, Ikin's portrayal of the *akhdām*, whom she called "Zabidi Arabs," was quite different than Lehmann's. She acknowledged Lehmann's hypothesis that the Zabidis are "of Veddoid origin" but then remarked that they "have acquired a lot of African genes" relative to other Arabs in the peninsula. She concluded that it was "an open question" whether the Zabidis had acquired the HbS gene "on its way from Asia to Africa" or "solely from Africa." With regard to the Eti-Turks, whom she described as "Arabic-speaking people living in Turkey," Ikin noted that their ABO and Rh blood group frequencies mostly resembled other Turks instead of "Arabians." However, Eti-Turks' MNS antigen frequencies were more like those of "other Arabic-speaking peoples" than the Turks.[44] She concluded her presentation with a call for more research in the Levant to bridge the geographic gaps in the tested populations, demonstrating her particular interest in drawing connections between the various Arabic-speaking populations studied at the BGRL.

Aksoy, on his own initiative, pursued just such a comparative study in the Levant. But having firmly denied any racial connection between Eti-Turks and neighboring Arabs, he aimed to undermine the presumed relationship between Arabic-speaking Alawites in Turkey, Lebanon, and Syria. In 1960, Aksoy coordinated with Lebanese physicians to compare the frequency of HbS among Eti-Turk schoolchildren to that among Alawites in

Tripoli, Lebanon.[45] He introduced his study with an acknowledgment of similarities between the two groups, commenting, "The Eti-Turks and the Allewits [sic] of Syria and Lebanon may have the same racial background."[46] Yet when publishing his results, he overexaggerated the genetic distinction between the two groups, explaining that 16.8 percent of Eti-Turks carried HbS, compared to "only 4 per cent" of Lebanese Alawites.[47] This summary is misleadingly oversimplified. His data from Lebanon came from a single village, but the percentage he offered for Turkey is an average calculated from the populations of five different villages, which, considered separately, reveal highly variable frequencies of sickling hemoglobin. In fact, James Neel, one of the peer reviewers of Aksoy's paper, pointed out that this data could well be interpreted as representing a genetic gradient and evidence of past emigration from the Levant with a serial founder effect.[48] However, in Aksoy's final draft, he obscured this potential hypothesis, leaving a clear implication that Turkish and Lebanese Alawites did not share a racial background after all.

Aksoy's rejection of shared ancestry between Alawites in Turkey and Lebanon was motivated at least in part by the local social context of Mersin, where Turkish discrimination against the Alawite community was expressed primarily through anti-Arab sentiment. Aksoy's biography contains extracts from an undated interview wherein he explains his decision to revive the term "Eti-Turk" for his sickle cell research:

> These patients all spoke Arabic along with Turkish and lived in the province's Garden District neighborhood. In Mersin and Adana, an expression like "Arab servant" [*Arap uşağı*] was used to describe the place of this community. Despite speaking Arabic, it was not known exactly from what origins they plausibly came. Great Atatürk, in a manner appropriate to his genius, gave the name "Eti-Turk" to this community during the Hatay crisis. I too opted to use this term.[49]

In other words, Aksoy claimed that his usage of the term "Eti-Turk" reflected not a discriminatory vestige of Turkish nationalism but rather a sympathetic gesture to his research subjects, a complimentary label for a community held in low regard by their ethnic Turkish neighbors. However, this does not explain why he should have been so invested in producing and publishing an anglophone scientific study to insist on an ethnic distinction between the "Fellah" of Turkey and of the Levant. I have found no evidence that the results of this study were ever published for a Turkish academic audience.[50]

Considering the descriptive shifts from "separate and distinct" to "Arabic-speaking" in Aksoy's coauthored papers with Lehmann and Ikin, his transregional collaborative network probably significantly shaped his motives for this study. Aksoy was responding to the British scientists' expectations of finding genetic relationships between Arab populations across the Middle East—an assumption surely influenced by contemporary Nasserism in Egypt and pan-Arabism in Syria. Indeed, Gamal Abdel Nasser's rising popularity across the Arab world was a severe blow to the pro-Western, anti-Soviet Turkish government's efforts to become a regional power during the 1950s. Syria's entrance into a mutual defense pact with Egypt and Saudi Arabia in late 1955, an agreement understood to be a neutralist counterweight to the Baghdad Pact, renewed Turkish-Syrian tensions to their highest level since the 1930s Hatay Crisis. Even as the delegates of the Abnormal Haemoglobins conference met in Istanbul in September 1957, Turkish troops had been mobilized along the Syrian border, conducting maneuvers in response to Syria's receipt of Soviet weapons shipments. The conflict stopped short of open warfare, but the political fusion of Syria and Egypt into the United Arab Republic (UAR) in February 1958 further exacerbated Turkish-Arab hostilities.[51] Although the UAR dissolved after only three years, the popular embrace of Arab identity by a wide range of Arabic speakers colored the subsequent discovery of sickle cell disease in the Levant, Egypt, and eastern Saudi Arabia.

WHICH ARABS HAVE SICKLE CELL DISEASE?

Medical researchers in other Arab countries responded to Lehmann's and Aksoy's work in different ways, reflecting different internal politics. The initial linkage of sickle cells to "Negro" ancestry meant that the trait was imagined to delineate the racial boundaries between black Africans and the "white" Berber and Arab inhabitants of the continent's Mediterranean fringes. In 1952, Anthony C. Allison tested several hundred Berbers and Arabs from French Algeria and the Hashemite Kingdom of Jordan for sickle cells. Finding only two individuals harboring sickle cells, Allison pronounced that the trait did not exist among "pure" Arabs and Berbers and that in the rare cases where sickling did appear, "there is undoubtedly an admixture of

Negro blood."[52] This apparent affirmation of Arab whiteness might have satisfied an earlier generation of Arab scientists. But by the 1960s, they were concerned less with international projections of Arab racial identity and more with the volatile internal politics of their individual nation-states. Major sectarian differences in rates of sickle cell disease reshaped their interpretations of ethnic origins and admixture. This, in turn, influenced how Arab physicians responded to Lehmann's Veddoid hypothesis and its promotion of an indigenous Arabian (if not Arab) origin for sickle cell disease.

Physicians working at the notoriously pro-Arab American University of Beirut (AUB) initially seemed most open to Lehmann's ideas. For example, in a 1962 report documenting the first discovery of sickle cell disease in Lebanon and Syria, AUB hematologists Munib Shahid and Najib Abu Haydar commented, "It appears that sickle cell anaemia is more prevalent in the northern confines of the Arabian peninsula than has hitherto been realized, possibly giving more strength to Lehman's assumption about the [Arabian] origin of the sickling gene."[53] However, the Lebanese often invoked Lehmann's hypothesis in a way that collapsed the careful racial distinctions Lehmann had constructed between the "Veddoid" *akhdām* and Yemeni Arabs. A pair of AUB pediatricians, Ibrahim Dabbous and Salim Firzli, noted that virtually all of their sickle cell patients belonged to Lebanon's Muslim communities. Dabbous and Firzli—a Sunni Muslim and a Greek Orthodox Christian, respectively—were concerned about the possible incorporation of this genetic trait into Phoenicianist nationalist discourses that posited Lebanese Christians as indigenous and Muslims as later invaders.

Noting that southern Lebanon "is known to have been populated by Arabian tribes from Southern Arabia and Yemen, since second century B.C.," they cited Lehmann to suggest that the sickle cell gene may have been "carried by the pre- and post-Islamic Arab settlers to Lebanon." They even included a map illustrating Lehmann's hypothesized spread of the HbS gene out from Arabia. However, Dabbous and Firzli warned against any simplistic interpretation that "the predominence of the sickle cell gene in the Lebanese Mohammedans is a reflection of the fact that the Mohammedans are predominantly of Southern Arab extraction while the Christians represent the 'original settlers,'" which would be "probably erroneous and at best impossible to substantiate."[54] In fact, they argued, the contemporary sickle cell distribution obscured the known historical facts that Christian and Muslim Lebanese share the same ancestral origins among southern Arabian tribes. In Dabbous and Firzli's interpretation, if Lehmann's hypothesis was correct, the sickling gene might have arrived during Lebanon's "early history" among the ancestors of both Christian and Muslim Lebanese.

To explain why only Muslims presently carried the sickle cell trait in Lebanon, they turned to malaria. Ultimately, they concluded that the current distribution of sickle cell disease dated neither to the Neolithic era nor to the original settlement of Lebanon. Rather, it was a product of more recent social and environmental history: the tumultuous period between the thirteenth and fifteenth centuries C.E., when Muslim groups from across the Middle East and North Africa "settled in the coastal towns outside Mount Lebanon proper."[55] While the local Christians and Druze remained in the malaria-free mountainous areas, the malarious coastal regions propagated the sickle cell gene within the newly amalgamated Muslim communities. Through this explanation, Dabbous and Firzli reconciled aspects of Lehmann's racial hypothesis with the environmental selective effects of malaria over a relatively recent timescale. They foregrounded Arab regional migrations and Lebanon's unique geography over a commitment to any one theory of genetic origin.

In contrast to the Lebanese, Iraqi and Egyptian physicians stridently opposed Lehmann's hypothesis. They argued that sickle cell disease in the Middle East was more likely to have African than Indian or Arabian Veddoid origins, pointing to various "negroid" physical features in their sickle cell patients from the Basra region and the oases of Egypt's Western Desert.[56] In particular, Karim Kamel of Cairo's Ain Shams University Medical School consistently rejected any indigenous Arabian source for the sickle cell trait he observed among patients from Egypt or the eastern coastline of the Arabian Peninsula. Instead, he argued that Arabs with the sickling gene owed it to the "great numbers of Negroes from Africa" brought to Egypt and the Trucial Coast through the slave trade, as well as through Omani imperial rule over East Africa.[57] Whereas the debate over the anthropological implications of sickle cells in Lebanon circled around internal political contestations of Arab identity, Iraqi and Egyptian scientists focused on the historical relationship between Arabs and Africans. This emphasis on recognizing the sizeable Arab role in the African slave trade deserves further examination, especially given the prominence of Egypt in the Non-Aligned and Pan-African movements in the 1960s.

Lehmann met similar opposition from American physicians working in Aramco's medical departments in eastern Saudi Arabia. As described in Chapter 3, in the Saudi blood samples sent to England by Aramco anthropologist George Maranjian, Lehmann found a high proportion of sickle cell carriers among the Shi'i communities of the al-Hasa and Qatif oases. In this case, he embraced the role of malaria: sectarian differences, he explained, reflected the longer residence of Shi'i communities in the oases, which had high endemic malaria rates. In contrast, Sunni communities had arrived more recently from the malaria-free deserts, wherein they had lost the sickle cell trait bequeathed by their "pre-Arab" ancestors.[58] But for Armand Gelpi, an Aramco physician who arrived in Dhahran in 1958, Lehmann's hypothesis overlooked "a convergence of historical and genetic evidence to support the theory that there was a great deal of African admixture into the populations of eastern Arabia."[59] The perpetuation of an East African slave trade near the oases into the early years of the twentieth century convinced Gelpi that Africans had brought the sickle cell gene to Arabia, rather than the other way around.[60]

In other words, Lehmann found his hypothesis contested not only by advocates of a malaria-driven, multicentric origin hypothesis but also by advocates of a single origin for the sickle cell gene—only in Africa, not in Arabia. In 1969, Gelpi left Aramco and took a position at Stanford University's medical school. As the Black Panthers active in neighboring Oakland launched a well-publicized campaign to screen African Americans for sickle cell disease, Gelpi continued to publish data he had collected from his Saudi patients.[61] In a 1973 review article, he synthesized historical and genetic evidence that the sickle cell gene found in India, the Middle East, and Southern Europe had evolved in Africa. He argued that the Arabian Peninsula was "pivotal in the dispersion of the sickle-cell gene" not because the gene had originated there but because the Arabs had readily intermarried with African slaves. Subsequently, this African gene had been carried by Muslim-Arab conquests and colonization across the Mediterranean and South Asia.[62]

In the 1980s, the development of restriction-enzyme genotyping technology confirmed that the sickle cell gene had, in fact, evolved independently in different locations. These genotyping techniques suggested that both Lehmann and Gelpi were correct about the importance of the Arabian Peninsula for understanding the global distribution of sickle cell disease, but not in the way that either one had imagined. The variety of sickle cell hemoglobin prevalent in southwest Arabia (i.e., Yemen) proved to be the same as the one dominant in Nigeria—one of at least four independent mutations across the African continent. Meanwhile, Gelpi's patients in eastern Arabia possessed their own HbS mutation independent of any of the African types, which they shared with the Indian tribal populations Lehmann had studied in the early 1950s.[63] In retirement, Gelpi admitted that he had overlooked clinical evidence that his Saudi patients had a form of the sickle cell gene differing from that of Africans, believing that the disease's different manifestations could be explained by coinciding medical conditions. However, he accepted the discovery of multicentric origins for HbS and its so-called "Arab-Indian haplotype" as a satisfactory explanation for both the clinical variation and geographic distribution of sickle cell disease.[64]

Gelpi also represented a trend of Saudi-based sickle-cell researchers who cited Aksoy's work yet challenged the latter's representation of the Eti-Turks. While Aksoy's Lebanese, Egyptian, and British counterparts had left unquestioned Aksoy's assertions about the Eti-Turks and their origins, Gelpi insisted that the "Eti-Turks are actually of Arab ancestry"[65] and that "ethnically, this is an Arab group, and likely to have descended from Syrian or Egyptian immigrants."[66] Similarly, clinicians at a Saudi military hospital in Jeddah found cases of a rare form of sickle cell disease in their Hijazi patients. Noting that Aksoy had reported identical cases among the Eti-Turks, they remarked that "Eti-Turk are Moslem Arabs [who] may be descended from the same tribe as our patients."[67] However, restriction enzyme analysis of the sickle cell gene seemed to buoy Aksoy's contentions that the Eti-Turks had little to no Arab ancestry: out of sixty-eight Eti-Turk sickle cell patients tested in 1985, only one carried a copy of the "Arab-Indian" haplotype, whereas the vast majority carried the "Benin" haplotype believed to have originated in West Africa. Aksoy and his (Dutch and Turkish) collaborators in the study concluded that "influence from the population of [the Hijaz] among the Eti-Turks is minimal," whereas "North African admixture to the Eti-Turk population during the nineteenth century may have played a most important role." However, they acknowledged that sickle cell genetics only told part of the story, because the "origin of [the Eti-Turks'] other genetic markers, such as blood groups, haptoglobins, transferrins, G6PD variants, appears to be Arabic."[68] The authors left the question of the Eti-Turks' own origins, whether ethnic or biogeographical, unresolved.

GENETIC DISORDERS AS RACIAL CATEGORIES

Aksoy's inconsistencies in his portrayal of the Alawites represent neither a solely individual bias nor a simplistic state-driven imposition of Turkish nationalist ideology. Instead, they are only another example of how human genetics, as a field, is rife with inconsistencies with regard to defining its subjects. The contrast between Lehmann's and Ikin's portrayals of the *akhdām* in Yemen also reflects this fundamental problem. Where Ikin describes them as "Zabidi Arabs who show some Veddoid features," for Lehmann, they were "Veddoid survivals" belonging to a "pre-Arab" indigenous population quite different from their Arab neighbors.[69] In either case, it was neither language, nor religion, nor even colorism, but only the caste-like features of Yemeni social structure that made the *akhdām* scientifically legible as a distinct population. For Ikin, who studied genetic traits found in all humans, the boundaries of this social category blurred at the biological level: their blood groups were "on the whole similar [. . .] to the Yemenite Arabs," with the addition of "a considerable Negroid component."[70] In other words, the *akhdām* or Zabidi Arabs were anthropologically unremarkable for the region they inhabited, and the aggregated genetic data could be interpreted as legitimating either of the local narratives of their origin. But the pathological trait of sickle cell disease in certain families provided the basis for Lehmann's racialization of the entire community as Veddoids, a narrative that detached them from Yemeni historical memory.

In the case of the Eti-Turks or Alawites, sickle cell disease similarly served as the diagnostic feature by which Aksoy could distance them from the Arabness suggested by their language. The British researchers evidently regarded the Alawites' language as more significant to determining their ancestry than did Aksoy, who personally communicated with the community's members in Turkish. Regardless, in contrast to Lehmann, Aksoy's anthropological determinations for the Eti-Turks were negative ones: he used sickle cells to argue that they were *not* Arab, that they were *not* the same as Levantine Alawites. But as for their "racial background" and the ultimate source of their sickle cells, he was content with inconclusive results. Looking to other kinds of blood proteins, Aksoy argued alternately that their sickle cell gene reflected Indian ancestry[71] or a history of African admixture.[72] The only consistent feature of his work on Eti-Turks was the erasure of any Arab ethnic relationship. Even when invoking Lehmann's hypothesis, Aksoy's publications in the 1960s dropped his colleague's insistent emphasis on Arabia and referred only to the "Veddoids of India."

A year before his death in 1985, Lehmann wrote a reminiscence of his early studies on sickle cell disease. By this point in time, new biochemical techniques seemed to support the multicentric hypothesis, wherein the sickle cell mutation had evolved independently in different geographical locations with high rates of malaria infection. Lehmann's Veddoid hypothesis languished in relative obscurity. Explaining his resistance to the multicentric hypothesis and its preoccupation with malaria, he complained:

> When [the relationship between sickling and malaria] became generally accepted there was a tendency to throw the anthropological value of the sickle cell trait into the waste basket. I consider this a mistake. Presumably, all human characteristics have selective value and will respond to environments. But does this permit us to remove any such characteristic from the arsenal of the anthropologist as soon as an environmental aspect has been clarified?[73]

This comment conveys some of Lehmann's motivations for the construction and longtime defense of his Veddoid hypothesis. It is difficult to take seriously Lehmann's claim to represent an anthropologically informed perspective, given that he was a product of colonial medical services who relied substantially on the racial classifications of the now-reviled Carleton Coon. Indeed, even as fellow geneticists and anthropologists disavowed Coon immediately following the 1962 publication of his *The Origin of Races*, Lehmann continued to cite Coon as an authority on Arabian population history as late as 1974.[74] However, Lehmann utilized Coon's work and rejected Yemeni beliefs that the *akhdām* had African origins in order to challenge other contemporary researchers who maintained that sickle cell disease was a "Negroid" trait. Furthermore, his attention to the "slave-like" condition of the *akhdām* revealed a desire to avoid environmental determinism and highlight social and cultural factors in human evolution. He emphasized how poverty and ethnic discrimination—that is, the effects of local forms of racialization—could maintain high rates of the sickle cell gene in specific classes of an urban population.

This social element links the *akhdām* of Yemen to the *Arap uşakları* of southern Turkey. Together, their transmutation into Veddoids and Eti-Turks pushed sickle cell research past a black-white racial dichotomy into an unstable realm, demanding the use of third racial categories and unverifiable histories of admixture. However, whereas Lehmann was driven to defend an evolutionary hypothesis before the international scientific community, Aksoy's motivations mostly reflected local concerns. An avowed Kemalist like most members of his professional class and age group, Aksoy expressly denied that his nationalist politics had affected his scientific work. He claimed that he used the term "Eti-Turk" to distance his Alawite patients from racialized Turkish discrimination against Arabs. He also pointed out that Westerners' publications on sickle cell disease (e.g., those of Lehmann) had also used this term.[75] By obliquely invoking the figure of the Western researcher, removed from local politics and thus above ideological suspicion, Aksoy legitimized his choice as scientifically neutral. This suggests that as a Turkish scientist, Aksoy perceived his authority as deriving from, but ultimately subordinate to, Western collaborators. In this way, he embodied the "paradoxes of position" that accompany other Middle Eastern scientists' efforts to establish both local and global legitimacy as "professionals participating in international scientific practices and debates."[76] But Aksoy did not acknowledge, or perhaps did not recognize, the extent of his own influence in constructing the Eti-Turk as a transnational scientific artifact. Aksoy appropriated and refashioned a narrative from 1930s Kemalist nationalism for the disembodied conceptual world of sickle cell genetics. In doing so, he imposed the modern borders of the Turkish state upon all attempts to locate a historical explanation for Alawite sickle cells.

Israeli clinicians, to a significant extent, had drawn the attention of sickle cell researchers worldwide toward the Middle East. The recent Israeli media scandal surrounding the early 1950s medical investigations demonstrates that, despite the fact that Yemenite Jews were eventually shown to not carry the disease, the search for Yemenite sickle cells reflected a broader medical racialization of Mizrahim. This medical discrimination enabled extensive research into another genetic disorder from which many Mizrahim did in fact suffer: favism. The presence of favism among a wider range of communities in the Middle East prompted geneticists to use the disorder not only to trace racial ancestry but also to substantiate more recent and specific histories of communal migration promoted by national origin myths. By pursuing such claims about local social history, these geneticists instigated further pushback on evolutionary hypotheses that suggested malaria and the environment had shaped human genetics more profoundly than social and cultural identities.

Chapter 5
GENES AGAINST BEANS

SHORTLY AFTER THE ANNEXATION OF AUSTRIA TO NAZI GERmany in 1938, the Jewish pediatrician Richard Lederer left the University of Vienna to take up a post at the Royal College of Medicine in Baghdad. Lederer, with his reputation as a dedicated physician and clinical investigator, was also appointed the personal pediatrician of young Faisal II, then the Crown Prince of Iraq. Despite this privileged position, Lederer had trouble adjusting to the Mesopotamian climate and diagnosing unfamiliar "tropical" diseases in his new patients. His Iraqi colleagues drew his attention to one particular condition, a seasonal phenomenon of acute anemia, jaundice, and dark urine in young children. After differentiating these symptoms from likely suspects, such as the "blackwater fever" caused by malaria and quinine treatment, Lederer began a concerted investigation to characterize what he believed to be a previously unknown disease, which he named "Baghdad spring anemia." In the course of his research, he discovered the Italian medical literature on *favismo* (favism), a condition brought on by the consumption of fava beans that allegedly occurred only among the inhabitants of Sicily and Sardinia. Lederer noted that favism had markedly similar symptoms to those of his own patients, and he found that many of them had eaten fava beans immediately before the onset of illness. However, he was unable to conclusively show that fava beans were the causative agent. Lederer therefore published a description of Baghdad spring anemia as a separate disorder with curious demographic features: all his patients were "light-complexioned" boys under age five, and thirteen out of fourteen were Iraqi Jews.[1]

After Lederer's untimely death by skin disease in early 1941, the task of confirming that Baghdad spring anemia and favism were the same disorder fell to a new group of researchers: European-trained Ashkenazi Jews, like Lederer, studying Iraqi Jewish immigrants to Israel in the late 1950s. By that time, this obscure and exotic disease of the Mediterranean and Fertile Crescent formed the basis for a transnational network of medical researchers extending from Seattle and Chicago to Naples, Tel Aviv, and Shiraz. Favism rapidly attracted this international attention due to the discovery of its cause: a hereditary deficiency of the enzyme glucose-6-phosphate dehydrogenase (G6PD). G6PD plays an important role in energy metabolism in the cells of all kinds of organisms, from bacteria to plants and animals. In humans, G6PD is crucial for the protection of red blood cells from oxidative stress. Mutations in the gene responsible for producing G6PD can result in abnormally low levels of the enzyme, i.e., G6PD deficiency (G6PDd). Certain medications, infections, and foods like fava beans may cause the rupture of red blood cells in individuals with this deficiency, producing anemia and jaundice. In mild cases, this may be experienced only as fatigue and shortness of breath, but severe cases may lead to organ failure and death. G6PDd is currently recognized as the most common inherited enzyme deficiency in the world, with recent estimates of 400 million people affected.[2]

However, the prevalence of the condition is not evenly distributed. Because the G6PD gene is located on the X chromosome, clinical symptoms of the deficiency manifest more often in males than in females. Furthermore, mutations causing the deficiency occur most frequently in African, Asian, and Mediterranean populations. The form of G6PDd most common in Africa is relatively mild, whereas the genetic variant common across the Mediterranean and Middle East—the one responsible for favism—is more severe. Favism had long been recognized as a potentially serious condition by local peoples, whose historical lore and folk medicine included warnings against the consumption of fava beans. However, its incorporation into a major international research agenda began only in the 1950s, in concert with postwar malaria treatment and eradication programs. Because certain antimalarial drugs trigger acute hemolytic anemia in G6PD-deficient patients, the discovery of G6PDd and the determination of its biochemical and genetic characteristics was a direct outgrowth of malaria research. Its relevance to malaria drove national health agencies to allocate funding specifically to G6PDd surveys in many countries throughout the 1960s, while the World Health Organization (WHO) convened expert committees to standardize terminology and methodology for the growing number of researchers working across the Americas, Southern Europe, Africa, and Asia.

Despite its close connection to antimalaria campaigns, from the moment of its discovery, G6PDd took on not only clinical but also anthropological significance. In the United States, University of Chicago researchers first characterized the condition during the Stateville Penitentiary malaria projects, which used prisoners to test a range of new antimalarial drugs.[3] They found that about 10 percent of African American prisoners, but not white ones, developed hemolytic anemia in response to primaquine; a further study published in 1956 demonstrated that a deficiency of G6PD caused this response.[4] Like the simultaneous research conducted on sickle cell disease, the U.S.-based studies of G6PDd initially framed the condition as a "Negro" disease indicative of African ancestry. But after studies in Italy and Israel independently confirmed that G6PD deficiency was also the cause of favism, geneticists influenced by the Chicago school, like Arno G. Motulsky, sought to explain the deficiency's prevalence among both Africans and the "Caucasians" of the Mediterranean and Middle East. After reviewing a compilation of available data from Central Africa, Asia, and the Mediterranean, Motulsky proposed that G6PDd, like sickle cell trait and thalassemia, could confer resistance to falciparum malaria parasites. Therefore, he argued, natural selection acted to maintain the defective gene within populations inhabiting regions of endemic malaria.[5]

Yet the data that Motulsky compiled from South and West Asia introduced a potentially confounding factor into his malaria selection hypothesis. Like sickle cell disease, the frequency of favism in these regions varied significantly between groups inhabiting the same environment but divided by social factors like religion, language, and tribal affiliation. These genetic variations therefore mapped onto major social contestations in the young nation-states of the Middle East. Focusing on the years between 1955 and 1970, this chapter traces how medical researchers in the Middle East used favism to create a medical-anthropological discourse that projected contemporary nationalist ideologies deep into the past. Scientists in Israel and Iran configured a metabolic response to both modern pharmaceuticals and traditional cuisine as living testimony to thousands of years of migration, war, slavery, religious conversion, and social transformation. Reduced quantities of the G6PD enzyme in a handful of individuals became the basis for dramatic assertions about the ancestry of ethnoreligious communities and, by extrapolation, about national origins.

I argue that by favoring locally specific narratives over universal explanations like the malaria hypothesis, favism research in the Middle East expressed ambivalence about the international public health regimes and biomedical research agendas directed by Europe and the United States during this period. As in the case of sickle cell disease, this ambivalence did not necessarily represent ideological rifts between Western and Middle Eastern researchers but rather produced factions of scientists who favored different kinds of evolutionary hypotheses. The distinguishing feature of favism lies in its entanglement with malaria not only as an abstract force of gradual natural selection but also as an immediate public health issue. Furthermore, the disorder's unique interaction with fava beans, a significant protein source in the Middle East, made it a concern for national and international management of agriculture and public nutrition. Medical researchers from these different fields all confronted the urgent need to address the ethnic and religious variability of favism rates. As a result, the faction that converged around Israeli and Iranian ethnonational explanations for genetic difference included a number of American and European scientists who collaborated with physicians in those countries.

MALARIA AND MIZRAHIM: FAVISM IN ISRAEL

Favism research brought several Middle Eastern scientists and institutions to international prominence. Chief among these was the Tel-Hashomer Government Hospital in Israel and its director from 1953 to 1971, Chaim Sheba. Sheba, who had served in leadership roles in the Israeli military and Ministry of Health before taking the reins at Tel-Hashomer, is now considered a quintessential representative of Zionist approaches to medicine and public health. Like the vast majority of Zionist settlers in Mandate Palestine, Sheba was an Ashkenazi Jew from Eastern Europe, and he regarded himself as a pioneer in an unfamiliar and even hostile landscape. In a 1971 keynote address in Tel Aviv just months before his death, he recalled his journey by ship from Europe to Palestine in 1932. Lamenting that he had not taken a course in "tropical medicine" while a medical student in Vienna, he spent the six-day voyage reading up on malaria. He noticed that physicians in Palestine had recorded a high number of "off-season" cases of blackwater fever occurring in spring rather than autumn. Upon arrival in the Mandate, he discussed this curiosity with a medical colleague who suggested the cause might be favism, a condition Sheba had "not even heard of" in Europe.[6]

According to Sheba's self-narrative, from that point onward, favism became the primary preoccupation of his medical career. In 1938, Sheba claimed, he struggled to correct his colleagues' misdiagnoses of anemia symptoms among Yemenite, Iraqi, and Kurdish Jews by pointing out that "fava bean is a favorite 'delicatesse' of the oriental youth." However, he found himself "doubted and even attacked by the more senior people who introduced into biology their social credo that all men are equal."[7] During the Second World War, Sheba worked for the British military medical services but found most of his superiors there equally skeptical of his warnings that soldiers of Indian and Mediterranean origin suffered hemolytic responses to sulfapyridine (an antibiotic) and pamaquine (an antimalarial) due to their ethnic background. Sheba claimed that his years of clinical observations, particularly of the "oriental youth" whose anemia Lederer had documented in Baghdad, were appreciated only after the first decade of Israeli statehood, when Mizrahi communities were compelled to immigrate en masse to Israel.

Israel's aggressive antimalaria practices figure prominently in Mizrahi memoirs about the discrimination they faced from Ashkenazi immigration officials. Having left their home countries under duress, they expected a warm and festive welcome upon their arrival in the Jewish state. Instead, they were shocked and humiliated by the experience of being marched off airplanes or ships and immediately sprayed with the pesticide DDT without explanation.[8] The extensive use of DDT in Israel was in line with international practices in the 1950s, and the policy of spraying new arrivals was adopted from the British mandate authorities' spraying of Holocaust refugees and survivors.[9] While Ashkenazim and Mizrahim alike suffered the indignity of DDT, for Mizrahim, this experience retrospectively seemed to foreshadow the many injustices perpetrated against them by the Ashkenazi-dominated Israeli medical establishment.[10] Indeed, government health officials unwittingly exposed many of these immigrants to further harm with prescriptions for primaquine and other drugs that triggered the dangerous effects of favism. By observing and studying the ensuing hemolytic reactions, Ashkenazi scientists like Sheba not only became leading figures within a growing international research network on inherited blood disorders. They also utilized the medical discourse on favism to biologically reify Ashkenazi perceptions of Mizrahi racial difference and Zionist visions of Jewish history.

Throughout the 1950s, Sheba and his colleagues at Tel-Hashomer Hospital identified a number of hereditary medical conditions, including favism, which appeared almost exclusively among Mizrahi patients originating from different countries (e.g., Iraq, Iran, Morocco, Yemen). Rather than follow the precedent established by other Israeli workers who focused their studies on single communities, the Tel-Hashomer team consistently invoked the term "non-Ashkenazi" (in Hebrew, *lo-ashkenazi*) as a biomedical category. In so doing, they collapsed the distinctions between the various "Oriental" communities, subsuming them all under a negatively defined category that constituted the Ashkenazi as the normative Jewish body. In their publications on favism, they also occasionally used terms like "Sephardi" and then "Oriental" to contrast their research subjects with the Ashkenazim. Regardless, their descriptions of difference fell back on a highly racialized language of "complexion." This rhetorical pattern traces its roots to Sheba's British military days, when he first published his observations on the differential incidence of blood diseases in "dark-skinned Mediterranean" as opposed to "lighter-skinned immigrant" Jewish populations.[11]

Between 1957 and 1958, Sheba and his colleagues published a series of highly cited papers that identified favism as a form of G6PD deficiency (independently of, but at the same time as, Italian researchers who published the same discovery). These papers all contained variations of this general description: "The Ashkenazic is the light-complexioned Jew originating from Eastern, Central or Western Europe, and the non-Ashkenazic usually is the dark-complexioned Jew of Oriental or Mediterranean origin. The chances of misplacements of subjects were negligible, as until now the distinction between the two groups is very clear."[12] The emphasis on a clear racial distribution for G6PDd mirrored not only local prejudices in Israel but also American and Italian reports on the condition, which stressed its uniqueness to African Americans, Greek Americans, Sardinians, and Sicilians.

The Israeli reports, framed in terms of complexion, called to mind the "black/white" racial discourse animating both Israeli and American society and provided a language for American researchers to understand the significance of the Israeli studies. Yet unlike American Ashkenazi hematologists like William Dameshek, who had so readily suggested that Yemenite Jews might be partly or wholly "Negro" (see Chapter 4), Israeli researchers like Sheba sought alternative explanations that preserved their faith in historical Jewish endogamy. Although his team's work highlighted a sense of racial difference between Jewish communities, the possibility of some ancestral connection between Jews and other Mediterraneans with "Negroes" appears to have troubled Sheba. Like Fritz Dreyfuss had done a few years earlier for sickle cell disease, Sheba proposed a "multicentric" hypothesis, dismissing the historical possibility of Mediterranean-Negro intermarriage: "Does [a similar incidence rate of G6PDd] indicate infiltration of Mediterraneans into the countries of origin of the Negroes in Africa? A more likely explanation would be that the trait is not purely Mediterranean and that a similar mutation occurred in various regions of the earth."[13] It did not take long to find international support and evidence for Sheba's contention that multiple mutation events were responsible for the global distribution of G6PDd. His research team collaborated with a pair of American physicians, Paul Marks and Ruth Gross, who synthesized the Israeli, Italian, and American data into a highly influential paper that proposed that different underlying genetic mechanisms could account for the different clinical manifestations of G6PDd between "Negroes and Caucasians."[14]

After this development, Sheba took no interest in the question of how and why G6PDd might have evolved in multiple places; rather, he focused on favism as an exclusively "Mediterranean" problem. In the process, Sheba gained local recognition, winning the Tel Aviv municipality's Henrietta Szold medical prize in 1959 for research on "hemolytic tendency among Oriental communities in Israel."[15] Sheba's rising star as a medical researcher, predicated largely upon favism, was thus indebted to a category of immigrants whose very arrival in Israel he had sought to block or delay while serving as minister of health.[16]

Despite his discriminatory attitudes toward Mizrahi Jews, Sheba ultimately perceived the variable incidence of favism as an opportunity to validate a biblical narrative of shared Jewish ancestry. Nurit Kirsh and Nadia Abu El-Haj have thoroughly analyzed how Sheba's personal commitment to Zionism guided his research interests, methods, and interpretations. For example, in scientific publications and conference presentations, he habitually referred to Jewish populations as direct descendants of "the tribes of Israel" (*shevaṭe yiśra'el*) and explained the shared genetic traits of Jewish communities with non-Jews in different regions as resulting from common ancestries traceable to Noah's sons, Shem, Ham, and Japheth.[17] These tropes reveal the apparent paradox between Israeli geneticists' fascination with the potential value of genetic anthropology for historical insight and their simultaneously fixed beliefs about Jewish origins, which genetic data could not shake.[18]

On the other hand, Abu El-Haj's suggestion that the work of Israeli geneticists "reads like an almost random quest for genetic signs that would prove that a single, historic Jewish population did indeed exist" filters out the international context of the discovery of genetic markers like those responsible for favism.[19] Furthermore, early Israeli medical research was conducted under strained material conditions. As a government hospital, Tel-Hashomer was strapped for cash throughout the 1950s and 1960s—indeed, it operated out of crumbling Mandate-era Quonset huts until the 1970s. Sheba had no stable research budget, let alone a directive from the Israeli state or other Zionist organization that would support such a "random quest." He routinely sought funds for research and even for hospital facility maintenance from American sources like the Blood Research Foundation (which funded Muzaffer Aksoy's work in Turkey), and the American Jewish Joint Distribution Committee (JDC). And, in fact, his passion for what he called "medical archaeology" or "medical anthropology" sometimes actually impeded his fundraising efforts.

For example, the JDC initially rejected Sheba's 1959 application to study "the wanderings of the Jews . . . through blood chemistry" on the basis that their philanthropic funds could not be used for "pure research with no medical significance."[20] Charles Jordan, an advocate for Sheba at the Geneva JDC office, intervened with a more practical description of the research at Tel-Hashomer:

> The work involves an investigation of a hemalytic [sic] trait possessed by certain Oriental Jews. Shortly after the Iraqi Jews arrived in Israel, it was found that anemia would result in these people during the course of treatment for tuberculosis with streptomycin and para-amino-salycilic [sic] acid. Also observed was the fact that certain Iraqi and other Oriental Jews, when exposed to industrial detergents and dies [sic], or when harboring certain virus diseases, or when receiving aspirin for fever, would also develop a severe hemalytic anemia. Strangely enough primaquine treatment for malaria would also produce this effect.[21]

Jordan conceded that while "the 'medical archaeology' that Sheba talk[ed] about" was quite interesting, Tel-Hashomer's proposed research primarily concerned the clinical aspects of G6PD deficiency. Given that its effects could "interfere with the absorption of a significant segment of the immigrant population," Jordan successfully convinced the JDC to fund Sheba's work. By the middle of 1959, Sheba received a $10,000 grant, appropriated from the JDC's Malben program for disabled Jewish immigrants.[22]

Despite the practical rationale used to secure JDC funds, later that year, Sheba used some of that money to supply a field expedition to Ethiopia. The official purpose of the expedition was to search for the presence of G6PD deficiency within six major Ethiopian tribes, given the condition's high frequency among both West Africans and Mediterraneans and its medical significance for the use of antimalarial and antibacterial drugs in the country. The field team, led by the geneticists Avinoam Adam and Mariassa Bat-Miriam and accompanied by two physicians and a nurse from Tel-Hashomer, took the opportunity to conduct a broad range of tests and measurements for other genetic traits in addition to collecting blood samples for G6PD assessment.[23] Of course, Sheba had additional motives for organizing the expedition and a decidedly nonmedical interest in Ethiopia: the presence of a Jewish Ethiopian tribe, known as the Falasha. According to his biographer, he "wanted to know how the Falasha had arrived in Ethiopia and whether they also belonged to the Jewish gene pool."[24]

Indeed, at the time, the Jewish status of the Falasha confronted the Israeli state with both internal social disputes and delicate foreign policy considerations. Although several prominent personalities in Israel at the time, including President Yitzhak Ben-Zvi, supported the Jewishness of the Falasha, the state did not declare them eligible to immigrate to Israel under the Law of Return, instead granting only individuals special permission to settle in Israel.[25] This ambivalence arose from a deference to the Chief Rabbinate's 1954 ruling that Falasha people must undergo rabbinical conversion ceremonies prior to marrying Israeli Jews, as well as a desire not to antagonize the Ethiopian government, a key regional ally against the Arab states. The Ethiopian government restricted emigration and opposed any encouragement of tribalism, including foreign aid directed preferentially to the Falasha.[26] Sheba's need to coordinate the expedition with the Ethiopian Ministry of Health, who provided extensive support to the field team, required him to downplay his special interest in the Falasha and frame the study as a malaria-related project—one of the many instances of general medical aid Israel provided to Ethiopia in the late 1950s.[27] A severe malaria epidemic in the Ethiopian highlands during the latter half of 1958 likely boosted Ethiopian health officials' enthusiasm for such a study.[28] In any case, the Tel-Hashomer team ultimately determined, on the basis of over fifteen genetic traits—including the absence of any cases of favism—that the Falasha were biologically indistinguishable from other Ethiopian tribes.[29] Sheba decided the Falasha must therefore be the descendants of indigenous Jewish converts rather than exiles from Judaea.[30]

Beyond Sheba's own interests, the thorough studies conducted by the Ethiopia expedition and its embrace by Ethiopian authorities exemplified the relative strength of Israeli medical genetics generally and its program on favism specifically. British researchers central to the postwar traffic in blood, including Hermann Lehmann and Arthur Mourant, participated as technical partners rather than as leading investigators in the expedition. This reveals the preeminent status attained by Sheba and his colleagues in studies of this condition. The very fact that the Falasha claimed a Jewish identity apparently further legitimized the Israeli dominance in every stage of the study, from organization and fieldwork to publication of the results. Although the results of the Ethiopian expedition defied Sheba's hopes of finding more evidence in support of his grander theories on Jewish historical migrations, he continued to identify the Falasha as Jews, as did Mourant in his compilations of Jewish genetic data.[31]

Favism research figured prominently in Israel's first international genetics conference, on "the genetics of migrant and isolate populations," held in Jerusalem in September 1961.[32] The conference took place immediately before the Second International Congress of Human Genetics in Rome, from which it drew many of its foreign attendees: nearly 150 participants, including about 40 from 14 other countries, attended the Jerusalem meeting to present their research. This conference was indeed a watershed moment for the international recognition of Israel's blossoming community of geneticists, who appointed J. B. S. Haldane as president of the meeting. Most of the foreign visitors hailed from Europe or North America, including such luminaries as Mourant, Lehmann, Curt Stern, and James V. Neel, who served as session chairs. Future Nobel Prize winner Baruch Blumberg and Italian phylogenist Luigi Luca Cavalli-Sforza were also present. Sheba, serving as the conference's vice president, humbly delineated the purposes of the conference as a gathering for foreign geneticists to help the Israelis by sharing their advice on local research, strengthening professional bonds, and promoting the emerging field of medical genetics.[33] Most of the other speakers conversely marketed the value of Israel as a research site for the international scientific community. Arno Motulsky rhapsodized that Israel's veritable menagerie of "well-defined" immigrant populations, with "well-known" histories of isolation, made the country "a population geneticist's dream."[34] Throughout the conference, Israel was described as a "unique laboratory" and a "unique opportunity" for genetic research, and plenty of foreign scientists expressed enthusiasm for the Israeli research programs.

The conference included more presentations and posters on G6PDd than on any other genetic trait and brought together many representatives from the American and Italian research networks studying the condition. The presentations and recorded discussions reflected the cutting edge of G6PDd research, including the recent discovery that the gene was sex-linked as well as intense debates surrounding Motulsky's malaria selection hypothesis. Although scientists did not yet know how G6PDd could provide a selective advantage against malaria, most of the American and Italian researchers demonstrated an inclination to accept this hypothesis because the geographical

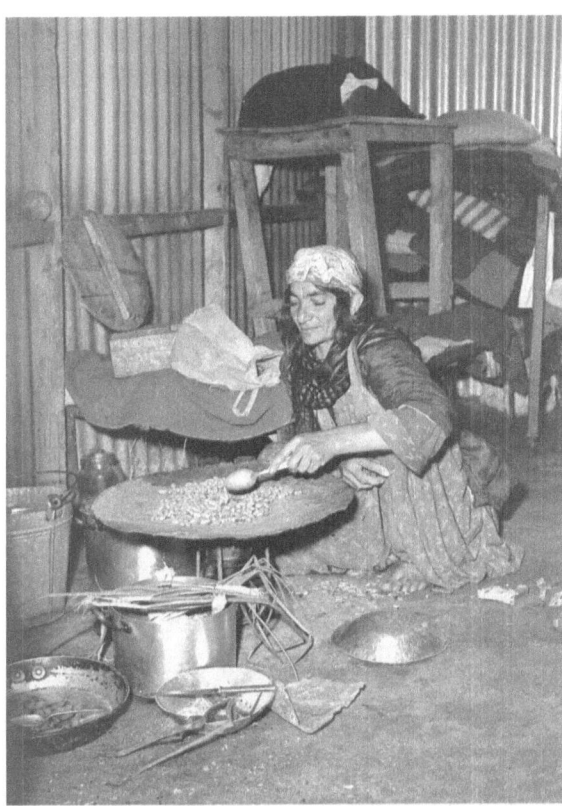

FIGURE 9. A Kurdish woman preparing food in her dwelling within the *ma'abarah* of Kiryat Shemonah, March 1956. Photograph by Moshe Pridan. Courtesy of the Government Press Office, Israel National Photo Collection.

distribution of G6PDd matched that of other blood disorders like sickle cell disease and thalassemia. In contrast, Sheba had profound doubts that natural selection could account for the favism distribution patterns in the Middle East, and Motulsky "challenged" him to present an alternative hypothesis.

Sheba responded to this challenge by pointing out that in Iraq, Jews and Muslims had lived alongside one another, "exposed to the same environment for about a hundred generations." Yet although Iraqi Jews had a relatively high frequency of favism, their Muslim neighbors almost never experienced it. Furthermore, the Jews of Kurdistan possessed the highest frequency of favism of any known community at the time—more than half of the tested Kurdish Jewish individuals carried the trait (Figure 9). However, the Kurdish Jews had formerly lived in high mountain villages completely free of malaria. Rather than by environment, Sheba argued, the distribution of favism could be explained by "ethnic origin, consanguineous marriage, and drift."

Essentially, his hypothesis relied on a series of founder effects through migrations that traced back to the ancient Levant: "The Kurdish Jews, a concentrate of the ancient Hebrews exiled to Babylon, and the Sardinians and other Mediterranean islanders who were Punes (Phoenicians) could have a common origin [...] on this assumption we would be able to explain at least why Sardinians should be so much more affected by favism than the Italians proper."[35] Sheba's romantic historical vision was effectively ignored in the subsequent discussions, which returned to the question of malaria.

In his final 1971 keynote address, after another ten years of research on favism, Sheba presented the following more-fleshed-out account: the defective gene was present among the ancient Judaeans and Phoenicians (that is, the Semites or sons of Shem), whose subsequent migrations across the Mediterranean and Southwest Asia explained why the Jews living in these regions, just like Sardinians, Sicilians, and Greek islanders, had relatively high rates of favism. The near absence of favism among Ashkenazim did not mean that they did not share origins with other Jews; rather, they were the descendants of a generation of Jewish men enslaved by Rome and forced to marry fair-complexioned "Japhethite" women, whose X chromosomes did not carry the gene for favism.[36] This basic narrative became increasingly convoluted as Sheba tried to account for differences between the various non-Ashkenazi communities and the geographic distribution of other inherited diseases.[37] Sheba also turned to favism to resolve the history of other religious communities in Palestine, such as the Samaritans. In contrast to previous anthropometric and serological studies that had supported the Samaritans' claims to be indigenous Israelites, Sheba accepted the Orthodox Jewish position that Samaritans had no ancestral relationship to Jews or the land of Palestine. After finding that Samaritans lacked any cases of favism, Sheba argued that genetics confirmed rabbinical accounts that the Samaritans actually belonged to "a foreign stock brought over by the Assyrians."[38]

Sheba's more fanciful hypotheses did not go uncontested by either his Israeli colleagues or their research subjects.[39] For example, the Samaritans reinterpreted Sheba's discovery that they lacked favism in a manner consonant with their beliefs. Pointing to the fact that favism in men is maternally inherited and also absent among Ashkenazi Jews, the Samaritan community leader argued that the Samaritans belong to the Tribe of Ephraim, while the Ashkenazim descend from the Tribe of Benjamin—the descendants of the biblical matriarch Rachel. He deduced that Rachel must not have possessed the gene for the enzyme deficiency, whereas the other wives of Jacob did and passed it down to their descendants among the Jewish population.[40] This seamless reversal of Sheba's interpretation demonstrates a Samaritan understanding that favism, or more specifically the genetic information attributed to it, carried an important epistemic weight for the community's ancestral claims vis-à-vis the Jews.

FAVISM AND RELIGION IN IRAN

The attachment of such detailed historical narratives to favism was not peculiar to the Israeli context. This phenomenon occurred across the entire international network that studied this enzyme deficiency. Furthermore, it was never simply the product of politically motivated local scientists imposing their nationalist views onto genetic data. Rather, these narratives were made possible by the convergence of national public health developments, international malaria eradication practices, global circulations of genetic expertise, and socioeconomic disparities between researchers and research subjects. The transnational dimensions of this convergence are readily visible in the favism research of James E. Bowman, an African American hematologist who began his career in Iran. He accepted the post of head of pathology at the Nemazee Hospital in Shiraz in 1955, seeking to escape segregation in the United States. Nemazee Hospital, privately funded by the wealthy Nemazee family, was a state-of-the-art medical facility—a far cry from the working conditions of Tel-Hashomer in Israel. Many of the other department heads, like Bowman, were Americans who had been recruited by the Iran Foundation, the joint Iranian-American organization created to manage and finance the hospital. Considered a "showpiece" of Iranian development, the hospital was regularly visited by Mohammad Reza Shah and other foreign dignitaries, whom Bowman had the opportunity to meet. Indeed, Bowman's social circle in Shiraz included the cream of the city's secular elite; he named as his closest friend Narcy Firouz, a wealthy land developer and scion of former Qajar royalty. In short, Bowman's position thrust him into a sheltered world wherein his knowledge of Iranian history was strongly shaped by the Pahlavi dynasty's secular nationalism, which exalted pre-Islamic Persian culture at the expense of other ethnicities and languages.

Bowman had no particular training in genetics, nor did he immediately recognize the symptoms of favism when he was confronted with afflicted patients in Shiraz. Yet once he made the correct diagnosis, he became fascinated by the condition and began corresponding with Israeli and American researchers studying G6PDd. Bowman worked with Deryck G. Walker, a British biochemist based at the neighboring University of Shiraz, to test enzyme levels using fava bean extracts. During 1959 and 1960, they traveled to Isfahan, Yazd, and tribal territories to take blood samples from different ethnic and religious communities. They found that favism appeared frequently within the normative population, the Persian-speaking Shi'i Muslim majority, as well as the nomadic tribal groups.[41] Meanwhile, religious minorities like Armenians and Zoroastrians had standard levels of the enzyme and seemed to have no cultural knowledge or folk medicine related to favism.[42]

Bowman and Walker traveled to New Julfa, the Armenian quarter of Isfahan, to add a further 153 Armenian blood samples to the 10 they had already collected from Armenian hospital staff in Shiraz. When publishing the data, they credited Caro Owen Minasian, the Isfahan-born physician and scholar of Iranian-Armenian history, for "making this survey possible" and thanked him for "his clarification of many points regarding the history of the Armenians."[43] These clarifications included a narrative of Armenian endogamy within the New Julfa community, which correlated with Bowman and Walker's genetic findings:

> The Armenians in New Julfa were brought to Iran as Christian captives in 1604 and 1605 by Shah Abbas the Great from Julfa in what is now Soviet Armenia. [. . .] While religious intolerance has not been permitted in Iran for many centuries, the Armenians were not encouraged to intermarry with the surrounding population. According to Dr. Caro Minasian, an Armenian physician and scholar, Armenians who have intermarried live outside of the area of New Julfa. This fact is supported by our finding of a virtual absence of [G6PDd] and by the blood group data.[44]

This account of the Armenians' arrival and enjoyment of "religious tolerance" in Isfahan represents an Iranian nationalist narrative that glosses over the total destruction of Julfa, the coerced relocation of the Armenian population, and the centuries of formal and informal discrimination they experienced in Iran. The fact that Bowman, who was quite active in the civil rights movement in his own country, could make such a distinction between captivity and slavery indicates his unquestioning acceptance of the dominant Iranian version of history, one to which Iranian Armenians also paid "homage" to secure their community's safety.[45]

Bowman took a special interest in Zoroastrianism, which had been the official religion of the Sassanian Empire of Persia prior to the Arab Muslim conquests of the seventh century C.E. Bowman's personal interactions with Zoroastrians had led him to expect biological evidence for their own claims to Iranian antiquity. The Zoroastrian chief of surgery at the Nemazee Hospital, Manouchehr Mavendad, not only helped Bowman collect blood samples but also "clarified Zoroastrian customs."[46] These customs allegedly involved maintaining strict endogamy within the community. Bowman was informed that "Zoroastrians have a strict unwritten code: they do not proselytize, outsiders are not accepted into the religion, and if a Zoroastrian marries into another group, neither he nor his children are considered Zoroastrian."[47] In a 2006 oral history interview, Bowman recalled, "[The Zoroastrians] knew that they were different and [were] very proud of it. They would say, 'We are the original Iranians.'"[48] Bowman himself would continue to refer to Zoroastrians as "the original Iranians," "the original Persians," and even "the original Aryan populations" for the rest of his career.

In 1961, Bowman left Iran to take up a research fellowship at the Galton Laboratory in London. Along the way, he traveled to Rome to present a summary of his favism research at the Second International Congress of Human Genetics. Rather than taking a clinical perspective, he focused on the anthropological significance of favism distribution. Bowman explained this pattern of distribution in a manner that aligned well with Pahlavi-era nationalist historiographies that cast Zoroastrians as the last living representatives of the "original" gene pool of the ancient Persians. He foregrounded Zoroastrianism in his description of Iranian ethnogenesis: "By the mid-sixth century B.C., a new people, the Persians, appeared on the world scene. They were Indo-Iranian in language, Zoroastrian in religion, and differed ethnically from Mesopotamians, Phoenicians, Hebrews and Egyptians," and, according to Bowman, they completely lacked the gene for favism.[49] Based on his population surveys, Bowman argued, the possible human sources of the condition who had introduced it into the ancient Persian population included the Muslim Arab forces who conquered Iran in the seventh century C.E.; Jews originally liberated by Cyrus in the fifth century B.C.E. who settled in Iran and eventually converted to Islam; and nomadic tribal groups who "migrated or were brought to Iran from bordering countries since the Islamic conquests to act as buffers against local dissident groups," many of whose descendants had gradually "left their peoples" and whose "progeny are lost in the Moslem population."[50] His argument was based primarily on the absence of favism within the Zoroastrian community as well as his acceptance of their claims that they had consistently maintained a total and permanent rejection of proselytism and outmarriage.

Bowman's convictions about the Zoroastrians' perfect preservation of their original gene pool are all the more striking given the skepticism he meted out to Israeli geneticists who made similar claims about Jewish history. In the course of his favism research in Iran, he had established a friendly professional correspondence with Chaim Sheba, Aryeh Szeinberg, and Bracha Ramot. In Bowman's view, the dramatic genetic variation observable between different Jewish communities was obvious evidence of past proselytization, even if Zionists had lately attempted to erase such incidents from their history:

> Going back to the history of Judaism, one of the things that I ran across that was very interesting was that Jews maintained they did not proselytize. I mean, they did *not* do it. I said, "But you do proselytize. Look at the Asian-Indian Jews. They're Asian Indians. [. . .] The Falasha—the Jews in Ethiopia—were there because you proselytized. That is part of your heritage, and it's a beautiful part. . . ." The Zoroastrians did not proselytize. They were a bunch to themselves.[51]

Bowman's conceptualization of Jewish history and the cultural and biological distinctiveness of various Jewish communities was probably also informed by his familiarity with Iranian Jews and some of their negative interactions with Ashkenazi-dominated Zionist organizations. The disparities in political and cultural ideology between Iranian and Ashkenazi Jews surely reinforced Bowman's perception of the latter as "Europeans," a perception further intensified by Bowman's eventual visit to his Ashkenazi geneticist colleagues in Israel. On the invitation of Sheba, Szeinberg, and Ramot, he and his family visited Tel Aviv and the nearby Tel-Hashomer Hospital for two weeks, during which he witnessed the stark inequities between Jewish and Palestinian residential areas. At a dinner party in his honor, his Israeli hosts asked him to describe his experiences living as an African American in the United States. Much to their consternation, he compared Black-white segregation in his own country to the inability of a Palestinian Arab to purchase a house in Tel Aviv. According to Bowman, this sent the party into an uproar. The Israelis admonished him for spending too much time living in "an Arab country," to which he pithily responded, "You better know your history. Iran is not an Arab country."[52]

Despite this moment of antagonism, Bowman's views on favism and its significance to anthropology aligned well with those of Sheba. Specifically, Bowman was not fully convinced by Motulsky's malaria selection hypothesis. During a formal discussion of Motulsky's work at a malaria workshop hosted by the Walter Reed Army Institute of Research in August 1963, Bowman marshaled his own studies on Iran as well as data from Mexico, New Guinea, and Thailand to point out the absence or low incidence of G6PDd in populations living in malaria-endemic regions. He conceded that this phenomenon did not disprove the hypothesis but also did not support it. He further elaborated on the history of the Armenians in Iran to point out that different strains and dosages of falciparum malaria might induce different effects.[53] Bowman argued that the malaria selection hypothesis was too simplistic and did not account for the many factors involved in the transmission and progression of the disease. At best, he argued, "the malarial selective effect might be operative in one area of the world and not in another."[54] Meanwhile, Bowman questioned the empirical data used to suggest that G6PDd conferred resistance to malaria, publishing a critique of arithmetical errors in the work of prominent geneticist and malaria-hypothesis advocate Anthony C. Allison.[55] Bowman therefore opposed malaria-based explanations for favism even more strongly than Hermann Lehmann did for sickle cell disease (see Chapter 4).

STUDYING FAVISM IN THE ARAB STATES

James Bowman had defended Iran's non-Arab identity in Israel, but he was directly responsible for the widespread interest in studying favism in Arab populations during the 1960s. His first ardent disciple was Armand Gelpi, the Aramco physician working in eastern Saudi Arabia. Like Bowman, Gelpi had arrived at his post with no knowledge of favism. He was also completely unaware of Israeli research on the condition. Only after hearing Bowman's presentation at a Shiraz medical conference did Gelpi begin tagging along on Aramco's annual malaria surveys to collect blood samples from Saudi oasis villages and test for favism.[56] Although falciparum malaria was highly endemic to these oases, Gelpi argued that Motulsky's malaria selection hypothesis could not account for his observation that Shi'i villagers, but not Sunni Muslims, experienced remarkably high rates of favism. He therefore suggested that contemporary patterns of favism in Eastern Arabia represented the effects of genetic drift rather than natural selection.[57] Kuwaiti pediatricians, too, cast doubt on Motulsky's hypothesis. They discovered a relatively high incidence of favism in their arid, malaria-free country, which offered compelling evidence against malaria selection. However, they argued, a more ethnographically informed study was needed to confirm or refute the malaria hypothesis because their data could potentially be explained by "the intermingling of the racial groups in the Gulf area over the ages and the fact that the present day inhabitants of Kuwait originally migrated [. . .] from other neighboring countries" like Saudi Arabia, Iraq, and Iran, which had endemic malaria. The Kuwaiti researchers' appeal to "racial groups" recalls Israeli approaches to favism distribution, and they suggested that the malaria question could be resolved only by accounting for "the racial origin of the Kuwaiti subjects studied."[58] However, for political reasons, the Kuwaiti publication in the *American Journal of Human Genetics* deliberately omitted any references to Israeli research and even erased Israel's name and borders from a map of the Middle East.[59]

Lebanese and Egyptian investigators cited both Bowman and Israeli researchers in their favism studies during the early to mid-1960s. Like Sheba and Bowman, they evaluated favism incidence according to ethnoreligious categories as well as to geographic distribution. In both countries, the authors found their results too ambiguous to weigh in on the debate over the malaria selection hypothesis. They conceded that either malaria endemicity or ethnic ancestry could equally account for favism distribution patterns. But their publications, like those in Israel, Iran, Saudi Arabia, and Kuwait, echoed the keen interest in correlating favism with histories of ethnic intermarriage and migration. For example, the Egyptians turned to historical invasions to explain a rural-urban divide in favism frequency: "From the historical point of view, the Egyptians are not a pure race [. . .]. But the greatest conglomeration of foreigners has always been in the metropolitan cities where intermarriage with Egyptians occurred. [...] The genetic make-up of the metropolitan cities is mixed, some of the people being of European descent, where the incidence of the G6PD trait is known to be low."[60]

These patterns of interpretation in Middle Eastern favism research reveal a certain dissatisfaction with international medical discourses that focused on global solutions and universal explanations for disease etiology, embodied by the persistent linkage of favism to malaria. Resistance to the Motulsky hypothesis by Sheba, Bowman, and others reflected not only the contemporary gaps in information about molecular mechanisms and vector prevalence. It also demonstrated a pushback against a reductionist story about human evolution—an example of natural selection applicable in every time and place—in favor of explanations that centered the importance of local histories and ways of life. However, this contrast did not reflect a simple opposition between Western and Middle Eastern scientists or the methodological biases of different medical specialties. Distinct kinds of international health interventions placed different valuations on local cultural knowledge. While malaria eradication campaigns readily imagined favism as an evolutionary response to malaria, campaigns to reduce malnutrition approached favism as a threat to traditional diets, which could be resolved through agricultural improvements and investigations of folk medicine.

INTERNATIONAL MEDICAL AGENDAS IN IRAN: NUTRITIOUS BEANS AND MALARIA ERADICATION

During his time in Iran, James Bowman had essentially been able to treat favism as a pet project. In the mid-1960s, however, favism transitioned from a relatively neglected, localized problem to a major consideration for national health services. In July 1965, the Iranian Food and Nutrition Institute convened a seminar on "Favism in Iran" in response to an "epidemic outbreak" of the condition in the Caspian littoral, with reports of 1,123 individuals suffering the effects of favism after a particularly fruitful bean crop.[61] The heightened incidence of the disease during the 1964 and 1965 fava seasons so inflamed the local newspapers that "false reports were frequently issued [...] sometimes given in the name of high government officials" to prohibit the consumption or planting of fava beans, and "a very strong group

FIGURE 10. Street vendor selling fava beans on Enghelab Street in Tehran, Iran, during public celebrations of the thirtieth anniversary of the 1979 revolution, on February 11, 2009. Copyright Shervin Malekzadeh, used with permission.

wanted to see the destruction of all bean fields and harvested beans."[62] Given the importance of fava beans as a cheap source of protein essential to the health of rice-eating northern Iranian populations, the Food and Nutrition Institute was eager to overcome the hysteria surrounding what was locally called *"bāqalā-zālah"* and funnel resources into systematic investigations of *"fāvīsm"* in concert with international experts on the topic.[63]

Israeli scientist Aryeh Szeinberg was among the most important of these experts. In his introductory lecture for the seminar, Szeinberg summarized all that was then known about the condition based on the investigations undertaken in Israel, Italy, and the United States (as well as those of Bowman in Shiraz), to an audience of over a dozen Iranian physicians and agricultural scientists, two German WHO biochemists, an Indian Food and Agriculture Organization representative, and Léon Lapeyssonnie, the French WHO representative for Iran. The seminar defined a research program on favism for the Caspian region, focusing primarily on botanical chemistry, seeking the causative mechanism by which fava beans induced hemolysis in individuals with G6PDd and by which the Caspian folk remedies for favism (the juice of wild persimmons and grapes) might work. The malaria selection hypothesis was referenced only in passing during the seminar, by Szeinberg and by Lapeyssonnie. Szeinberg commented, "This theory has not been yet proven with certainty, and many investigators doubt its validity," while Lapeyssonnie noted that "as malaria is [...] less important in the Caspian Area than in most parts of Iran this hypothesis does not give much satisfaction."[64] Participants in the nutrition seminar, with its emphasis on local foodways and plant chemistry in contrast to hematology and epidemiology, showed no interest in the possible relationship of favism to malaria.

It was malaria itself, not favism, that drew in researchers arguing that favism distribution in Iran could support Motulsky's hypothesis. In 1966 the U.S.-born British physician Peter Beaconsfield collaborated with Iranian scientists at the University of Tehran's newly founded Institute of Public Health Research to conduct a survey of favism in different regions of Iran. The WHO-sponsored malaria eradication program for Iran, which relied heavily on the use of fourteen-day courses of primaquine and therefore caused outbreaks of hemolytic anemia, spurred a number of such favism surveys to stave off these incidents.[65] Unlike Bowman, Beaconsfield and his Iranian colleagues largely analyzed populations according to administrative geography rather than seeking out particular ethnoreligious groups. Only in a few areas, like Tehran, Yazd, and Kurdistan, did they follow Bowman's protocol of sorting samples by religious communities (e.g., separating out Jews and Armenians in Tehran, Zoroastrians in Yazd).[66] Generally, they found only minor differences in favism rates between such communities living in a particular area. For example, Beaconsfield found that Zoroastrians did, in fact, possess the gene for favism (albeit at very low levels), challenging the basis for Bowman's narrative of Iranian history. According to Beaconsfield and his colleagues, both Zoroastrians and Muslims living in Yazd experienced low rates of favism because this region had no endemic malaria.[67] Indeed, Beaconsfield believed all favism data collected by his Iranian coworkers supported the malaria selection hypothesis.[68]

The different sampling practices and interpretations of favism in Iran did not consistently correlate with medical specializations. However, they do reflect the divergent interests of two transnational constituencies: (a) medical scientists involved in malaria eradication programs, alongside geneticists who believed that natural selection had played a major role in human evolution; and (b) nutritionists and agricultural scientists, alongside geneticists who believed that human social customs, especially religious self-segregation, were more important to shaping human genetic variation. As examples of the second constituency, when Arthur Mourant and Hermann Lehmann conducted independent research in Kurdistan, they tended to reinforce the views of Sheba and Bowman. Mourant and Lehmann were interested in surveying genetic traits in Iran's Kurdish population due to the remarkably high frequency of favism Israeli researchers had discovered in Jewish Kurds. In 1969, Lehmann traveled to Iran with a team of British physicians; like Beaconsfield before them, they collaborated extensively with Tehran's Institute for Public Health Research. Lehmann relied on Iranian public health officials and clinicians to facilitate his fieldwork in Kurdistan province among Kurdish villagers who were "believed to be almost completely unmixed genetically with other populations."[69] In addition to collecting 184 blood samples that were shipped to the United Kingdom, Lehmann's team also field-tested more than 300 Kurds for PTC-tasting ability and more than 500 for hereditary color blindness.[70] Ultimately, they found that the most significant genetic difference between Jewish and non-Jewish Kurds was the Jews' much higher rate of favism (about 40 percent, compared with less than 7 percent).

Mourant and Lehmann took an ambivalent stance on Arno Motulsky's malaria hypothesis, invoking Sheba's old contention that mountainous Kurdistan was not highly malarious and that, in any case, the hypothesis could not explain the variable frequency between Jewish and non-Jewish Kurds: "If the distribution of this deficiency is an evolutionary response to malaria, there is no reason to think that the frequencies found among the Iranian Kurds are far from the equilibrium frequencies corresponding to the incidence of falciparum malaria in recent times. It is the frequencies among the Kurdish Jews which require a special explanation."[71] This explanation, they argued, must relate to "some circumstance peculiar to their own ancestry or mode of life. The gene itself may have accompanied the Jewish community, which presumably migrated a very long time ago from the region of Baghdad, or perhaps directly from northern Israel." But Mourant and Lehmann did not fully commit to Sheba's explanations of ancestry and inbreeding either: "It is unlikely, however, that it is solely through inbreeding that the frequency found in this community has persisted and even increased [. . .] in the absence of some special environmental influence favouring this normally slightly deleterious gene. [. . .] One is led to look for some cultural peculiarity."[72] Mourant and Lehmann thus deemed both Motulsky's and Sheba's explanations inadequate on their own. Their turn to "culture" shows how favism research reinforced a fascination with the overbearing presence of ethnoreligious difference in Middle Eastern societies. This fascination is evident as early as Richard Lederer's 1941 claim that his "Baghdad spring anemia," in part because it occurred almost exclusively in Iraqi Jews, was a distinct local disease and not the same as Italian "favismo."

GLOBAL PUBLIC HEALTH BECOMES EVOLUTIONARY ANTHROPOLOGY

By the 1960s, later favism research in Israel and Iran overturned specific results obtained by Sheba and Bowman. However, their approach to understanding this disorder would be replicated both for its management through genetic counseling and for the study of other hereditary conditions in the region. Furthermore, they established a paradigm wherein favism channeled public health resources and funds into a form of "medical archaeology" that pitted the condition's relevance to malaria against questions of ethnic ancestry. This approach proved influential to G6PDd researchers in other parts of the world, such as the hematologist Rubén Lisker, who studied indigenous peoples across Mexico. As described by Edna Suárez-Díaz, Lisker gained Mexican and U.S. government support for his existing interests in the anthropological significance of sickle cell anemia and G6PDd through the relevance of these conditions to the national malaria eradication program. Using state infrastructure created to manage the indigenous populations of rural Mexico, Lisker sampled the blood of dozens of indigenous communities in the early 1960s. Similar to Sheba and Bowman, he demonstrated that the frequency of G6PDd in his samples did not correlate with malaria incidence and explained elevated rates of G6PDd in certain communities as the historical legacy of the African slave trade and "Negro admixture."[73] Bowman himself visited Mexico to study G6PDd in the Lacandon people of the Yucatan peninsula and, with Lisker's assistance, the Sephardic Jewish community of Mexico City.[74]

From his new base at the University of Chicago, Bowman compiled genetic data on various tribal populations of West and East Africa as well as Vietnam. Into the 1970s, his publications on this material continued to express doubts about the global applicability of the malaria hypothesis and consistently returned to the examples of Iranian and Israeli populations.[75] In contrast, after Sheba's death in 1971, resistance to the malaria selection hypothesis gradually dissipated among his successors in the Israeli biomedical community. At an international symposium in Sheba's honor held in March 1973, Aryeh Szeinberg recalled, "In the early studies of this trait, it seemed to us that it might constitute a good tracer for the ancient Hebrew gene pool," but due to the accumulation of new data showing the high frequency of favism in non-Jewish Middle Eastern populations, "this view became untenable." Conceding that "most investigators agree" that this distribution can be explained by the malaria selection hypothesis, Szeinberg explained that favism had now "lost much of its appeal as a tracer for ethnic relationships."[76] However, Israeli acceptance of the malaria hypothesis did not preclude a continued focus on collecting "ethnic incidence" rates for favism.[77]

In a trajectory that coincides with sickle cell disease, the advent of restriction-enzyme genotyping definitively showed that G6PDd was caused by multiple independent mutations of the G6PD gene. By the 1980s, Italian hematologist Lucio Luzzatto's comprehensive review of G6PDd research stated forcefully that malaria's role as a selective agent had been established by the preponderance of the evidence, without discounting the additional effects of migration and drift.[78] The triumph of the malaria selection hypothesis, however, ultimately did little to dampen Middle Eastern enthusiasm for G6PDd as a marker of ethnic boundaries and migration history. Luzzatto himself went on to study the distribution of G6PD-Mediterranean, the mutation responsible for favism, in the Middle East in collaboration with Israeli and Saudi researchers. This team decided that the widespread appearance of G6PD-Mediterranean throughout the region signaled its utility as a population genetics tool confirming shared ancestry for all Middle Eastern populations: "This mutation may be quite ancient and may have spread by migrations that have taken place perhaps over millennia."[79] Although Luzzatto proffered Archaic Greek civilization as a migration pathway, his Israeli coauthors highlighted the similar antiquity of the Kurdish Jews, "who must have taken the two mutations with them some 27 centuries ago, when they migrated to Kurdistan."⁸⁰

Luzzatto and Israeli scientists further investigated the case of the Kurdish Jews, who prompted a novel synthesis of the malaria hypothesis with the assumption of Jewish common origins in Israel. They suggested that malarial selection in Kurdistan had increased the frequency of G6PD-Mediterranean in the Kurdish-Jewish population but also argued, "We think it is more likely that the Jews who were exiled from the Jordan river valley already had a high frequency of G6PD deficiency. Since this frequency is much higher than in other Jewish communities, we postulate that they came from a region of marked malaria transmission within Israel, such as the notoriously malarious Hulla valley."⁸¹ Although the paper contains no references to Sheba's work, it encapsulates all the basic features of Israeli favism research in the 1950s and 1960s: Ashkenazi and European scientists, working on a genetic disorder of Mizrahi Jews, crafted a narrative of Jewish exile from the Levant and subsequent reproductive isolation. Here, the malaria selection hypothesis was used not to challenge Sheba's visions of a shared Jewish gene pool fractured by founder effects and genetic drift but instead to substantiate them—by rooting the Kurdish Jews' original encounter with malaria firmly within the present-day territory of the Israeli state.

This apparent fusion of the two dominant hypotheses explaining the distribution of G6PDd demonstrates that they had never been mutually exclusive and in fact could be easily reconciled. It is for epistemological, not empirical, reasons that certain researchers had positioned the two as dichotomous and firmly stood against the view that privileged malaria, at a time when malaria eradication campaigns remained a major public health priority in the Middle East. These campaigns represented major opportunities for local researchers to secure material resources and recognition on the international stage of biomedicine. However, the campaigns also represented a centralized agenda emanating from American- and European-dominated international bodies that had little interest in or patience for the national particularisms that animated many within the professional classes in the Middle East during this period. From the perspective of scientists embedded in the region, differential incidence of favism or malaria was simply another diagnostic feature of the ethnic, linguistic, and religious boundaries that characterized Israeli and Iranian social and political life. To them, the idea that such culturally significant boundaries could be primarily or solely responsible for favism distribution was intuitive and compelling in a way that the malaria hypothesis, with its implications of environmental determinism, was not. The debates over favism therefore reflect larger tensions between the ideologies and practices of international public health and its aspirations to universality and the sociopolitical interests of national scientific communities, who instrumentalized nationalist approaches to history and anthropology.

However, given the intrinsic violence of nationalist state-building, the historical and cultural narratives offered to explain the distribution of favism in Jewish and Iranian groups were neither neutral nor innocent. Rather, they were iterations of hegemonic discourses of national identity endorsed by state authorities. The "medical archaeology" produced through favism research was therefore an opportunity *not* to excavate new, unknown facts about Middle Eastern groups' ancestry and history but instead to justify existing facts of social and medical discrimination within nation-states. As the chapters in Part II have shown, these anthropological and historical claims were produced by medical practitioners drawing on national and international medical infrastructures designed to circulate blood samples for transfusion and diagnosis. The social inequalities embedded in these infrastructures determined who became a professional scientist, who acquired the necessary research funding and equipment to collect blood samples and interpret genetic data, and whose blood was circulated and instrumentalized for the careers of other people. Part III highlights the relationship between these inequalities, the increasing role of anthropologists in genetic research, and the ongoing territorial contestations between and within Middle Eastern national entities. The next chapters examine how the violence of the Arab-Israeli wars, the end of British colonialism in southern Arabia, the 1979 Iranian revolution, and the Kurdish conflict in Turkey obstructed, but also enabled, opportunities for research on human genetics. These conflicts, from the late 1960s to the early 1990s, reveal how geneticists have leveraged national militaries and the conditions of unrest to expand their access to new research subjects.

Part III

COLONIAL AND ETHNIC VIOLENCE

Chapter 6
COLLECTION AGENTS

IN NOVEMBER 1966, JUST OVER TEN YEARS SINCE DOUGLAS BOT-ting's Oxford expedition to Socotra, Arthur Mourant saw another opportunity to obtain blood samples from the remote island. At the seventy-fifth-anniversary celebrations of the Lister Institute in London, he met Air Vice-Marshal William P. Stamm, a pathologist for the Royal Air Force (RAF). Through Stamm, Mourant connected with two RAF physicians about to be deployed to Aden. One of them, Squadron Leader M. A. Pallister, would accompany an expedition to Socotra between March and May 1967. Pallister, initially willing to collect Socotran blood for Mourant, quickly became disenchanted by Mourant's inability to provide adequate material or financial support from his laboratory budget. Like Botting's expedition, Pallister faced the major problem of refrigerating blood for storage and transport. However, lacking the resources of the Oxford team, Pallister had to scavenge "an old battered oil burning refrigerator from an Army dump in Aden." Once he arrived in Socotra, he found it easy to convince his patients to donate blood as a routine part of medical examinations. But to his dismay, the "refrigerator would not work. With this simple piece of equipment failure the whole system failed." After his return to England, Pallister recounted his frustrations to Mourant with a pessimistic forecast: "I do not think a further opportunity to get samples out of Socotra will occur to judge by the way the Middle East is developing, but if it does you must be prepared to provide your field workers with considerable support."[1]

Pallister's experience reflected the fragility of Mourant's blood supply chains in a decolonizing Middle East. His concluding comment highlights the immense significance of 1967 as a violent turning point in the region's politics. On November 30, British forces withdrew completely from Aden and the rest of the southern Arabian protectorates, ending more than 100 years of colonial rule in the area. During the preceding four years, a period known as the Aden Emergency, the British had struggled to suppress an insurgency of leftist and nationalist guerrilla groups, some supported by Gamal Abdel Nasser's regime in Egypt. Pallister left Socotra shortly before a peak of violence in June 1967, when the Aden Armed Police and Arab soldiers in the South Arabian Army mutinied against their British officers. The mutiny was partly a response to rumors that the British had assisted Israel two weeks earlier in its stunning victory over Egypt, Jordan, and Syria in the Six-Day War. To the shock of an Arab world inspired by Nasser's bravado, Israel not only crushed the military forces of all three countries but also became an occupying power. By the time a cease-fire was signed on June 11, Israel had captured the Golan Heights from Syria, East Jerusalem and the West Bank from Jordan, and Gaza and the entire Sinai Peninsula from Egypt. In less than a week, Israel had more than tripled its territory; within these new borders, more than one million people now fell under Israeli military rule. Amid the final collapse of Britain's Middle Eastern empire, Israel expanded its control over Arab populations, with profound effects on their political rights and access to social and medical services.

This chapter explores how the Israeli-Arab conflict shaped the logistics and methodology of Middle Eastern genetic research in the 1960s and 1970s. In doing so, I follow Jess Bier's analytical insights regarding the role of territorial occupation in knowledge production—that is, "how segregated landscapes help produce different scientists."[2] Specifically, I examine how the shifting borders and mobilities caused by warfare and military administration strained various collaborative relationships between "privileged internationals" who could cross borders, the local geneticists who could not, and the research subjects who struggled simply to survive in place.[3] Prior to 1967, a network of British, American, and Israeli geneticists had worked together in relative harmony to manage the collection of blood samples from several Arabic-speaking communities. Israeli citizens, restricted from crossing borders into neighboring countries, primarily studied displaced Arab Jewish communities and relied on British and American colonial and corporate infrastructures to provide comparative data on non-Jewish Arabs from across the region. However, the conditions of access to these Arab populations as enforced by territorial borders and military regimes changed dramatically in the wake of the 1967 war. Israeli researchers were suddenly left with the upper hand, able to directly study coveted populations like the Samaritans of Nablus and the Sinai Bedouin tribes, while their British and American counterparts were confronted by anticolonial unrest. However, the shifting control over Arab genetics in this period did not represent a simple replacement of one colonial power with another, because the politics of scientific labor did not shift in concert with Israeli military expansion. Israeli geneticists had to demand increased scientific recognition from British and American collaborators who perceived the Israelis as mere "collection agents" serving to ship blood samples from the Middle East to Europe. The question of who could access Arab blood became a dispute over who had the greater right to benefit professionally from its scientific circulation, analysis, and publication. This dispute, not coincidentally, ignored the professional and human rights of Arab scientists and research subjects.

Below, I trace the early career of Israeli geneticist Batsheva Bonné through her quest to study the Samaritans, Jews from Habban (Yemen), and the Bedouins of southern Sinai. In order to obtain data from the Nablus Samaritans and Yemeni Arabs, she initially needed to collaborate, generally to her disadvantage, with Mourant and his protégés based in Lebanon and Aden. The tables turned in 1967. With Mourant now dependent on Bonné to reach the Samaritans and benefiting from her Israeli military connections to access Sinai Bedouins, she assertively challenged his assumptions about the positionality of Middle Eastern field-workers vis-à-vis Euro-American laboratories. She claimed that her permanent residence in the region and ethnographic approach to blood collection gave her more scientific priority than Mourant to manage and publish the data produced from what she considered "Middle Eastern ethnic groups." This claim represents a certain paradox: on the one hand, it challenged patterns of Eurocentric coloniality in the international scientific infrastructure; on the other, it relied on the coloniality of Israeli military occupation, reflecting a co-constitutional reframing

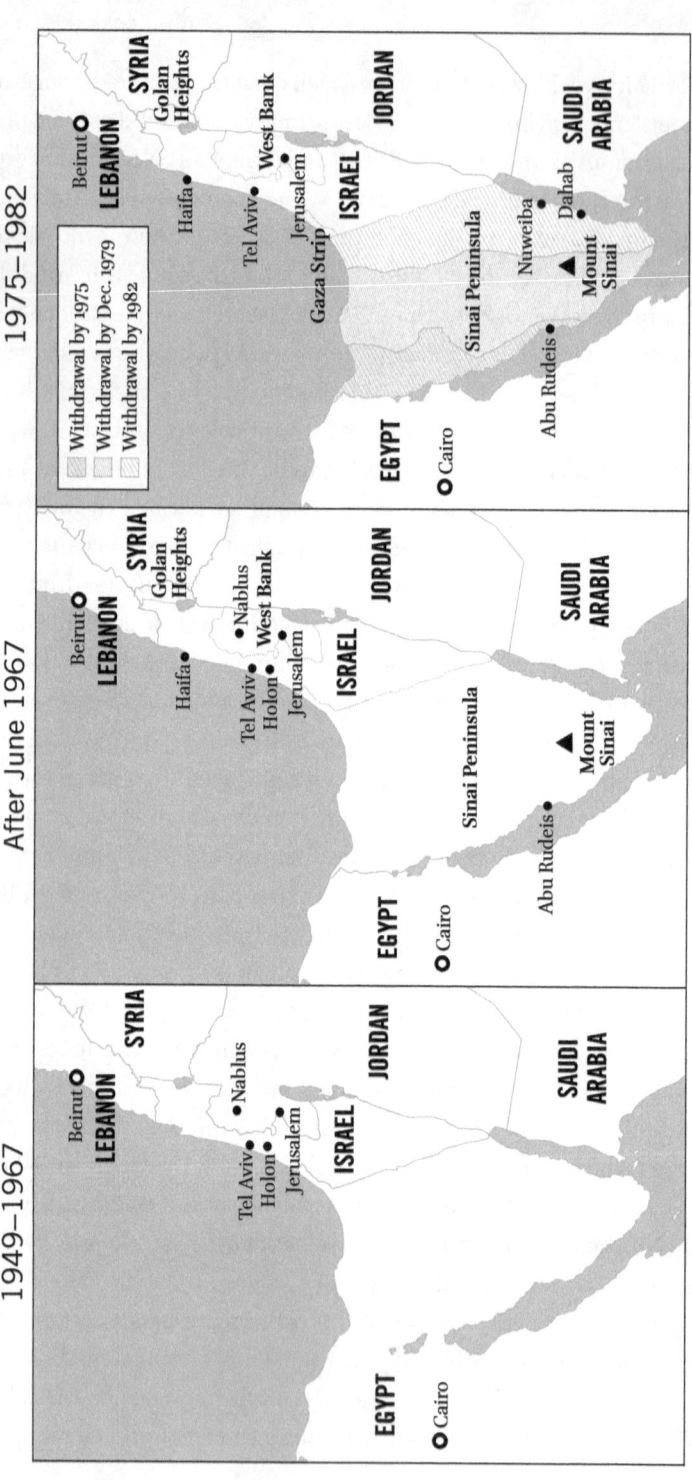

MAP 2. Map showing Israeli state borders after 1948, Israeli territorial occupations following the 1967 war, and the staged Israeli withdrawal from the Sinai Peninsula between 1975 and 1982.

of the Israeli state as a Middle Eastern power. Bonné's activities inspired her compatriots to study more Palestinian communities within Israel proper in the early 1970s. However, increasing unrest among Palestinian citizens of Israel highlighted the dangers involved in the scientific, and sometimes political, "collaboration" of Arab research subjects. The communities desired by geneticists were politically and socially precarious, and in the context of colonial violence during the Aden Emergency and Israeli-Arab wars, they demanded increased compensation from scientists or withheld their cooperation entirely to offset the potential fallout of associating with the "enemy" in the event of regime change. The shifting territorial borders of the late 1960s therefore transformed the traffic in Arab blood at all stages and scales of genetic research.

SCIENTIFIC TERRITORIALISM IN NABLUS, 1961–1967

In 1962, the American University of Beirut (AUB) established the Anthropological Blood Grouping Laboratory (ABGL) within the hospital's clinical pathology department. Created by American biologist Harry Madison Smith with funding from the United States' National Institutes of Health (NIH), AUB administrators hoped that the ABGL might contribute to their ambitions to educate Middle Eastern research scientists. Instead, the lab essentially functioned as an experimental pied-à-terre in preparation for the International Biological Program (IBP), a ten-year initiative that from 1964 to 1974 channeled vast quantities of human blood samples from developing countries to laboratory freezers across Western Europe and North America.[4] Mourant served as the IBP's general coordinator for worldwide human genetics surveys, drawing on fifteen years' accumulation of contacts with hematologists, anthropologists, and medical practitioners across the globe. However, Mourant did not regard all of his correspondents as having equal professional status. When considering "essential collaborators" to organize projects in the Middle East, he thought primarily of British, American, and French scientists working in the region. He took for granted that Arab, Israeli, Iranian, and other local researchers would assist Westerners with the collection of blood, and he had little interest in supporting their professional development.[5] These attitudes reflected not only race and gender biases but also the uneven distribution of research funds and passport privileges that enabled Western researchers to travel and cross borders in ways that Middle

Eastern scientists could not. The Israeli-Arab conflict, which prevented direct collaboration between Israeli scientists and their counterparts in neighboring Arab states, reinforced the axes of sample collection and genetic analysis that funneled Arab blood out of the region to laboratories in Europe and North America. For his part, Mourant took Harry Smith under his wing and threw the weight of his support behind Smith's Beirut laboratory, suggesting that it "could be a very valuable outpost of American serological genetics for the whole of western Asia and N.E. Africa."[6]

Smith first joined AUB in 1954 and spent four years on the faculty as a biology professor.[7] In 1958, Smith received a two-year NIH-funded postdoctoral fellowship at Columbia University to study the geographic distribution of human blood groups. As part of this work, he traveled to the BGRL in London, where he met Mourant. Though not his direct supervisor, Mourant formed a personal attachment to Smith because the latter's father hailed from Mourant's home island of Jersey.[8] In 1960, Smith joined a BGRL expedition to Ibadan, Nigeria, where he collected blood from Yoruba villagers, and he flew the specimens back to London for lab testing.[9] His understanding of the norms of fieldwork in postcolonial regions was thus forged by his experiences with Mourant's lab. Subsequently, Smith decided to return to Beirut to initiate his own blood grouping program in the Levant. Late in 1961, he won an NIH grant to set up a laboratory affiliated with AUB's medical school. Shortly thereafter, he was offered a faculty position in the biology department of Springfield College, a small university in western Massachusetts. Springfield's dean, Cummins E. Speakman, had met Smith at AUB in 1956 and eagerly promised that Springfield could serve as the U.S.-based institutional partner for Smith's NIH grant.[10]

When Smith arrived at the AUB hospital in June 1962, he hired only one Lebanese technician, Fares Aftimos Ghareeb. For the position of senior research assistant, who would have the authority to oversee the laboratory in Smith's absence, he asked Mourant to hire a qualified British researcher. Smith's preference for foreign over local staff would backfire in several ways, the first being that Mourant's choice, Victor Alan Clarke, faced long delays in obtaining a Lebanese work permit and did not arrive in Beirut until February 1963.[11] In the meantime, Smith and Ghareeb began testing blood from Lebanese AUB affiliates, sorting their genetic data into the following categories: "Druzes, Shi'a, Alawi, Assyrian Orthodox, Maronites, Latins, and other

Christian sects inhabiting Mount Lebanon and adjacent regions in northern Israel."[12] In mid-August, Smith and Ghareeb flew to the Jordanian side of Jerusalem on a flight chartered by the United Nations Relief and Works Agency. They visited hospitals in Jerusalem, Nablus, Ramallah, and Amman, collecting blood from fifty Palestinians and twenty Circassians and Bedouins. Two weeks later, they made a similar trip to "Gaza strip territory controlled by United Nations" to collect more Palestinian samples.[13]

Smith positioned his lab as an effective satellite of Mourant's BGRL in London and considered it a major point of pride that Mourant and William C. Boyd, "the world's two leading blood groupers in the field of genetics and anthropology," visited the ABGL in January 1963.[14] But the circumstances behind their visit signaled a looming clash between the self-conscious Western "outpost" in Beirut and the regional aspirations of another national scientific community: the new generation of Israeli geneticists affiliated with Tel-Hashomer Government Hospital. Smith had long planned for Mourant to come to Beirut with his top two technicians, Elizabeth Ikin and Patricia Brooks, to assist his expedition to survey the blood of the Samaritans of Nablus. However, Boyd was a late addition to the team, representing his Israeli doctoral student Batsheva Bonné, who had begun independently studying the Samaritans two years earlier. As a young anthropologist, Bonné had already forged strong ties to the branch of the Samaritan community that had resettled in the Israeli town of Holon during the 1950s, and she felt that her ethnographic experience and familiarity with life in the Middle East lent her a scientific priority to study the Nablus Samaritans. Bonné's interactions with Smith and Mourant would expose a range of methodological and ethical problems behind their center-and-outpost approach to anthropological genetics.

Bonné was born and raised in Mandate Palestine by a family of secular German Jewish academics.[15] After completing her mandatory military service, she joined Kibbutz Tzorah, where she spent several years balancing the manual labor of kibbutz life with extension courses from the Hebrew University of Jerusalem. She finally left the kibbutz to finish a degree in sociology, alongside courses on genetics, and wrote a bachelor's thesis on the marriage patterns of kibbutz members. In 1958, she entered a master's program in physical anthropology at the University of Chicago on a full scholarship. She first learned of the Samaritans while looking for a suitable subject for her master's thesis, and she quickly established contact with Yisra'el Tsedaka, who became her longtime liaison with the Samaritan community of Holon.[16] Based on his information, she analyzed the community's demography for her thesis (which was eventually published in the journal *Human Biology*).[17] Bonné met Tsedaka in person during the summer of 1961, when she initiated a genetic survey of the community with the assistance of researchers at Tel-Hashomer. She planned to continue studying the Samaritans for her doctorate in human genetics at Boston University under the supervision of Boyd. However, as an Israeli citizen, she could not personally visit the rest of the Samaritans in Jordan. Boyd decided to go to Nablus on her behalf, applying to the Wenner-Gren Foundation for travel funding in November 1962. Meanwhile, Bonné wrote to her Samaritan contacts in both cities to arrange his visit.[18]

Their preparations were interrupted by Mourant, who had been assigned to review their Wenner-Gren application. With much chagrin, Mourant wrote that he and Smith had firm plans to conduct the very same research in January 1963. He explained that "testing of the Samaritans has had first priority with Harry Smith ever since he planned his Beirut laboratory, and he has for years been in personal contact with the community" and that Mourant was therefore in no position to "relinquish the Samaritan blood grouping." However, he proposed a "joint scheme," whereby Smith's lab would share blood samples with Bonné.[19] Severely disappointed, Bonné felt she had no choice other than to agree to Mourant's terms.[20] Smith and Mourant took the leading role in planning the now joint expedition and generally disregarded Bonné's advice, particularly on the question of whether and how the Nablus Samaritans should be financially compensated for donating their blood. Mourant, deeply invested in Smith's long-term ability to sample local Arab populations from his Beirut outpost, supported Smith's opinions in a letter to Boyd:

> I understand from Miss Bonné that you are planning to pay your donors. Harry Smith thinks that this practice should be discouraged, but, as he says, 'The Samaritans have made it clear to me in private conversation that they will not give their blood voluntarily. I have not paid a single donor thus far and do not intend to resort to bribery because it might make further work impossible in this area. The Samaritans, however, have been pampered by tourists, and expect money for the slightest consideration.'[21]

Bonné, who was more concerned with the immediate success of the Samaritan study, responded by emphasizing the information she had been given by the Israeli Samaritans:

> Although it was very clearly indicated by the Samaritans that they will agree to be bled only if they will be paid (and they even quoted a definite sum) I have never promised to do so. [...] I certainly understand Dr. Smith's attitude, and I never had any problems with the group in Israel. However, I was told by the Samaritans there themselves, that their fellow-brothers in Jordan are different and that I cannot expect any cooperation unless they will be paid.[22]

To Bonné's consternation, Smith and Mourant did not seem to value either her personal relationships with the Samaritan communities or the advice she passed on, instead treating her prior correspondence with the Nablus Samaritans as a liability to their fieldwork plans. Indeed, conflicting expectations of compensation between the Nablus Samaritans and the Anglo-American research team ultimately derailed the expedition.

After Boyd departed in early January 1963, Bonné anxiously waited to hear from the men at Nablus. After weeks with no word and no blood samples forthcoming, she finally received a telegram from Boyd that read, "Reason for delay is lack of Samaritan cooperation. Only 82 samples sent to Beirut." In a community of more than 200, Mourant and Smith had not sampled even half of its members. A few days later, Bonné received a letter from Boyd explaining that only the priestly families who had previously been in contact with Bonné had given blood. The rest of the community refused, due to unfounded rumors that the priests had received substantial sums from the researchers (which were ostensibly not being shared with the other Samaritans). The team enlisted one Dr. Masud, a Jordanian physician, to wheedle blood samples from the disagreeable faction, then tried financial negotiations, offering 50 cents and then $3 per sample, all to no avail. The team then gave up and left Nablus. Boyd added, "You may think that you would have successfully collected the bloods if you had come yourself. Let me say that if Harry Smith, Arthur Mourant and William Boyd failed to do so, it seems that the task is indeed impossible."[23] Bonné clearly believed otherwise. Invoking her greater familiarity with the region, she wrote to Mourant, "I was naturally disappointed to hear that the Samaritans did not cooperate as expected, though growing up in the Middle East, I can perceive quite clearly what has happened, and thus I should be grateful for what has been accomplished by you."[24]

Not ready to give up on the prospect of achieving a full data set for her dissertation, Bonné wrote to Smith in spring 1963 to find out if Dr. Masud was able to pursue further sample collection with the Samaritans. Smith strongly discouraged her from resuming contact with the Samaritans, emphasizing the dependence of his laboratory on the cooperation of Arab governments and the danger she posed to the Samaritan project as an Israeli citizen. He mentioned that Dr. Masud's life had been threatened due to her involvement and that Boyd's own visit to Beirut (after which he had visited Israel) had prompted protests from his Arab colleagues at AUB against collaborative work with Israelis.[25] Mourant, too, cautioned Bonné that if she contacted the Nablus Samaritans, she could endanger both the community and Smith's future research plans: "An unwise step could therefore have unfortunate personal [consequences] for the Samaritans (e.g. in preventing the free movement of those from Israel to join in the Passover—or even in more unpleasant ways). It could also prejudice future blood group work in the Arab countries."[26]

Despite these protestations, Smith himself visited Israel for a week in November 1963, where he met with "geneticists, doctors, biochemists [...] with whom he discussed at length possibilities for future projects."[27] Despite her suspicions about Smith's opportunism, Bonné also introduced Smith to the Holon Samaritans, who had recently welcomed Bonné's team of Israeli workers from Tel-Hashomer Hospital for weekly visits to the community. Describing her first collection visit as a remarkable achievement, Bonné wrote to Mourant: "The Samaritans were indeed so cooperative that they almost stood in line to give blood. There was no reward or money involved. . . ."[28] Within two months, Bonné's team had collected blood and saliva samples from 90 percent of the community members, who also submitted to more tedious procedures such as color blindness testing and anthropometric measurements.[29] Bonné attributed her success to the nature of her relationship with the Holon Samaritans, which she consistently portrayed as one of sustained friendship and mutual respect, in contrast to Mourant and Smith's extractive approach to the Nablus Samaritans.

Regardless, Mourant remained committed to Smith's Beirut laboratory and helped to expand its territorial remit to the Arabian Peninsula. In February 1963, shortly after the debacle in Nablus, Smith, Ghareeb, the freshly arrived Clarke, and two other AUB physicians traveled to Saudi Arabia and Kuwait. The team arrived in Dhahran at the invitation of Aramco. After consulting with the Aramco physicians about best practices, the AUB team headed to Kuwait, where the Ministry of Health funded their expenses for a week while they collected blood samples from Kuwaitis, Omanis, Saudis, "and other inhabitants of Eastern Arabia." Mourant further promoted Smith's lab as a strategically located way station for particularly remote field surveys, ensuring that blood samples from British expeditions to Afghanistan and Iran were shipped first to Beirut.[30] Although the ABGL was not as well equipped as the IBP reference laboratories in Europe and North America, its staff conducted the more-time-sensitive blood grouping tests before forwarding sera from these samples to such labs.

Despite this apparent productivity, Smith's relationship with the AUB administration deteriorated after 1964, and he faced increasing pressure to return to Springfield College. As a result, he planned to downscale the facilities in Beirut and turn the local ABGL staff into a roving unit that would simply collect blood and ship it to him in Massachusetts. In letters to Mourant, Clarke argued that Smith's plans were unethical and proposed that AUB "should serve as a reference center for the Middle East," under the supervision of AUB hematologist Munib Shahid.[31] Mourant sympathized with Clarke but hesitated to support his and Shahid's plans to set up a "rival establishment" before confirming the final demise of Smith's laboratory in September 1966.[32] Only following the full withdrawal of his own protégé did Mourant suggest that Shahid attempt to secure funds for "a Lebanon contribution" to the IBP.[33]

Yet a "rival establishment" already existed—at AUB's traditional institutional rival, the francophone Université Saint-Joseph (USJ), which had its own laboratory for biochemical hematology. USJ physician Najib Taleb, who earned a PhD in anthropology under the supervision of Jacques Ruffié at the Collège de France, had initiated an independent axis of collaboration between Beirut and Toulouse. Between 1961 and 1964, Taleb "rigorously identified" the ethnic origins of over 3,000 Lebanese research subjects, whom he sorted into five Christian and three Muslim communities.[34] Blood samples from these subjects, tested first at USJ and then forwarded to Ruffié's Center for Hemotypology in Toulouse, were tested for six antigen systems, sickle cell hemoglobin, and G6PD deficiency. Taleb also traveled with Ruffié to Jordan, where they collected blood from Jordanian and Palestinian soldiers as well as Bedouin tribes for direct shipment to Toulouse.[35]

Unlike Smith, who never managed to publish any of the ABGL's data on Lebanese and Palestinian communities, Taleb and Ruffié produced striking interpretations of Levantine population history as part of a French IBP project on "human adaptation to arid areas."[36] Contradicting the 1930s research performed out of AUB, they argued that most genetic variation in Lebanon, Jordan, and Palestine corresponded not to environmental distribution but rather to sectarian and tribal divisions. For example, they proposed that Greek Orthodox Christians—the community to which Taleb himself belonged—represented "the purest descendants" of Lebanon's first indigenous inhabitants.[37] Meanwhile, they found no significant differences between Palestinians and sedentary Jordanians, claiming that their frequencies of chromosome Rh_0 (or cDe) indicated African admixture, whereas nomadic Bedouins in Jordan more closely resembled the "pure" Arabs (i.e., those of the Arabian Peninsula).[38] Virtually all of the Levantine Arab communities Mourant had hoped to access through Smith were therefore studied through Franco-Lebanese channels, bypassing his laboratory entirely.

Mourant's attempt to circumvent the politics of the Israeli-Arab conflict by strategically investing in Smith at the expense of local researchers backfired with Smith's disengagement from Lebanon. Furthermore, within a year of the Beirut lab's closure, the June 1967 war drastically changed the politics of Arab population sampling. The blood shed by the fighting along the Syrian, Jordanian, and Egyptian borders heralded the territorial occupation through which Israeli geneticists collected blood for their own research. Eventually, the study of the Nablus Samaritans was successfully completed not through changing blood collection practices or collaborative infrastructure but through military conquest.

HABBANITE JEWS AND THE ADEN EMERGENCY, 1965–1967

After graduating from Boston University in 1965, Batsheva Bonné moved back to Israel permanently, taking a faculty position at the new medical school at Tel Aviv University with plans to build her own program in anthropological genetics. Still unable to reach the Samaritans in Nablus, she approached a new community: the Jews of Habban, a rural village in the Hadhramaut region of Yemen, who had emigrated en masse and settled in the moshav of Bareket in Israel. According to Bonné, the Habbanites were an appealing group for genetic study because they were "very inbred," with about 800 individuals belonging to only 5 clans. Also, they were "known to have lived in very primitive conditions for many centuries," isolated from both Muslim and Jewish neighbors in Yemen.[39] In February 1966, Bonné contacted Mourant to ask if his new establishment in London, the Serological Population Genetics Laboratory (SPGL), could assist her with blood group tests. Meanwhile, she would directly supervise tests for hemoglobin variants and G6PD deficiency in the laboratories at Tel-Hashomer. With Mourant's agreement, she started sending him regular shipments of Habbanite samples in May, beginning with the community's schoolchildren.

Their renewed collaboration fell into the same pattern as the Samaritan study. Before the blood survey, Bonné and her students regularly visited Bareket, establishing a rapport with community members. Over the summer, she reported how cooperatively and even eagerly the Habbanites volunteered to donate their blood, enthusing to Mourant, "I wish you could see the people; they are most pleasant and intelligent folks."[40] Some duplicate samples were even erroneously collected when children presented themselves to be bled again, enjoying the fuss of the experience.[41] Bonné asked Mourant to send her results as quickly as possible so that she could inform the Habbanites of their ABO and Rh types. Mourant, for his part, was most excited about the Habbanites' major genetic differences from all other Arab and Jewish populations his lab had previously tested, such as the Saudi Arabs sampled by George Maranjian at Aramco. Compared to other groups, Mourant explained to Bonné, the Habbanites' blood contained many more "African" traits, indicating a "very high Negroid component" in their ancestry.[42]

Mourant capitalized on Bonné's ready access to such fascinating research subjects to raise his own lab's profile. In October, he listed five ongoing international collaborative projects in an application to have his SPGL designated an official IBP center. Among these projects was Bonné's Habbanite study, which Mourant described as "eligible for inclusion" in the IBP.[43] By November, Mourant was cultivating his contacts in the RAF to collect Arab blood from the Aden protectorates. While the aforementioned M. A. Pallister was supposed to supplement Botting's older specimens from Socotra,

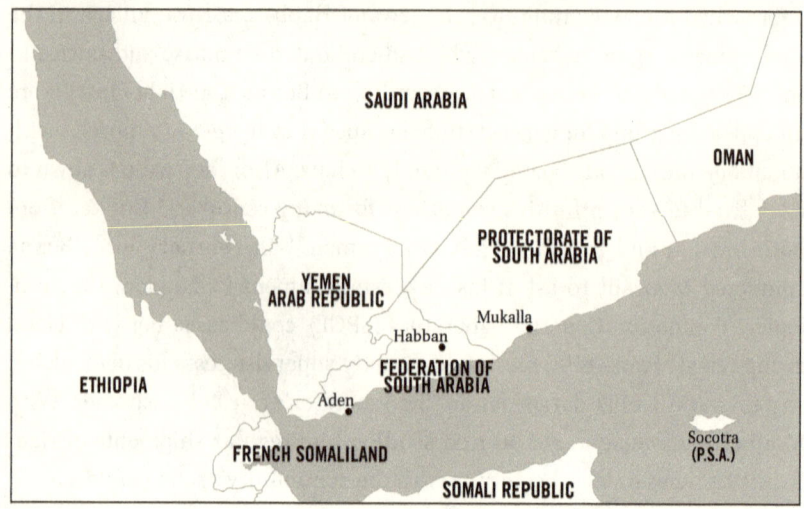

MAP 3. Map showing political divisions of southwestern Arabia during the Aden Emergency, 1963–1967.

Mourant urged Alain Marengo-Rowe, a pathologist posted to the Khormaksar Beach RAF Hospital in Aden, to obtain samples from previously untested tribes on the Arabian mainland. These included Arabs of Mahra and Dhofar, hypothesized to be ancestors of the Socotrans, and, more importantly, Arabs of the Hadhramaut (Hadhramis) living in the vicinity of Habban.

The Habbanite Jews took center stage in Marengo-Rowe's January report to his commanding officers at the Khormaksar hospital. Because the Habbanites were "totally different from any other known Jewish or Arab population," the ongoing investigation of their genes provided the "immediate reason" for testing Hadhramis. Reflecting Mourant's own concerns, Marengo-Rowe cautioned that "it must not, of course, be said publicly that we wish to examine these Arabs in order to compare them with Jews!"[44] He outlined his plans to collect about 300 samples from Hadhramaut, as well as 200 samples from Mahra and Dhofar, with the assistance of a network of British medical officers and commanders of Arab military units. Unfortunately, securing the permission of commanding officers and other local authorities was insufficient to convince Arab soldiers, such as the Omani Dhofar Forces, to donate their blood. By the end of January, Marengo-Rowe managed to ship Mourant 104 samples from a hodgepodge of hospital and clinic patients hailing from Hadhramaut, Mahra, Dhofar, and other regions of Muscat and Oman, with the admonition that "you cannot imagine the difficulties we have had to overcome."[45] Meanwhile, Mourant kept Bonné in the dark about these machinations in the Hadhramaut. His instructions to Marengo-Rowe to avoid public discussions about the Habbanite Jews suggest that Mourant believed the successful collection of Arab blood was simply a matter of concealing their underlying interests in Jewish groups—and, by extension, excluding Israeli scientists.

At the same time, Bonné found that the Habbanites in Bareket were losing interest in her project and becoming less cooperative. When she tried to work with Habbanite families who had moved outside the village, she noted that they were even more opposed to participating, being "extremely sensitive and obstinate."[46] Feeling that she had managed to obtain a representative sampling of about 90 percent of the community's members, she expressed interest in turning to a different Yemenite Jewish community, whose origins lay outside the Hadhramaut, to serve "as a control group."[47] Mourant agreed with her plans but asked Bonné to hold off on sending any new samples from Israel until the end of April. The SPGL, he explained, was anticipating a high volume of shipments from Bhutan and Arabia that would keep the staff too busy to assist Bonné. He did not mention that the "Arabian" samples would be the Hadhramis collected by Marengo-Rowe, who themselves would serve as a non-Jewish control group for the Habbanites. None the wiser, Bonné agreed to wait until May to send any Yemenite samples to the SPGL, noting that in any case she needed time to build up trust with the community.

Mourant's expectations for what Marengo-Rowe could achieve amidst increasing violence in southern Arabia turned out to be overly optimistic. By the end of February, Marengo-Rowe reported that his contacts in Hadhramaut and Dhofar had stopped responding to his messages. For "political reasons," he could not work at the RAF Hospital at Mukalla, in the Hadhramaut. Furthermore, Colonel Eric Johnson, the commander of the Hadhrami Bedouin Legion, had "withdrawn permission to bleed his men" in light of the recent assassination of his predecessor. Thus unable to access the peoples and regions Mourant had requested, Marengo-Rowe could supply only small batches of blood belonging to men from Aden and tribes residing in the broader Federation of South Arabia.[48] In April, Donald Tills, Mourant's technician and deputy director, encouraged Bonné to resume sending samples to the SPGL, commenting, "Several of our expeditions seem to have dried up [due] to unpleasant political situations."[49]

"Unpleasant," of course, was an understatement. Marengo-Rowe made his final breakthroughs in May, sending blood from eighty Dhofari Arabs posted in Dubai and nearly fifty precious samples from Hadhramis living in "Habban and immediate surroundings." In addition to his own difficulties convincing research subjects to donate blood, Marengo-Rowe warned that "disaster has struck my organization." Of his two chief contacts in the Hadhramaut, one had been shot and killed, and the other had committed suicide. All Europeans had been evacuated from the British Residency at Mukalla, with Marengo-Rowe's own family due to depart within weeks, and all airlines had ceased flying to the interior regions of Yemen. Marengo-Rowe's hospital in Aden was swamped with growing numbers of casualties.[50] Even under these circumstances, he still hoped to make one final venture into the Hadhramaut, with Mourant's support and encouragement. In parallel, Bonné sent samples from her "control group" community of Yemenites, whom she found to be thoroughly cooperative. However, the brewing military storm on Israel's borders soon interrupted her work as well. On May 23, Egypt blockaded the Straits of Tiran, turning away Iranian oil tankers bound for the Israeli port of Eilat. In response, Israel mobilized for war; Bonné's colleagues at Tel-Hashomer set up emergency blood banks across Tel Aviv, requesting urgent supplies of ABO and Rh antisera from the United Kingdom.

On June 5, 1967, Israel launched a surprise attack on Egypt. Within hours, the Israeli military managed to destroy almost the entirety of the Egyptian, Jordanian, and Syrian air forces. Within three days, Israel had captured East Jerusalem, the West Bank, and the whole of the Sinai Peninsula. The elation of the Israeli public at their military's swift success cannot be overstated. On June 9, even as the fighting raged on in the Golan Heights, Bonné wrote to Mourant proposing to revive their study of the Samaritans in Nablus, "as well as other neighbouring populations which were previously inaccessible to us."[51] Meanwhile, Marengo-Rowe's research prospects in Aden could hardly have been more different. Two weeks after Bonné's letter, Mourant received Marengo-Rowe's apologetic message that the "situation is now impossible and our soldiers are being killed every day," including one of his principal contacts in the civil armed police, who was shot during the mutiny.⁵² Giving up any further attempts to provide Mourant with blood, he left Aden in July. This last gasp of British colonialism in Aden, in contrast to Israel's sudden territorial conquests, transformed the nature of Bonné and Mourant's collaborative relationship. Over the rest of the summer, Mourant and Tills expressed interest in expanding their work with Bonné and finally shared with her their results from Marengo-Rowe's hard-won Hadhrami samples.

THE SINAI BEDOUIN AND THE POLITICS OF SCIENTIFIC LABOR AFTER 1967

Within days after the conclusion of the June 1967 war, the Holon Samaritans prepared to visit their kin in Nablus. Batsheva Bonné joined the visit, at long last stepping across the Green Line. She fondly recalled that the family of the Nablus high priest hosted her "like a queen."⁵³ This personal satisfaction heralded new expectations of professional power. Arthur Mourant quickly became a supplicant to Bonné, clearly hoping she might take over the role of Harry Smith's abandoned Beirut laboratory. "I am delighted that you have been to see the Samaritans at Nablus," he wrote to her. "Do you think there is any hope of resuming a study of their blood. As you know, we tested only 82 and then for some reason they became hostile and we could get no more."⁵⁴ Within six weeks, Bonné began collecting and shipping Samaritan samples to the SPGL. Mourant responded, "I am touched to see that [the Samaritans] seem to trust you better than they did me and my team. Even though I did all I could [to help them] I suppose I was, in the final analysis, associated in their minds with their Arab masters, and so had to pay for the blood and put up with considerable obstruction."⁵⁵ Evidently, it had never crossed Mourant's mind that his own approach to blood collection might be flawed. Rather, he blamed "Arabs" generally and the Israeli-Arab conflict specifically for his fieldwork problems and the ensuing collaborative tensions with Bonné. He and Smith had constantly pointed to the turbulence of Middle Eastern politics as a justification for the continued Anglo-American dominance over scientific activity in the region. Now Mourant's complacent reliance on British imperialism to sustain his protégés' outposts in the Arab world had become a liability.

In this context, Mourant wrote that he was "delighted at Israel's successful resistance" in the June War.⁵⁶ The tone of his correspondence suggested that he now considered Israel his ideal outpost for collecting Arab blood samples. Meanwhile, Donald Tills, who had already proposed visiting Bonné in Israel to help her set up new enzyme testing techniques, now planned to use the trip as a launchpad for both an IBP project and his own master's thesis. Unfortunately for Mourant and Tills, Bonné was no placid replacement for Smith. Although she helped arrange Tills's journey, she expressed reservations about the idea that Tills's thesis would focus on Middle Eastern populations, pointing out that this would amount to a duplication of her own planned research rather than an original project. Bonné regarded the collaboration as a temporary necessity stemming from her laboratory's lingering infrastructural issues rather than a permanent arrangement for the benefit of the SPGL. She had been given a space at Tel-Hashomer that had once served as the research laboratory of the blood bank; it contained "basic equipment" like centrifuges and power units but lacked any electrophoresis apparatus. Her technician ran tests for abnormal hemoglobins and G6PD deficiency in the hospital's biochemistry laboratory.[57] Reflecting on the ups and downs of Tills's experience working with her in Israel, she commented: "We are still characterized here by the unusual combination of on the one hand many technical difficulties (electricity, equipment, availability of materials etc.) and on the other hand ease of approach to both 'ethnic' subjects and also to institutions and colleagues."[58]

By the time Tills arrived in Israel in mid-December 1967, Bonné had been presented with an irresistible opportunity to explore a new group of "ethnic subjects": the Bedouin tribes of southern Sinai, now living under Israeli military occupation. Travel to the peninsula required permission from its military administration, which proved easy for Bonné to secure. Tel-Hashomer's directors, Chaim Sheba and Baruch Padeh, had personal connections throughout the Israeli military establishment dating from their days in the Haganah, clearing the way for the army's support. Tel-Hashomer was allocated responsibility for the Bedouins' health care, and Sheba, with his personal interest in genetic anthropology, proposed that Bonné travel with the hospital physicians, who were carrying out medical surveys. On November 5, she flew on an Israeli Air Force plane to Abu Rudeis, the Israeli administrative center for managing Bedouin affairs, to describe her research to the military governor. The Bedouins of southern Sinai, she explained, were of particular interest to geneticists because of their geographic isolation from the rest of the peninsula, which meant that they might represent descendants of the region's oldest inhabitants. However, the most exciting prospect was the Jebeliya tribe, which had completely avoided intermarriage with the neighboring tribes. According to Jebeliya oral tradition, their ancestors had been Bosnian and Romanian serfs who were sent by the Byzantine emperor Justinian I to tend Saint Catherine's monastery on Mount Sinai in the sixth century B.C.E. Lending support to this narrative was the occasional birth of blond, green-eyed tribe members. Otherwise, they were indistinguishable from the other Bedouins in terms of their dress, language, and Muslim religion. Like the Samaritans and the Habbanites, they offered an opportunity to examine the genetic effects of inbreeding as well as to correlate biological data with historical ancestry claims.[59]

The governor approved the project, and Bonné returned to Sinai less than three weeks later with her Tel-Hashomer colleague Avinoam Adam and several of her graduate students. From Abu Rudeis, they were driven in army jeeps and accompanied by Israeli navigators and interpreters to find Bedouins of seven different tribes. During their first expedition, they collected blood, anthropometric measurements, and demographic data from about 250 men. In her autobiography, Bonné claims that no Bedouin refused to participate in the study, either because they perceived the researchers as representatives of the "new government" or because the accompanying physician offered free medical examinations for all participants.[60] However, her interactions with the Bedouins were not as consistently smooth as this account implies. Even on the first expedition, not all participants agreed to donate blood, and during her eight subsequent visits to Sinai, Bonné encountered additional forms of resistance. She reported to Mourant that the Bedouin men she sampled were "not eager" to allow her to bleed their families; "they themselves agreed, I suppose, only because I paid them well."[61] When Tills and Mourant requested that she send a higher volume of blood from each individual to improve their test results, she complained that this was impossible because "all family members, including the tribe's Sheick [sic] stand around and give direct instruction that 2–3cc are more than enough. By then, the subject pulls his arm away. . . . All this after hours of convincing. . . ."[62] The South Sinai Bedouins had had little experience with modern medicine because Egyptian administrations had never set up any clinics in the region. The Israeli military government quickly built up a basic health-care system, hoping to establish a rapport with the tribes as well as a humanitarian reputation.[63] However, at the time of Bonné's visits, Israeli plans for Sinai were still shrouded in uncertainty, in light of the ongoing War of Attrition along the Suez Canal front. The Bedouins rightfully responded to their sudden encounter with Israeli physicians and the demands for blood donations with the same wariness they had applied to interactions with Egyptian officials.

At first, the Sinai project seemed to represent the peak of Bonné's successful collaboration with the SPGL. Tills traveled to Sinai with Bonné in January 1968, and she briefly visited the London laboratory that summer. In their correspondence, Mourant and Tills regularly expressed enthusiasm for the fascinating results they obtained with the Habbanite and Bedouin samples, urging Bonné to send more specimens and publishing some of their findings in *Nature* (choosing not to credit her by name, with her permission).[64] However, Bonné increasingly resented that the SPGL took many months to send her serological test results, which interfered with her own ability to present and publish her research. To ameliorate her reliance on overseas collaboration, she sought to accumulate more equipment and build up her own serological laboratory. She confided to Tills in March 1968 that she had ordered "centrifuges, water-bath, incubator, pH-meter, spectrophotometer, analytical balance, freezer, refrigerator, vacuum-pump, stirrers, arrangement for distilled water, and a variety of glassware and chemicals," aiming to be able to do routine blood typing in her own lab.[65] However, the equipment took months to arrive, and in the meantime she remained substantially dependent on the SPGL. At the end of 1968, despite difficulties obtaining grant money to fund her trips to Sinai, she assured Mourant that she would be able to continue sending him samples of the Jebeliya and other tribes in light of her "excellent relations with the Army."[66] In contrast, her relations with the SPGL were severely tested in March 1969, a few weeks after she sent a final shipment of Jebeliya blood to London.

In the summer of 1968, Bonné had presented some data from the in-progress work on the South Sinai Bedouins to the staff of Tel-Hashomer, and the text of her talk was printed in a special issue of the hospital's internal proceedings, dedicated to Sheba.[67] When Mourant discovered that this data had been published without his consultation, he angrily wrote to Sheba, complaining that Bonné had failed to seek his approval and downplayed the SPGL's contribution. She had also drawn conclusions from the data with which he did not agree. Bonné responded to Mourant directly with a tirade of her own:

> I do not regard myself only as an agent for collecting bloods and sending them to labs abroad to be typed and classified. I have my deep anthropological and genetic interest in these populations, hence I spend many hours visiting them; becoming acquainted with them, and observing their way of life and thus collecting as many genetic markers and traits as I can. [...] It seems to me, for unknown reasons that the fact that we are typing *here* the bloods, screening the sera, etc. is not only not encouraged or supported by you and your colleagues but rather the opposite.[68]

This exchange shows that Mourant's perception of Bonné's role in the research remained fundamentally unchanged since the 1963 Samaritan expedition. In his mind, she was a necessary intermediary, with a convenient geographic location and set of connections to subject populations and Israeli government authorities—an ideal "collection agent." Yet Bonné rejected this subordinate status and resisted Mourant's expectations that her own research program would remain a mere satellite of his own, asserting herself as an autonomous scientist capable of analyzing, interpreting, and publishing her own results. Her suggestion that Mourant's behavior reflected fears about the increasing technological capabilities of her own laboratory shows a keen insight that his scientific prominence emerged from, and perpetuated, the dependency of the international genetics community on Western-based reference laboratories. Mourant, however, did not acknowledge the potential for Bonné's lab to become substantial professional competition, and he regarded her complaints as a product of her "difficult personality."[69]

In the aftermath of this incident, Bonné stopped sending blood samples to the SPGL, instead running all new tests in her own laboratory. However, their correspondence continued, as Bonné urged Mourant and Tills to send results from previously sent specimens and approve her manuscripts for their joint publications on the Habbanites and Sinai Bedouins. Hostilities surged anew in April 1970, when Bonné wrote to the editor of the journal *Human Heredity* to delay the imminent publication of two articles by Tills and Mourant that included unpublished data on "ethnic groups in the Middle East," whose blood had been supplied by Bonné. She claimed that Tills and Mourant had never shown her the drafts and, accordingly, many of their population figures included "duplicate specimens, closely related individuals and incorrect division of population groups all of which distort the representation of these populations and their gene frequency."[70] She forwarded this letter to Mourant, explaining her actions in terms of Israeli ownership over the data: "I think there has been a basic misunderstanding [. . .] with regards to our collaborative efforts. Since the group here at Tel-Hashomer and myself have and will continue to devote all of our research efforts to the study of the Middle Eastern populations, it would seem only logical that we should have the right to publish first and that we should be properly consulted on all matters regarding the use of this data."[71]

Unsurprisingly, Mourant and Tills vehemently disagreed. Tills argued, "In your letter to [the editor of *Human Heredity*] you state that these papers include your work; this as you well know is not true. All the data included in the two papers was on work performed in this laboratory, and *your connection with it was in collecting and sending the samples*."[72] His lengthy letter continued with an itemized list of grievances, which ultimately centered on Bonné's resistance to the use of her lab as a Middle Eastern outpost for the SPGL. He complained, "You appear to completely fail to understand that if you send samples to outside laboratories for testing, it is not unreasonable for them to use these results in general review articles [. . .]. You are the only person who has ever complained of such a procedure!"[73] Disregarding Tills's ire, Bonné continued to insist on the importance of her sustained relationships with research subjects as she and Mourant drafted joint publications on the Habbanites and Sinai Bedouins.[74] She wrote:

> I have often the feeling, Dr. Mourant, that you completely disregard or underestimate my own ability to sort the results of tests obtained by you, according to number, sex, tribal affiliation, family relationship etc. [. . .] I know many of [the Bedouins] personally (as I do the Habbanites, Yemenites and the Samaritans) as a result of more trips to the Sinai and my constant contact with them. And all this information is not always at hand immediately upon blood letting; often only after some time we realize that we bled the same person twice or that he didn't tell the truth about his tribe. This is the very reason why I have emphasized so often that it is important for me to see the results and prepare them *myself* for publication.[75]

By expressing her anxieties about data sharing not only in terms of professional courtesy but also in terms of scientific accuracy, Bonné upended the logic of the knowledge-control regime through which Mourant and Tills had published data gathered by others with impunity.[76] Directors of centralized laboratories assumed that truly scientific work happened only in Western facilities using equipment unavailable in developing nations; they regarded the labor that moved samples from outpost to center as merely logistical. In contrast, Bonné highlighted the significance of both logistics and ethnography to the production of scientific knowledge. Finally, her insistence on controlling the interpretation and publication of Middle Eastern population data asserted a newly distinct professional status for Israeli geneticists. This status emphasized a territorial propriety, girded by Israeli military strength, which legitimized Israeli control over both access to and data obtained from Middle Eastern subject populations. For example, in her autobiography, Bonné referred to the Samaritans, Habbanites, Jebeliya Bedouins, and Karaites together as "'our' four isolates" (*arba'at ha-izolaṭim "shelanu"*).[77] The equivocation indicated by placing "our" in quotation marks downplays her own role, and that of the Israeli occupation, in elevating Israeli geneticists from "collection agents" for European and American laboratories to a self-sufficient research community pursuing independent interests.

PALESTINIAN COMMUNITIES AS "ISRAELI ETHNIC GROUPS," 1971–1978

In March 1971, Arthur Mourant was obliged to retire by the Medical Research Council, who also shut down the testing facilities of the Serological Population Genetics Laboratory. However, Batsheva Bonné soon resumed shipping blood samples to the United Kingdom. Now her primary collaborators were Walter and Julia Bodmer, a husband-and-wife team of immunogeneticists based at Oxford University. The Bodmers had first visited Israel early in 1971, where they met Chaim Sheba just months before his death, and proposed a joint Israeli-British research project to study the human leukocyte antigen (HLA) system in Israeli populations.[78] The HLA system is a highly diverse series of cell-surface proteins that allows the immune system to distinguish between the body's own cells and foreign matter. Discovered in the 1950s, these proteins are responsible for causing the rejection of human organ and tissue transplants and are therefore immensely important in surgical medicine. Like the ABO and Rh antigen systems and inherited blood diseases, the HLA system was also readily co-opted for anthropological purposes. The seemingly unlimited variability of the HLA proteins, and the genes that encoded them, suggested that they might be able to reveal events in human evolutionary history at a much finer scale than classical genetic markers could detect. By the 1970s, geneticists argued that HLA tests could provide definite answers to questions about family and population genealogies for which other blood antigen systems might give ambiguous results. They could definitively determine an individual's paternity, and they could help scientists reconstruct the genetic makeup of deceased individuals—including common ancestors of long-segregated populations.

Bonné enthusiastically embraced the Bodmers' proposed HLA study, which dovetailed with her own long-standing ambitions to direct "a large scale project to investigate a variety of population groups in Israel."[79] In April and May 1971, Bonné and her students collected blood from Yemenite Jews in Rosh Ha'ayin, then airmailed the samples to the Bodmers' laboratory.[80] The HLA study therefore began as only the latest incarnation in a long line of genetic studies performed upon the Rosh Ha'ayin Yemenites since 1950. As the project developed, Israeli collaborators planned to survey five additional Mizrahi communities. However, Bonné's intensive fieldwork on the Samaritans and Sinai Bedouins during the 1960s had helped to reorient Israeli population genetics away from a formerly near-exclusive interest in Jewish populations. While Sheba and his Tel-Hashomer colleagues had routinely utilized non-Jewish blood samples in their research, they did not deliberately seek out non-Jews as research subjects. Rather, they employed non-Jewish hospital patients and blood donors (designated "Arabs" or "Israeli Arabs," sometimes divided into "Moslem Arabs" and "Druze" after 1957) to serve as comparative "outgroups" for the Jewish populations. In the first decades of Israeli statehood, the notion of "ethnicity" primarily described differences among Jewish Israelis, specifically the cultural and physical differences of Mizrahim from Ashkenazim, and it was in this sense that the first generation of Israeli geneticists had investigated the biological relationships of Jewish "ethnic groups" (i.e., 'edot). Bonné, however, had planted the seeds for a more expansive concept of "Israeli ethnic groups," including categories of non-Jews subject to Israeli military authority.

This new ethos permeated the HLA project in its early years. Another Israeli collaborator—Sarah Nevo, an MD at the University of Haifa—planned to study non-Jewish Israeli citizens, specifically Muslim and Druze Palestinians. Nevo reflected idealistically on her work, writing to Walter Bodmer, "You are contributing, indirectly, to develop [sic] communication between

Arabs and Jews here as I am beginning to socialize with Arab doctors [for the purposes of the study]."[81] Through this socializing, Nevo eventually collected Palestinian blood samples from a maternity hospital and a genetic counseling clinic, both near the northern coastal town of Acre.[82] In 1972, Bonné turned her attention to the Armenian community living in the Old City of Jerusalem. After securing the cooperation of teachers at the Armenian Theological Seminary and Sts. Tarkmanchatz school, Bonné led weekly trips with Tel Aviv medical students to the Armenian Quarter to collect blood. Cyril Levene, an employee of the Israeli Ministry of Health, also occasionally joined Bonné in these visits.[83]

The optimistic engagement with non-Jewish Israelis ground to a halt in October 1973. Nevo's enthusiasm for promoting social relations between Jewish and Arab citizens was abruptly derailed by the death of her nineteen-year-old son, an Israel Defense Forces soldier, in the Yom Kippur War. While the Nevo family mourned, Bonné forged ahead with studies of Jewish populations for the joint project. This exacerbated a professional rivalry between the two Israeli collaborators and ultimately proved illusory the notion that "Israeli ethnic groups" could offer a conceptually neutral basis for the genetic comparison of Jewish and non-Jewish citizens. By March 1974, Bonné was able to resume her sampling of Libyan, Moroccan, and Iraqi Jews near Tel Aviv after the disruption of the war; Nevo could not even make satisfactory arrangements to sample the Druze of Isfiya village (outside Haifa) until June 1975.[84] This meant that Nevo was embarking on new fieldwork at a time when a significant proportion of "Israeli Arabs" was undergoing a process of "Palestinization" through their deteriorating relationships with the Israeli state and renewed contact with Palestinians in the occupied West Bank.

The success of the Palestine Liberation Organization (PLO) in attaining UN recognition in 1974 appealed to Palestinians who rightly observed that cooperation with Israeli authorities—and, in the case of the Druze, their service in the Israeli military—had not secured their equal status with Jewish Israeli citizens. By the end of the year, a harsh campaign of land expropriations by the Israeli government transformed the simmering discontent among the Muslim, Christian, and Druze Arabs of northern Israel into political action against the "collaborationist elite" of their communities.[85] Sakhnin, one of the towns from which Nevo had collected Arab Muslim blood samples, was particularly affected by land expropriation. In 1972, the Druze Initiative Committee formed to demand an end to the compulsory conscription of the Druze, calling it a "blood tax" on the community, which suffered from socioeconomic discrimination, neglected town infrastructure, and land confiscation, just like other Arab citizens who did not serve in the military.[86] The Committee's activities intensified into a violent demonstration against foreign minister Yitzhak Rabin in April 1974 and a campaign to support the inhabitants of Kisra, a Druze town on whose agricultural fields the Israel Land Administration encroached in the summer of 1975. Meanwhile, Christian and Muslim leaders met in Nazareth to form the "National Committee for the Protection of Lands" (*al-lajna al-waṭaniyya li-ḥimāyat al-arāḍī*) in October 1975, which organized a national day of strikes and demonstrations on March 30, 1976. Druze, Muslims, and Christians alike participated in the events of Land Day, which drew a brutal response from the Israeli Army and ended with the death of six Palestinian citizens, half of them from Sakhnin.[87]

In her correspondence with the Bodmers, Nevo mentioned that she had been temporarily forced to stop collecting Druze samples due to "some objections to being bled among the villagers" and did not achieve her desired sample size of 200 people until April 1977.[88] Although she did not offer a reason for the villagers' resistance, the aftermath of the Land Day protests and the polarization of Druze communities into pro- and anti-Israel camps surely played some role. Nevo's fieldwork difficulties occasionally fueled resentment against her colleague Batsheva Bonné-Tamir (married name),[89] whom Nevo perceived as unfairly advantaged because she occupied a more prestigious post at Tel Aviv University and studied more "friendly" populations. In the summer of 1976, when Bonné-Tamir requested Nevo's data on Muslim and Druze Arabs to incorporate into her own statistical analysis, Nevo lashed out: "The Arab Moslems and the Druzes are 'my baby.' I spent a 'fortune' in time, effort and money, to get this done. I need not tell you how difficult it is to collect blood samples from Arabs, in villages, here. [...] I simply cannot afford to sell for so cheap, so much hard work of mine to just be mentioned as 'unpublished data of Nevo S.'"[90] In fact, Bonné-Tamir's request was not particularly untoward, as the entire joint British-Israeli team was wrapping up work and preparing to publish the results at this time. Nevo herself had hoped to be able to include her Druze data in the publication drafts, but laboratory results were not ready in time.[91] Ultimately, completion of HLA testing on the Druze samples took nearly two more years, and the first data to be published from Nevo's samples (on a different genetic marker) was buried in a paper that did not compare the Druze to other Israeli populations at all.[92]

After the difficulties working with "Israeli Arabs," it is hardly a surprise that Nevo rejected Bodmer's 1978 suggestion to conduct a study "amongst the refugee camps in Gaza," where Israeli researchers could not expect willing participants.[93] In the aftermath of the 1979 Egyptian peace treaty with Israel, which involved evacuating Israeli settlements in the Sinai Peninsula, Nevo expressed a decided ambivalence (likely shared by many Israelis) toward her relationships with different categories of Arab neighbors. In a 1980 letter to the Bodmers, Nevo enthusiastically insisted that she would buy an "authentic Bedouin embroidered dress" for Julia Bodmer; lamented the loss of Dahab, a Sinai tourist resort; mentioned her desire to visit Neviot (Nuwayba), a similar resort town, before it passed from Israeli to Egyptian control; and, finally, proposed a new study of local Arab villagers. She commented that she had "already secured 'Arab collaboration'—a very rare thing these days in Israel when we find at the university such lovely messages written by Arab students [in the restrooms] reading: we are all PLO, when the time comes we will slaughter you all [. . . following] the achievement of the free Palestine. Great, is it not? The era of peace in the Middle East has started." On a less sarcastic note, she concluded, "As you see Arab collaboration with me in one of the most hostile villages in the country (this is where your samples came from), is a real historical event."[94]

The increasing disenchantment with non-Jewish research subjects is also evident in Bonné's attitude to the Armenians, as displayed in her autobiography and scientific publication on the community. Compared with the effusive descriptions of her relationships with the Samaritans and even the Sinai Bedouins, she maintained a striking detachment from the Armenians. She explained her interest in them as due only to their status as "one of the ancient communities of the area" and to Sheba's fascination with "the resemblance between them and the ancestors of Sephardi Jews."[95] Perhaps this social distance reflects the extent to which Armenians were culturally and economically integrated with Arab Palestinians under British Mandate and Jordanian rule as well as the shared experience of expulsion by Zionist forces during the 1948 war, which had driven thriving communities of Armenians in Haifa and Jaffa across the borders to Lebanon and the Jordanian-held Old City of Jerusalem.[96] After 1967, the few thousand Armenians who remained in Jerusalem and the West Bank under Israeli occupation mostly retained their Jordanian citizenship as well as what Bonné characterized as "close contact with Armenians in Lebanon and Jordan."[97] Despite acknowledging these contacts, Bonné treated the Old City Armenian Quarter as an isolated community somehow immune to the effects of Israeli-Palestinian violence. She mentioned that only "about half of them are descendents [sic] of the original Armenians and the rest are later 'joiners,'" with no further elaboration or analytical acknowledgment of either this original-latecomer distinction or of the recent historical events, ranging from Ottoman genocide to Israeli state policies, that had produced this population structure.[98] Rather, Bonné and her coworkers chose to flatten the Armenians' complex demographic history by treating them, like the Arabs, as a single undifferentiated "Middle Eastern population" against which the various Jewish categories were measured.

The grand plans to integrate non-Jewish groups into the genetic study of "Israeli ethnicities" had thus faded by the publication time of the HLA project, whose title went through several telling revisions. Chaim Brautbar (of the Hebrew University) first labeled it "HLA Polymorphism in Jewish Ethnic Groups"; Bonné-Tamir amended it to "Israeli Ethnic Groups" to acknowledge the Armenian and Arab data; and Walter Bodmer finally reduced it to "HLA Polymorphism in Israel," as he said, "for brevity."[99] A number of factors had led to the elision of Mizrahi and non-Jewish material from the project, which was finally published as a series of nine short papers: four on Ashkenazi Jewish groups, three on Mizrahi Jewish groups, one on the Armenians, and a final comparative analysis that mentioned Armenians and Arabs only in passing. In fact, the unintended innovation of the HLA project turned out to be its significant emphasis on Ashkenazi populations, particularly their consideration by the specific country of origin (Russia, Poland, Germany, and Romania), rather than the undifferentiated "block" portrayed by an earlier generation of Israeli geneticists.

GENETICS AND MILITARY OCCUPATION

The traffic in blood underlying postwar genetic research was built on a foundation of technological and financial disparities between Western laboratories and non-Western field sites. In the Middle East, British and American colonial, missionary, and military infrastructures provided ready access to Arab populations, whereas the funds and equipment to study them remained largely in Western laboratories. The solutions applied by Mourant to this distribution problem—either creating "outpost" facilities like Smith's lab in Beirut or relying on "collection agents" like Alain Marengo-Rowe or Batsheva Bonné-Tamir—aimed to keep the flow of genetic material and information into his own hands at the expense of local collaborators.[100] Tensions of gender, race, and nationalism further marginalized the contributions of Middle Eastern scientific actors, as demonstrated by Smith and Mourant's dismissal of Bonné-Tamir's legitimate concerns as naïve misunderstandings. These forms of discrimination were exacerbated by the Israeli-Arab conflict, which in turn provided Westerners like Smith and Mourant, who had the privilege of crossing contested state borders, with a practical justification for their administrative control over Middle Eastern genetic research.

The 1967 June War reversed these fortunes, laying the foundations for the ascendancy of Bonné-Tamir's own laboratory and of the Israeli genetics community more generally. The current prominence of Israeli anthropological genetics is now conceptualized as a validation of Zionist founding ideals that the Jewish state and its culture would represent an outpost of Western Enlightenment values in the Middle East. However, it actually stems from Israel's self-repositioning as a regional military power ruling over Middle Eastern populations and the increased willingness of Israeli researchers to market themselves as interpreters of local Arab cultures. However, the nature of the Israeli-Arab conflict, particularly the Israeli occupation of the West Bank and Sinai, belies the moral contrast Bonné-Tamir posed between herself and her British collaborators. In 1975, she wrote to Mourant, describing a recent visit to the Nablus Samaritans, whom she hoped to add into the HLA project with the Bodmers. She confessed that "they are still not very easy about blood-letting" and that she had been able to collect only 125 samples, "after some persuasion."[101] However genuine her personal relationships with research subjects were, she still represented Israeli governance for communities whose legal statuses were fundamentally precarious. She visited the Armenians in East Jerusalem with workers from the Ministry of Health and drove around the Sinai in Israeli Army jeeps, appearing alongside an Interior Ministry team assigned to register all Bedouin families.[102] Their cooperation with Bonné-Tamir and other Israeli geneticists was therefore always, to some extent, an instrumental strategy of group survival in the aftermath of violent conquest. Even the Habbanite Jews, once the novelty of Bonné-Tamir's attention wore off, grew tired of indulging the intrusions of an

Ashkenazi scientist whose very presence confirmed their otherness within the Israeli state. The Muslim and Druze citizens of Israel had even more reason to react to Sarah Nevo's overtures with skepticism or outright hostility. "Collaboration" with Israelis, scientists or otherwise, had not protected the Palestinians' political rights, and it became particularly dangerous in the context of resistance to the 1970s land expropriations.

If Bonné-Tamir's arguments were ultimately self-serving, they also raised the question of how Arab geneticists responded, sometimes opportunistically, to Israel's territorial conquests. The displacement of people from northern Sinai into Egypt's Tahrir province provided convenient new fodder for Karim Kamel's research group in the Ain Shams University clinical pathology department. Funded by a grant from the U.S. Office of Naval Research, Kamel and his colleagues had been studying the blood of the inhabitants of the far-flung desert oases of Kharga, Dakhla, and Siwa. They also took a keen interest in the Nubian communities displaced by the construction of the Aswan High Dam and resettled north of Aswan city. Blood samples from these distant locales had to be airlifted back to Cairo for analysis, with all the usual expense and logistical headaches.[103] But within a few hours by car, Kamel's team could now reach hundreds of evacuees from Sinai, including two Bedouin tribes (Bayadia and Akharsa) and city dwellers from El-Arish and East Qantara. They analyzed blood samples from refugee schoolchildren, comparing their results intently with Bonné-Tamir's published findings on the southern Sinai Bedouin tribes as well as other Egyptian data.[104] Even as the 1967 war had cut off Egyptians' access to some populations, it had made others more readily available. Bonné-Tamir's work also made it possible for Kamel's colleagues to complete more quickly their picture of genetic diversity across Egypt and Sinai. Socially and professionally, Egyptian scientists had more in common with their Israeli counterparts than with the Sinai Bedouins, displaced Arabs, and other marginalized groups that they studied. The exploitative nature of Anglo-American and Israeli genetic anthropology on Arabs, therefore, is not unique to the violence of foreign colonialism and military occupation. The next chapter engages with the concept of "internal colonialism" to explore how national surveys of genetic diversity relate to civil strife and fears of ethnic separatism in Iran and Turkey.

Chapter 7

DOMESTICATING DIVERSITY

WHEN MOHAMMAD REZA SHAH ARRIVED AT THE WEST GERMAN capital of Bonn for a state visit on May 27, 1967, a crowd of his angry compatriots gathered for a spontaneous demonstration. Nearly 150 Iranian students protested against the Shah's regime that day, including Mansur Bayatzadeh, a doctoral candidate at the Institute of Anthropology at Johannes Gutenberg University in Mainz. Hundreds of police officers surrounded the protestors, and one allegedly dislocated Bayatzadeh's shoulder in the process of arresting him.[1] Over the following months, he was plagued by the legal fallout from his political activity, but this did not prevent his dissertation on Persian population genetics from being approved in July 1967. In fact, after standing trial for civil disorder in late 1968, Bayatzadeh remained in Mainz for at least another five years, during which he continued to publish genetic research with his doctoral adviser, Hubert Walter.

The Mainz Institute of Anthropology, founded in 1946, was first directed by Egon Freiherr von Eickstedt, who had been deeply involved in the theory and implementation of Nazi racial policy. Following his retirement in 1961, his longtime assistant Ilse Schwidetzky took over, largely carrying on the institute's research emphasis on anthropometry. In contrast, Walter, who joined the institute in 1962, hoped to rehabilitate German anthropology by rejecting racist scientific concepts and discriminatory practices. Walter soon developed a strong program in serological genetics, attracting a number of students who had been engaged in medical studies.[2] Bayatzadeh, Walter's first Iranian protégé, started work on his thesis in 1964. Bayatzadeh collected 48 blood samples himself from fellow Iranian students in Mainz and received a further 972 samples delivered by air from the Tehran Reference Laboratories of the Iranian Ministry of Health.[3]

Bayatzadeh had no trouble studying the local Iranian community, due to his leadership roles in cultural organizations and his anti-Shah activism. By 1966, he was a board member of Johannes Gutenberg University's Iranian Student Union, and he participated in anti-Shah demonstrations organized by the Confederation of Iranian Students–National Union (CISNU).[4] CISNU, since its founding in the early 1960s, connected pro–National Front Iranian students in Tehran with their compatriots studying abroad in Europe and the United States, bringing together a range of opposition ideologies. Bayatzadeh's decision to cancel the Mainz Iranian Student Union's planned Nowruz (Iranian New Year) event in March 1967 to honor the recent death of the former Iranian prime minister Mohammad Mosaddegh suggests that his sympathies aligned with Mosaddegh's vision of Iranian nationalism.[5] As for Bayatzadeh's relationship with the Tehran blood banks, he felt that their collection practices hindered the anthropological value of his work. As mentioned in Chapter 3, these institutions did not collect ethnic or religious data from blood donors, restricting Iranian geneticists to comparisons of populations on a provincial or regional basis or comparisons of national averages with those of other countries. Bayatzadeh repeatedly argued that regional differences could potentially be attributed to the "ethnic structure of Iran," but due to lack of "suitable information," this could not be confirmed.[6]

Yet even when Bayatzadeh collected his own samples from Iranians in West Germany, he claimed similar difficulties in obtaining information on the "ethnic structure" of his blood donors. Ultimately, he maintained the same geographical categorizations (six broad regions) in all of his work for the sake of comparative consistency. He argued that if significant genetic differences existed between ethnic groups, this regional division should be able to expose it, correlating as he did certain ethnic groups with Iranian geography—for example, "North Iran is mainly settled by ancient Iranians . . . Northwest Iran by Turks, Kurds and Armenians. . . ." But after years of surveying about a dozen genetic markers, Bayatzadeh discovered that his data was characterized by "striking" homogeneity. After reexamining his own work alongside that of foreign geneticists in Iran, he concluded that "on the basis of the genetic data from Iran published up to now... there is no clear evidence for the assumption of marked regional or ethnic genetic differences within this country."[7] Nevertheless, he called for further genetic research in Iran, "which should be based on representative samples of the various ethnic groups."[8] However, Bayatzadeh himself would not be able to pursue the question of Iranian ethnic diversity. After his trial in 1968, Bayatzadeh's own research relied exclusively on Iranian students and residents within West Germany to supply Iranian blood samples, indicating that he was politically barred from traveling to Iran or making use of government institutions like the Ministry of Health laboratories.[9] By 1972, his career as a geneticist had sputtered to a halt.

Bayatzadeh's experiences demonstrate, in an incidental yet conspicuous way, the subversive potential of genetic anthropology within Middle Eastern political contexts. The medical orientation of genetic research in the first two decades of the postwar era had enabled anthropological investigations to emerge as a subsidiary activity of blood transfusion facilities, disease inheritance studies, and public health interventions. In the early 1970s, the renewed push for an anthropologically based genetics highlighting ethnic differences presented particular challenges for geneticists working under regimes deeply invested in narratives of national homogeneity, like Iran and Turkey. The revived specter of Kurdish ethnic separatism and the outbreak of widespread political violence in both countries further complicated the ability of Iranian and Turkish geneticists to participate in this emerging international field of research without running afoul of their governments. The "ethnic factor" of these civil conflicts made the notion of diversity deeply contentious, and new Iranian and Turkish regimes adapted and intensified previous nationalist policies to either co-opt or outright deny the recognition of ethnic minorities and their grievances. This chapter examines the strategies used by Iranian and Turkish geneticists to "domesticate" ethnic diversity as a pattern of genetic variation that did not undermine the essential unity of the nation-state.

GENETIC ANTHROPOLOGY AMID THE THREAT OF ETHNIC POLITICS

Physicians and hematologists like William C. Boyd, Arthur Mourant, and Hermann Lehmann had never wavered in their belief that their work on inherited blood traits represented an enlightened form of anthropology, although most physical anthropologists remained skeptical of the utility of serological genetics through the 1960s. A new generation of biochemists, led by Emile Zuckerkandl and Morris Goodman, tried again at a 1962 Wenner-Gren symposium to promote a "molecular anthropology," this time one that would "privilege the point of view of hemoglobin." Again, most of their colleagues working in paleontology, anatomy, and taxonomy were initially unenthusiastic.[10] However, around this time certain academic centers did begin to integrate genetics and physical anthropology, such as the Institute for Anthropology in Mainz. In the Middle East, anthropologists like Batsheva Bonné-Tamir had also worked their way into the medical establishments that had largely dictated genetic research practices. In 1970s Iran, the growing involvement of individuals with anthropological training significantly changed local sampling strategies for human genetics. Instead of compiling blood bank statistics, they conducted field surveys that targeted specific ethnic, religious, and tribal communities. In contrast, Turkish anthropologists continued to rely on anonymous blood donor records until the 1990s. The divergent paths toward genetic anthropology in the two countries were all the more noteworthy given that they both endured major political and social upheaval throughout the 1970s, culminating in the 1979 Iranian Revolution and the 1980 Turkish coup d'état.

The transformation from the imperial Pahlavi state to the Islamic Republic of Iran entailed a drastic reorientation of political culture. However, it must be remembered that the Iranian Revolution coalesced around popular resistance to the autocratic rule of Mohammad Reza Shah, not around any particular ideology. The hundreds of thousands of Iranians who participated in the snowballing protests between October 1977 and the departure of the Shah and his family in January 1979 represented a broad coalition of liberal nationalists, leftists, Islamists, and everyone in between, including a number of activists for ethnic minority rights. During the late 1960s and 1970s, the Shah's government had violently subdued peasant uprisings in Kurdistan and "low-scale guerrilla war" in Baluchistan.[11] Leftists and Islamists alike had appealed for the support of disaffected Kurds and other ethnolinguistic minorities who hoped for increased regional autonomy; a reversal of government resettlement policies; and rights to property, free movement, and education and media in their native languages. Ayatollah

Ruhollah Khomeini and his immediate supporters were strongly anti-nationalist, and many outside his circle also believed that the hypothetical Islamic Republic, based on a religious program of social justice, would end the ethnocentric Persian-oriented nationalism that had defined the Pahlavi dynasty. But after Khomeini returned to Iran and began consolidating his personal authority and vision for clerical rule, he abandoned his tenuous alliances with secular leftist groups and ethnic activists. The new constitution approved at the end of 1979 contained several provisions for the protection of minority religious and cultural rights (including the public use of "ethnic languages") and prohibited all forms of discrimination. However, autonomist movements in Kurdistan and other border provinces that had emerged in the heat of the revolution were quickly labeled communist dissidents and suppressed by Khomeini's supporters.

In a phenomenon that attests to the highly contingent formation of the Islamic Republic, Saddam Hussein can be partially credited for preserving Iranian nationalist sentiments. The Iraqi invasion of western Iran in September 1980 began a war that would drag on for eight bloody years. It also proved to be a boon to Khomeini's new regime as it constructed a hybridized religious-nationalist discourse, lending a patriotic facade to its insistence on Iranian national unity and the embargo of all requests for increased local self-government.[12] The public campaigns of Islamization, such as the Cultural Revolution, which shut down the country's universities from 1980 to 1983, belied the extent to which the Islamic Republic co-opted much of the state structures and nationalist mythologies the Pahlavis had left behind. This included not only the central government's privileging of Persian language and culture as the normative national identity but also its approach to managing and representing other ethnic identities.

In 1968, the Pahlavi regime had funded a major expansion of activities at the Center for Iranian Anthropology, charging it with the task of conducting research in ethnology, folklore, and cultural and physical anthropology. During the early 1970s, most Iranian anthropologists worked in rural and tribal areas, where the completion of each study was celebrated with organized exhibitions "to familiarize the local people with their local culture."[13] Medical and, in turn, genetic research mirrored this pattern, with physicians and scientists dispatched from Tehran and other urban universities to measure and interpret the characteristics and health problems of marginalized ethnic and tribal groups. The Islamic Republic tacitly approved and perpetuated this approach to "domesticating" ethnic difference, ensuring that all of its expressions fundamentally identified with the Iranian nation-state.

In Turkey, the 1970s marked a period of severe political polarization, with government repression against left-wing activists accompanied by the local resurgence of the Kurdish nationalist movement. Kurdish activists had long affiliated themselves with the Turkish radical left, but after 1975, they embraced a new language for their plight, describing Kurdistan as a colony of Turkey. This concept, although first formulated by a Turkish sociologist sympathetic to the Kurdish struggle, "came as a sudden shock to Turkish intellectuals who considered Kemalist Turkey an anti-colonialist and anti-imperialist country [. . .] now, in the hands of Kurdish intellectuals. . . . Kemalist Turkey had become a colonialist and racist state."[14] New militant parties, such as the Kurdistan Workers' Party (PKK), mobilized under the contention that only violence could end "Turkish colonialism" and secure a free, independent Kurdistan.[15]

Combined with the violence of militant left- and right-wing groups across the country, more than five thousand Turkish citizens lost their lives between 1975 and 1980. Citing the breakdown of law and order, the military dissolved the civilian government and parliament, suspended the constitution, and declared martial law on September 12, 1980. Over the next three years, the ruling generals harshly suppressed all expressions of left-wing and ethnic separatist ideologies, arresting, torturing, and imprisoning hundreds of thousands of suspected dissidents. This included a thorough purge of "politically unreliable" staff from the universities, which now came under strict state supervision.[16] Finally, in addition to rewriting the constitution, the military government endorsed a new formulation of national identity, the "Turkish-Islamic synthesis," intended to undermine the threat of communism. A military representative charged that communist sympathizers had "not only fostered class conflict but, even worse, also accentuated differences of identity among the people" and "fomented conflicts among the different Muslim sects and between the 'so-called Kurdish citizens' and the rest of the population."[17] Even as it extended the ban on using the Kurdish language, the junta enabled the expansion of religious education and the construction of hundreds of mosques, believing that the ascendancy of a new Islamic public culture and a reduced emphasis on ethnic nationalism could provide an appealing alternative to communism and a more stable social unity.[18] By the time a new civilian government established itself, in 1984, the PKK had renewed its guerrilla warfare. In response, although by 1987 the military had lifted martial law over most of Turkey, the southeastern provinces home to most of Turkey's Kurds remained under a "state of emergency" until 2002.

Despite this extraordinary turmoil, Iranian and Turkish geneticists suffered only short-term setbacks to their research rather than long-term infrastructural or technological deprivation. In fact, this same time period witnessed the increased development of local scientific facilities in both countries, which all became less dependent on foreign laboratories for testing most genetic markers. Local investigators in Iran and Turkey accordingly took on more leadership roles in the organization and dissemination of their own research. Iranian and Turkish scientists also hosted a higher number of international conferences and symposia than in prior decades.[19] However, political constraints on access to and the identification of ethnic groups profoundly shaped the development of genetic anthropology in both countries. As part of a professional class subject to political scrutiny, the expanding Turkish scientific community remained quite conservative in its approaches toward studying the country's genetic diversity. Iranian geneticists increasingly conducted research on the communities most marginalized by the dominant national culture, but no such change could be observed in Turkey. Turkish researchers recognized ethnic differences—even, in some cases, that of the Kurds—but only within the strictly bounded frameworks of hereditary disease rates and classical blood markers.

REVOLUTIONIZING IRANIAN BIOLOGY

During the late 1960s and 1970s, a nascent conglomerate of Iranian geneticists built several new institutions that revolutionized the practices of human genetic research in Iran and ensured the survival of a nationalist approach to Iranian biological identity even after the downfall of the Pahlavi regime. The Genetics Society of Iran (*anjuman-i zhinitīk-i īrān*), founded in 1968, was followed in 1974 by the Genetic Research Center at Jundishapur University, the Iranian National Blood Transfusion Service (INBTS), and a research group on "Human Genetics and Anthropology" at Tehran University. Each of these organizations emerged separately, largely at the behest of individual scientists with disparate research agendas, professional ambitions, and relationships with the Pahlavi government. Yet, taken together, they provided the infrastructure for a major shift in Iranian human genetics, wherein Iranian scientists (in concert with foreign investigators) led more and more field surveys targeting ethnic, linguistic, and religious subpopulations across the country.

The institutional pioneer was Pezeshkpour Mostashfi, a biology professor and later the vice chancellor of Tehran University, who organized the first meeting of the Genetics Society of Iran in June 1968. Ten years later, the Genetics Society had grown from a group of seven Tehran University scientists to more than two hundred members across Iran.[20] Meanwhile, Habibollah Fakhrai, a native of the southwestern city of Dezful, earned his doctorate in human genetics at Michigan State University in the 1960s. In 1974, Fakhrai accepted a teaching position at Jundishapur University in Ahvaz, on the condition that he could establish a genetics research center. Under Fakhrai's leadership, the Jundishapur Genetics Research Center administered multiple research programs in cytogenetics, hemoglobin abnormalities, blood groups and disease, and even a seminar series on the origins of life.[21] Fakhrai's center hosted the final pre-revolution meeting of the Genetics Society in 1978.

The long-term effort to reform Iran's blood banking system fell to Fereydoun Ala, the son of the Shah's trusted adviser, ambassador, and sometime prime minister Hossein Ala. Educated at Harvard and Edinburgh, the young Ala founded Tehran University's hematology department in 1965. In 1972, he proposed a national administration for blood transfusion that would take over the hitherto uncoordinated efforts of blood collection, testing, and storage from individual hospitals and universities. By 1978, his INBTS had largely taken over the independent transfusion services of the Iranian military as well as the Red Lion and Sun Society.[22] Although genetic research was never a primary function of the INBTS, the institution ultimately played a significant role in Iranian population genetics due to its ready access to blood samples and laboratory facilities. For example, Victor Alan Clarke, the former technician at Harry Smith's ill-fated Beirut laboratory, became the "Principal Scientific Officer" for the INBTS in 1975, overseeing the serology laboratory and tracking population-genetic markers in donated blood samples.[23]

Mostashfi, Fakhrai, and Ala, along with the institutions they created, represented the strong medical roots of human genetic research in Iran. For these investigators, population genetic research on ethnic and religious minorities in Iran was interesting primarily if it could be correlated with disease incidence. The medical priorities of Iran's emerging genetics community were demonstrated by the array of papers presented at the 1978 Genetics Society meeting in Ahvaz. Out of the twenty-three papers dealing with human genetics, seventeen are directly related to health and disease, while only six are concerned with population genetics. Even most of those six were simply side projects derived from medical research on underserved communities. For example, Fakhrai and his colleagues contributed a paper comparing the ABO frequencies of Arabic- and Persian-speaking residents of Khuzistan province.[24] This research was an outgrowth of the Genetics Research Center's project linking blood groups with disease resistance (e.g., correlating group B with higher resistance to smallpox). Fakhrai's colleagues discovered that an inordinate amount of their type-B hospital patients came from one small, Arabic-speaking tribal population. With the help of Fakhrai's brother, who by chance was the governor of the rural area where the tribe resided, the Jundishapur researchers visited the tribe three times to collect blood samples.[25]

Dariush Daneshvar Farhud, the figure who claims to have singlehandedly introduced the field of genetic anthropology to Iran, can at least be credited with promoting an anthropologically oriented research agenda for Iranian population genetics distinct from medical genetics. This stemmed from his education in Germany, where he spent ten years successively obtaining a bachelor's degree in psychology, a medical degree, and finally a doctorate in anthropology from the Institute for Anthropology in Mainz. Farhud joined the institute in 1964, the same year as his compatriot Mansur Bayatzadeh, to study under Hubert Walter. In contrast to Bayatzadeh's activism with Iranian students in Germany, Farhud seems to have devoted himself exclusively to his studies and forged stronger ties to Tehran, specifically with Ahmad Azhir. In the early 1970s, Azhir headed the blood transfusion services of the University Clinics of Tehran, and he granted Farhud access to the clinic's blood donors to obtain samples for his own research.[26]

Like Bayatzadeh, Farhud was strongly influenced by Walter's use of the World Health Organization–sanctioned racial categories "Caucasoid,"

"Mongoloid," and "Negroid" and subsequently incorporated this terminology into Persian-language scientific journals as "white," "yellow," and "black" races (*nizhād-i safīd/zard/siyāh*, or *safīd-pūstān, zard-pūstān, siyāh-pūstān*).[27] Also like Bayatzadeh, Farhud's early publications acknowledged regional genetic diversity within Iran. However, Bayatzadeh's work emphasized the analysis of Iranian regional variation far more than did Farhud's, which consistently treated Iranians as a single unit in contrast to other nationalities arranged in racial supercategories.[28] Only after Farhud returned to Iran did he begin to direct his attention to ethnic groups within Iran, effectively taking up Bayatzadeh's call for more "representative" ethnic sampling practices. Although Bayatzadeh's career in genetics appears to have been stymied by his political activities, Farhud's own connections back home set him up for success. Immediately after completing his doctorate in 1972, Farhud returned to Iran to join the faculty of Tehran University. Based at the School of Public Health, he soon formed the first research group on "Human Genetics and Anthropology."[29] By 1974, he had also begun serving as the secretary of the Genetics Society, a post he would hold until 1981.

Members of the Society had been interested in anthropological applications since before Farhud's return from Germany. The proceedings of the Second National Genetics Congress, held in Tehran in 1970, included an abstract on the "anthropological features" of five villages of the Fars region. These "features" were primarily demographic—e.g., data on household structure and marriage patterns with a particular emphasis on the incidence of consanguineous marriages, a topic of interest to geneticists worldwide at the time.[30] The proceedings also contained an introductory article on the genetic basis of "human races" that discussed differential geographic distribution of human blood groups as well as "visible racial differences"— namely, traditional anthropometric measurements like skin and eye colors, as well as shapes of skulls, lips, and noses, etc.[31] Regardless, the creation of Farhud's special research unit explicitly integrated genetic anthropology into the existing structures that had fostered Iranian medical genetics.

In 1973, Farhud participated in Tehran University's health survey of the Qashqa'i nomads in southern Iran, which collected a vast amount of sociological, demographic, and anthropometric data alongside blood samples.[32] Farhud's colleague in the School of Public Health, Kambiz Montazemi, tested the Qashqa'i blood and eventually published a detailed comparison with thousands of other Iranian samples collected by various regional health institutions across the northern provinces as well as the Persian Gulf coast.[33] The School's study of the Qashqa'i coincided with the Iranian government's expanded support for cultural anthropology on tribal groups across the country. Back in the late 1950s and early 1960s, when James Bowman wanted to search for favism among the Qashqa'i, Mamasani Lur, and Basseri tribes, he had first been interrogated by SAVAK, the Shah's secret police, in keeping with the state's heavy surveillance of tribal groups.[34] By the early 1970s, recalcitrant khans had been killed or exiled and the tribes forcibly disarmed, neutralizing their potential threat to the central government. Accordingly, the Iranian state began advertising the "exotic nature" and "touristic appeal" of the Qashqa'i and other nomadic groups, encouraging foreign and domestic audiences alike to appreciate Iran's "cultural diversity." Years of government repression of minority groups and propaganda about the unity of the Iranian nation-state culminated in a new trend of domestic tourism, with a steady stream of middle-class, urban, Persian-speaking Iranians visiting nomads' campsites.[35] By exploiting Qashqa'i traditions of hospitality without intending to reciprocate the favor, these Persians were unwittingly participating in a long tradition of internal colonialism that dictated the subordinate status of minority communities within Iran.

Since the foundation of the Pahlavi state, this internal colonialism had been framed in the ethnolinguistic terms of Aryanism. A revival of Aryanism in Iranian public discourse emerged in 1967, when the Iranian Parliament bestowed upon Mohammad Reza Shah the new title of *"Aryamehr"* ("Light of the Aryans") on the occasion of his formal coronation ceremony. In the ensuing decade, the usage of the term *"āryā"* ("Aryan") proliferated in the names of new neighborhoods and public structures in Tehran (e.g., Aryashahr district, Aryamehr University of Technology, Aryamehr Stadium). The renewed prevalence of Aryanism found its way into the work of Farhud and his colleagues in the School of Public Health who joined him in conducting anthropological genetic studies. For example, Nahid Mohagheghpour and Hamideh Tabatabai took a special interest in Tehrani religious minorities as windows into Iran's "ethnic history." Tabatabai completed a master's thesis comparing human leukocyte antigen (HLA) types in Iranian Armenians and Jews in 1977, and Mohagheghpour directed a similar study on Zoroastrians. They relied on each community's leaders, physicians, and medical facilities to collect blood. Mohagheghpour thanked two members of the Zoroastrian Society (Mr. K. Keshavarz and Mr. H. Hormozdiari) and a Zoroastrian priest (Moubed Rustam Shahrzad) for "consultation" and "cooperation"; Tabatabai credited a Mr. Karandian for collecting samples from the Jewish Kurush-i Kabir (Cyrus the Great) Hospital, and Dr. Haratonian and Mr. Salari for samples from the Armenian Avedissian Polyclinic.[36]

The work on Zoroastrians invoked the familiar claim of total reproductive isolation that had driven genetic research on the community for decades. As a result, Mohagheghpour explained, "the small size and isolation of the Zoroastrian population through the ages makes it a very important population for genetic studies," in contrast to Iranian Muslims, who "represent a racially heterogeneous group of people with varying degrees of Arabic, Turkish, and Mongoloid influences."[37] Like Bowman before them, Mohagheghpour and her assistants concluded that "the HLA profile of the present-day Zoroastrians is quite distinct" from that of their sample of Iranian Muslims.[38] Tabatabai's study on Jews and Armenians deals more explicitly with the question of how these two minorities, who occupied a more unstable position within the Aryanist national narrative than Zoroastrians, were related to other Iranians. In Tabatabai's narrative, Aryans are the "original inhabitants" of Iran, in contrast to subsequent "invaders" like the Jewish and Armenian communities, who arrived by migration at precise times:

> The ethnic history of Iran, originally inhabited by Aryans, is marked by invasions of Assyrians, Medes, Greeks, Arabs, Turks and Mongols. In addition to these major invasions, other occasional migrations have occurred, such as the settlement by Jews in 538 B.C. after the conquest of Babylon by Cyrus, and by the Armenians in the 17th Century. [...] Because of religious restriction the Jews and the Armenians have remained relatively isolated from the surrounding populations....[39]

Tabatabai's introduction thus marked Jews and Armenians as ethnically distinct from the rest of the "assimilated" (i.e., mixed) urban Muslim populations of Iran.[40] Most importantly, Tabatabai and Mohagheghpour demonstrated the model for how Farhud and other Iranian researchers subsequently framed their genetic studies of minorities. All categories of difference (religion, language, geography, etc.) came to be subsumed by "ethnicity" (*qawmiyyat*), and, furthermore, all communities became "Iranian populations,"

"Iranian groups," "Iranian ethnicities": variations on an essential Iranianness defined by nation-state boundaries. In her Persian-language thesis, Tabatabai used the terms *"īrāniyān-i yahūdī"* and *"īrāniyān-i armanī"*—that is, "Jewish Iranians" and "Armenian Iranians"—rather than "Iranian Jews" or "Iranian Armenians." This language testified to an underlying conceptualization of minority identities as a superstrate over Iranian subjecthood, rather than the other way around.[41]

PRESERVING "IRANIAN ETHNICITIES"

Although the 1979 revolution and the Iran-Iraq War were major disruptions in Iranian politics and society, from the perspective of scientific development, these events were not turning points so much as accelerants for processes that had already begun. For example, the Iranian government had been fretting since the 1960s about the "brain drain" that sent many of Iran's physicians and biologists abroad for good.[42] Habibollah Fakhrai, who had planned a 1979 sabbatical in the United States, chose to remain in California indefinitely.[43] Fereydoun Ala, after a harrowing arrest that would have brought him before the ruthless Revolutionary Tribunals, managed to smuggle himself out of Iran and settled in the United Kingdom in 1981.[44] Pezeshkpour Mostashfi retired from Tehran University in 1979 and relocated first to France, then to Canada. The activities of the Genetics Society of Iran were interrupted for twenty-two years.[45]

Dariush Farhud, who remained in Iran, swiftly became the country's preeminent geneticist. Amid the political turmoil surrounding the downfall of the Shah's regime and the subsequent Cultural Revolution, which purged foreigners and potential dissidents from Iranian universities and government institutions, Farhud managed to strengthen his ties with British institutions with strong genetic anthropology programs.[46] The British scientific officer of the INBTS, Victor Alan Clarke, described Farhud as a close friend and a "workaholic" dedicated to securing the necessary funds and laboratory connections to make good on his "enthusiastic" assessment of Iran's genetic research potential.[47] Between 1978 and 1982, even as the revolution and Iran-Iraq War were raging, Farhud helped orchestrate field surveys to collect and test blood from more than a dozen different populations in Iran. Three of his students carried the samples to Britain, where they earned doctorates in anthropology and human genetics.

Yousuf Seyedna and Parivash Amirshahi went to Durham University, where they studied under anthropologist Eric Sunderland. Seyedna pursued a study of Iranian Zoroastrians, relying on contacts through the Zoroastrian Association to facilitate his collection of blood samples and demographic information from Tehran, Yazd, and Kerman during three field trips between 1979 and 1981. Some of these Zoroastrian samples were shared with Amirshahi, who also collected blood from Muslim Tehranis and Kermanis in 1979. In three subsequent field surveys between 1980 and 1982, Amirshahi traveled to Iran's provincial fringes, working with "Turks" and Kurds from two different regions: the northwest (Urumiya, formerly Rezaiyeh) and northeast (Shirvan County, Khorasan); Lors from Khorramabad; and Baluchis and Zabolis from Zahedan.[48] Meanwhile, Mohammad Taghi Akbari carried blood samples from several hitherto neglected populations to the human genetics department at Newcastle University. In 1979, Akbari journeyed across the country from Bandar-e Turkoman to Bandar-e Abbas to gather blood from "Mongoloid" Turkmens from the northeast and the Bandaris, a "morphologically ... rather negroid" group dwelling in Hormozgan province on the Persian Gulf coastline. In 1981, Akbari targeted Iranian Christian communities, sampling Assyrians from three urban and rural localities and Armenians from four.[49]

Neither in their doctoral theses nor in the related publications do Seyedna, Amirshahi, and Akbari explain in any detail the circumstances under which they encountered their research subjects and convinced them to part with their blood. This omission is troubling given the timing of the field surveys in an atmosphere of uncertainty and fear, particularly for religious and ethnic minorities. All Iranians faced the danger of imprisonment and execution from accusations of collaborating with the Shah's administration or undermining the revolutionary regime. But as Khomeini and his supporters began to shape Iran into a theocracy, leftist groups and religious minorities, including Sunni Muslims, had reason to be anxious. In March 1979, Kurdish and Turkmen factions seized control of provincial resources to implement leftist policy reforms, demanding autonomy from the central state. The Revolutionary Guards were deployed to violently suppress these uprisings, deemed to be "counter-revolutionary" and "separatist," and many Kurds and Turkmens lost their lives. The Kurdish representative elected to serve on the Assembly of Experts (charged with drafting the Islamic

Republic's new constitution) was subsequently excluded from participating, on Khomeini's orders.⁵⁰

The Assembly of Experts, convened in August 1979, also included representatives from the four recognized religious minorities—Zoroastrians, Jews, and Armenian and Assyrian Christians—who attempted to protect the religious and cultural rights of their communities within the constitution. During the proceedings, these representatives struggled against the prejudices of their Muslim colleagues by performatively expressing their communities' Iranian patriotism and loyalty to Khomeini.⁵¹ Meanwhile, Baluchi representatives voiced concerns that Sunni Muslims like themselves were being marginalized under Shi'i governance. These tensions escalated into a Baluchi boycott of the constitutional referendum in December 1979 as well as armed clashes between Baluchis and Revolutionary Guards in Zahedan that left several dozen killed and wounded. Guerrilla skirmishes in Kurdish, Turkmen, and Baluch areas continued into the early 1980s.⁵² It was in this context that Seyedna, Amirshahi, and Akbari appeared in provincial cities, identified research subjects according to their ethnicity and religion, and collected their blood.

In the abstract of his dissertation, Seyedna explained that his research aimed to determine whether the Zoroastrian communities in Tehran, Yazd, and Kerman were genetically different from one another or from the "host population" of Iran. After analyzing more than a dozen blood antigen systems, serum proteins, and red cell enzymes, he concluded that Zoroastrians were genetically distinct from "other indigenous" Iranian groups. Seyedna also dedicated a chapter to discussing the biological relationship between Iranian Zoroastrians and Indian Parsis, even combining the groups together to speculate on the genetic profile of the "ancestral Zoroastrians" and, by extension, the ancestors of all Iranians.⁵³ However, Farhud's Human Genetics and Anthropology unit was not particularly invested in this kind of historical reconstruction. As clarified in the work of Amirshahi and Akbari, the primary aim of their wide-ranging population surveys was merely a descriptive cataloging of ethnic groups' genetic profiles "to appreciate the genetic structure of [the] Iranian population as a whole."⁵⁴

Iran "as a whole" is visually portrayed through several maps Amirshahi reproduced as front matter to her dissertation, which she dedicated "to my country." Political and geographical maps of Iran were accompanied by an

FIGURE 11. The 1983 doctoral thesis of Parivash Amirshahi included a photocopy of an "Ethno Composition" map of Iran, published in 1978 by Sahab Geographical and Drafting Institute. The more recent version shown here (acquired by the author in 2019) has Persian-language title and captions rather than English but is otherwise identical.

"Ethno Composition" map, produced in 1978 by the Tehran-based Sahab Geographic and Drafting Institute (see Figure 11). This map featured fifty-five photos and illustrations of smiling, colorful individuals displaying "local costume," handicrafts, and agricultural products. A numbered key identified the figures as representative types: "Kurdish man from Javanrood area," "Young Bakhtiary mother carries her baby," "Lovely girl from southern Khuzestan." This appealing kaleidoscope of cultural diversity was sharply outlined by Iran's land and sea borders. The Caspian Sea and Persian Gulf were dotted with images of fish and boats, but only empty space stood in for the dry land beyond Iran's boundaries. The map's apparent celebration of ethnic difference was therefore contingent on literally containing these groups within the Iranian nation-state, rather than incorporating Iran's peoples into the broader Middle Eastern cultural and geographical landscape.

Amirshahi's dissertation and subsequent publications emphasize that the "Iranian population is among the most heterogeneous in the world," with invaders, traders, and pastoral tribes from the Aryans onward contributing to its "great diversity."[55] However, this ethnocultural diversity did not always correlate with genetic difference. For example, Amirshahi and her colleagues discovered that the Turks and Kurds of Khorasan differed little from Tehranis and Kermanis. These four "eastern" populations "did not show any conclusive genetic heterogeneity" among themselves and therefore did not "warrant any further analysis" of their genetic "infrastructure."[56] In contrast, a different combination of Amirshahi's sampled populations—Lurs, Zoroastrians, Baluchis, Zabolis, and the Turks and Kurds of Urumiya—offered significant degrees of genetic heterogeneity. She concluded that "the genetic composition of the present day populations of Iran is like a complex tapestry," although the differing threads of that tapestry seemed to represent geographical rather than ethnic isolation. The northwestern Kurds and Turks, although "ethnologically distinct," were genetically quite similar to one another, as were the Baluchis and Zabolis of the southeast.[57]

Meanwhile, Akbari overtly framed his work as a comparison of religious and ethnic minorities with Persians, "the predominant group in Iran[;] in racial origin they are thought to derive from admixture of the ancient inhabitants of the Iranian Plateau with Arabs and Aryans."[58] He found that the frequencies of Persian blood groups, serum proteins, and enzyme variations resembled those of Armenians and Assyrians, especially those living in Tehran. On the other hand, the Turkmens and Bandaris sharply diverged from each other and all other Iranian communities. Akbari explained that the Turkmens "resemble Mongols and historically [descend from] the Oghoz tribes who mixed with the indigenous settled and nomadic peoples," whereas "ethnically and linguistically [the Bandaris] represent admixture of Portuguese and Arabs with the negroes brought into the region in the 15th century."[59] These historical origins explained why the Bandaris' genetic profile demonstrated "their ethnic affinity to Negro populations" as well as the "mongoloid influence" on the Turkmen gene pool.[60] Meanwhile, he argued that his similar results for the geographically separated populations of Assyrians and Armenians "indicate[d] the relatively closed nature of the Christian community as a whole," although "moderate differentiation has already occurred among the local groups."[61] Overall, he concluded, the genetic variation detected by previous regional studies of Iran could indeed be attributed to the existence of distinct ethnic communities. As Akbari put it, "the genetic constitution of these populations reflects their ethnohistory."[62]

In the early 1990s, Farhud published two comprehensive reviews of Iranian serological genetics with Hubert Walter, his former adviser at Mainz, to detail exactly what this ethnohistory was and to align it with the countrywide distribution of genetic markers. Farhud sorted all existing genetic data on Iran into twenty-one population categories, which included not only the standard religious minorities and ethnolinguistic groups but also subdivided Shiʻi Persians according to province of origin. The authors determined that the ABO frequencies of the latter groups, which they called "'true' Iranians," clustered together, whereas "ethnic groups . . . with a different origin and history" deviated from these frequencies.[63] Patterns of enzyme variations, too, seemed to "link up" to Iran's "ethnic structure," in which "geographical, linguistic, and religious factors apparently prevented an extensive intermixture of the entire population of Iran and thus contributed decisively to the maintenance of their genetic diversity."[64] Just over twenty years since Mansur Bayatzadeh had been logistically frustrated in his attempts to genetically substantiate Iran's ethnic diversity, Farhud and Walter successfully consolidated a "diversity catalog" of Iran's ethnic groups. The ability of Farhud and his students to collect blood samples from ethnically defined populations in the immediate aftermath of the revolution demonstrates how the Islamic Republic perpetuated ethnic policies of the late

Pahlavi state, wherein minorities were politically neutralized and anthropologically museumized in the name of the country's "territorial integrity."[65]

Ultimately, postrevolutionary structural and ideological changes had little direct impact on the late-1970s methodological and discursive transformation of Iranian human genetics. In fact, since the Islamic Republic was slow to create any systematic policy toward human genetic research at the state level, Farhud and his network of students and collaborators predominantly shaped the agenda of Iranian population genetics through the 1980s and early 1990s. This agenda largely involved analyzing minority groups according to ethnic and religious categories in order to situate them within a national biological narrative. Farhud's priorities reflect his international connections to German and British trends in genetic anthropology as well as his own proud identification as an Iranian nationalist. In 2013, he published a Persian-language book called *Cultural Iran* for general audiences. Although the book makes little reference to his genetic research, it strikingly illuminates his vision of Iran as a national entity dating back to antiquity. The tome is essentially a color-illustrated, chronological encyclopedia of Iranian history, archaeology, and ethnography recapitulated from the early-twentieth-century Iranian scholarly tradition of Hassan Pirniya (see Chapter 1). Farhud's book is distinguished by its preoccupation with maps: sixty-five color maps of Iran illustrate the territorial domains of all ancient civilizations, past ruling dynasties, and extant religious and ethnic groups that happen to overlap the present-day borders of Iran. Paradoxically, although Farhud's maps are intended to demonstrate Iran's endurance as a civilizational entity over time, they actually highlight the arbitrariness of its current nation-state borders, featured in bold lines. No civilization or dynasty before the Pahlavi period occupied a geographic space matching that of present-day Iran. The distributions of all the religious and ethnic communities either reside in small and apparently well-defined pockets within the country or spill over its borders into neighboring territories.

Aryanism features prominently in Farhud's narrative reconciliation of geography and ethnic diversity into a unitary nation-state. For example, Farhud's map of Armenians displays their communities as residing neatly within the boundaries of present-day Armenia and in patches surrounding major Iranian urban centers. Although he acknowledges that Armenians were forcibly brought to Iran by the Safavids, Farhud emphasizes a deeper and longer-term connection between Persians and Armenians, calling them "Christians of Aryan ancestry" (*āryāyī-tabār*), as opposed to the "Semitic ancestry" of the Assyrians.[66] As for Iran's "Turkish peoples" (*aqvām-i turk*), Farhud's map subdivides them into Khorasani, Azeri, Gagauz, Anatolian, and Crimean communities. He depicts "Azeri Turks" as dwelling beyond as well as within Iran's state borders, conflating this ethnic label with the tribal groups of southern Iran as well as Turkic speakers in the Caucasus, Anatolia, Iraq, and even Palestine. On the other hand, Farhud does not highlight a single group in Central Asia, the oft-cited homeland of the Turks, and this is no accident. In the 1920s, on the basis of dubious linguistic evidence, the Iranian Azeri historian Ahmad Kasravi argued that Azeris, despite their Turkic language, were not ethnic Turks. They were actually Aryans like Persian-speaking Iranians, who had simply adopted a Turkic dialect after the Mongol invasions.[67] Since then, other Iranian nationalist intellectuals, who feared pan-Turkist and Soviet irredentism that might dislodge the Azerbaijani provinces from Iran, adopted this argument and attempted to document other Azeri linguistic features to support Kasravi's case.[68] Farhud updates this discourse for the DNA era, mentioning that genetic studies in Turkey support the hypothesis that Azeris and Anatolian Turks are not really "racially" Turkish—that is, they are a mixed people who have very little ancestry from "Mongols" (*zardpūstān-i mughūlistān*), but rather have primarily Iranian, Caucasian, and Greek "racial" origins.[69] Farhud's leadership in genetic anthropology therefore enabled the continuity of certain threads of Pahlavi-era nationalism within Iranian genetics, particularly its conflation of Aryanist discourses with the maintenance of Iran's historical identity and territorial integrity.

TURKEY: STALLED HOMOGENIZATION

Nationalist Turkish scholars from Afet İnan onward were troubled by their detection of genetic variation across different regions of the country. They consistently attributed this evidence of internal diversity to a legacy of Ottoman imperial decentralization and a lack of efficient mobility between regions. This explanation avoided any discussion of ethnic differences, which would require acknowledging the presence of non-Turkish ethnicities with competing nationalist movements. Accordingly, most Turkish hematologists did not seek out shrinking non-Turkish minorities to catalog their distinct genetic characteristics. Rather, they analyzed Turkish blood groups through large, "random" data sets of blood donors, all the while welcoming the prospect of future nationwide genetic homogeneity.[70] Like early Iranian blood bank research (see Chapter 3), Turkish blood banks collected data only on donors' birthplaces, not their ethnicity, religion, and language.

Turkish scientists did perform targeted field studies of ethnically defined subpopulations and even did so with state funding—but exclusively within the context of medical research, such as Muzaffer Aksoy's analyses of abnormal hemoglobins among the "Eti-Turks" (see Chapter 4). In the late 1960s and early 1970s, Aksoy and his students and colleagues received funding from the state agency TÜBİTAK (Turkish Science and Technology Research Council) for their surveys on both Eti-Turks and the Greek population living on Turkey's Aegean island of İmroz (Imbros).[71] Circumstances of medical emergency, such as the 1977 malaria epidemic in Turkey's eastern Mediterranean region, also provided opportunities for "ethnic" genetics. With the aid of the WHO, the Turkish government mounted an intense campaign to control the epidemic, which included the widespread use of the antimalarial drugs chloroquine and primaquine.[72] Accordingly, the Turkish Malaria Eradication Organization investigated eighteen villages for incidence of G6PD deficiency, designating their inhabitants as members of four ethnic groups: "Arabic speaking Eti-Turks," "ethnically unmixed Turcomen," Armenians, and "mixed" villagers.[73] This study of Armenians is particularly notable; in contrast to the profusion of Iranian genetic research on Armenians, Turkish scientists rarely acknowledged the existence of Armenians in their country.

Outside the medical sphere, the ongoing "problem" of genetic diversity within the Turkish population, as well as the increasing visibility of ethnic separatism, induced a small number of Turkish anthropologists to dabble in genetic anthropology during the mid- to late 1970s. However, unlike the simultaneous developments in Iran, in Turkey this amounted to a short-term revival of interest in sero-anthropology. Turkish anthropologists neither conducted any ethnic-based field surveys nor made use of the dozen-plus genetic marker systems employed by the Iranians, instead relying exclusively on ABO and Rh blood group frequencies. The first example of this sero-anthropological revival was undertaken by the students of Istanbul University's social anthropology and ethnology division, directed at this time by Nephan Saran, during the 1969–1970 academic year. Saran's project involved an ABO and Rh survey of 1,421 Istanbul University students from multiple campuses but offered no methodological or analytical innovations to existing blood bank studies.[74]

In contrast, members of Ankara University's Language and History-Geography Faculty (DTCF) used blood group data to address the issue of ethnic minorities in Turkey. Between 1975 and 1978, physical anthropologist Armağan Saatçioğlu analyzed ABO frequencies across Turkey, aiming to historicize geographic variation with reference to long-term population movements. In some respects, her work was radical: she sharply critiqued earlier sero-anthropological works for their geographically limited sampling procedures and poor statistical methods.[75] She also framed the importance of her work in the language of salvage genetics: "If any genetic variation exists, then it is important to record it while there is yet time, because increased mobility and intermixing will cause any remaining genetic heterogeneity to disappear rapidly by the beginning of the twenty-first century."[76] Most significantly, hers was the first Turkish genetic study to explicitly discuss the existence of the Kurds as a distinct ethnic entity in eastern Turkey and consider their genetic relationship to Turks as well as to Kurds living in Iraq and Iran.

On the other hand, these seemingly major departures from previous Turkish scholarship belie the fact that Saatçioğlu's work was ultimately a conservative project, rooted in the nationalist foundations of Turkish anthropology. She had been trained, from baccalaureate to doctorate between 1961 and 1970, by the old hands leading Ankara University's physical anthropology division: Şevket Aziz Kansu and Seniha Tunakan. The division's research methods still leaned heavily on anthropometry and racial paleoanthropology, which formed the basis of most of Saatçioğlu's own scholarly oeuvre. Kansu and Tunakan both retired in the early 1970s, leaving Saatçioğlu as the only physical anthropologist at Ankara University still actively engaged in research by 1975.[77] Meanwhile, Nermin Aygen Erdentuğ—the scholar who had effectively shut the door on Turkish sero-anthropology in 1946—held the university chair in ethnology. Despite, or perhaps because of, these factors, Saatçioğlu pursued a sero-anthropology project for submission to her promotion committee (which included Erdentuğ as well as Saran).[78]

The foreword to the Turkish publication of Saatçioğlu's study described her work as a national advancement in Turkish population genetics: "The determination of the current genetic condition [*bugünkü gensel durumu*] of the people of Turkey, from both a historical and a biometric approach, is an unavoidable [*kaçınılmaz*] task for contemporary anthropologists. To fulfill this task and to make a contribution to the emerging 'population genetics' [*topluluk genetiği*] science from the perspective of our country, I set out on such a study." She also expressed the hope that others would further her research by investigating other blood systems.[79] Her actual methods can be best described as a fusion of the blood bank surveys produced by hematologists and the anthropometrically based racial histories constructed by her mentors at the Ankara DTCF. The process of data collection was facilitated by the blood bank at the Gülhane Military Medical Academy in Ankara, who provided information on "3881 Turkish persons" sampled in 1975. Just like many hematologists before her, Saatçioğlu sorted the donors' birthplaces into regions and compared the frequency averages of the pooled regional data. But her assessment of the genetic differences indicated by the blood data was preceded by a seventy-seven-page section titled "The Arrival of the Genes" (*genlerin gelişi*), actually a detailed history of racial groups migrating through Turkey, from the first arrival of *Homo sapiens* in Anatolia to the post-Ottoman period. The paleoanthropological research of DTCF and European anthropometrists constituted the bulk of the evidence for Saatçioğlu's narrative, which coincides significantly with the "Race History of Turkey" that her mentor Kansu had recently published.[80] Her discussion of the racial origins of Armenians and Kurds, leaning on the speculations of European racial anthropologists, stands out as the major distinguishing feature.

Much of her analysis of the blood data, like that of earlier seroanthropologists, focused on Turkish racial classification. Saatçioğlu made detailed statistical comparisons of average Turkish ABO frequencies to those of various European, Asian, and Middle Eastern countries. She then compared different Turkic groups to East Asian averages to determine the Turkic populations' relationship to "the Mongol race" (*Mongol ırkı*).[81] She decided that the Kyrgyz alone could be classified as "Mongols," while the other Turkic groups descended from an original "white race" that had lost its distinguishing characteristics under the pressure of Mongol admixture. As for the descendants of the Oghuz Turks who now populated present-day

Turkey, she suggested that they had either been a "purer Turkish nation" (*daha arı bir Türk budunu*) to begin with or had biologically diverged from other Turkic groups since their arrival in Anatolia.[82] Her work echoes 1930s Kemalist anxieties and ambivalences about the relationship between the Turks of Turkey, the Turkic peoples of Central Asia, and the "Mongol race." However, it may also represent a more-timely rebuke to the ultra-right-wing Nationalist Action Party (*Milliyetçi Hareket Partisi*, MHP), whose brand of nationalism emphasized the Turkish racial connection to Central Asia and whose youth branches (known as *ülkücüler*, or "idealists") notoriously contributed to the public violence of the 1970s.

The most novel analytical section, and the only part of Saatçioğlu's serological research published in English, experimented with the geographical divisions of Turkey. Most researchers before her had presented and analyzed their data according to the seven broad geographical regions of the country delineated by Turkey's First Geography Congress in 1940. Saatçioğlu intervened by subdividing the length of the Black Sea region into three parts, creating nine total regions, which she believed would provide a clearer picture of frequency variation between East and West Anatolia. Furthermore, she cross-checked those results against a slightly different scheme of geographical divisions: the eight "population regions" outlined by the State Statistical Institute in 1973. In the process, her analysis departed sharply from the seroanthropologists of the Atatürk era, who had narrowly focused on the east-to-west frequency increase of group A as proof that the Turkic migrations had brought the "A allele" from Central Asia to Europe. Rather, she attributed the low B frequencies of the eastern Black Sea region to the gene pool of the medieval Greek Trebizond Empire.[83]

Saatçioğlu reserved her most remarkable comments for the region of Southeast Anatolia, which varied conspicuously from the national frequency averages. To explain this heterogeneity, she noted state statistics that the majority of the region's population had a "mother tongue" other than Turkish and that "[İsmail] Beşikçi writes that they are Kurds."[84] The publications of Beşikçi, a leftist Turkish sociologist who abandoned his academic career to campaign for Kurdish rights, had roused the ire of both the Turkish government and much of the Turkish intellectual establishment. At the time of Saatçioğlu's writing, Beşikçi had been arrested for the second time (of many), soon to be convicted of "undermining [Turkish] national feeling."[85]

Saatçioğlu, despite distancing herself through the reference to Beşikçi, effectively accepted the Kurds' existence as a distinct ethnolinguistic group and their likely influence on the genetic variation displayed by the population of Southeast Anatolia. For example, she pointed out the similarity between her Southeast Anatolia ABO frequencies and those that Iraqi researchers had calculated for Iraqi Kurds.

However, she then emphasized the extensive genetic diversity within the Kurdish population as evidence of its historical admixture with many other regional ethnicities, citing work by Hermann Lehmann and Arthur Mourant suggesting that Iranian Kurds had probably intermixed with Turkish people. Finally, she argued that the figures for Southeast Anatolia more closely resembled those for Arabs residing in Aleppo and Baghdad and, "therefore, it should be wrong to think that the Kurds alone are responsible for the population of southeast Anatolia being one of the sources of heterogeneity in Turkey."[86] Saatçioğlu's revolutionary acknowledgment of a distinct Kurdish population thus ultimately served to undermine the potential "national feeling" of Kurds in Southeast Anatolia by highlighting their status as an admixed people, their differences with Kurds in Iraq and Iran, and even their inability to appear as an incontestable genetic bloc within Turkey.

Shortly after she completed her own work on Turkish ABO frequencies, Saatçioğlu supervised a master's thesis by Can Şentuna that analyzed Rh frequencies. Şentuna's methods and interpretations borrowed heavily from his mentor, although he used an independent sample population. In 1978, he obtained Rh data on 2,096 Turkish blood donors tested over the previous four years by Ankara's Specialized Medical Training Center and Istanbul's Red Crescent Blood Center. Unlike Saatçioğlu, he did not analyze his data according to Turkish geography. In fact, he neglected any discussion of internal diversity within Turkey, instead simply comparing Turkish national average frequencies of Rh genotypes with those of other national and ethnic groups. In this regard, he echoed the most conservative features of Saatçioğlu's research—namely, its concern with racial classification. Şentuna used his results to distance Turks from the "Mongol race" and emphasize their close relationship to Central Europeans. He also repeated Saatçioğlu's historical interpretations regarding Ottoman-era admixture between Turks, Kurds, and Arabs. Finally, Şentuna tried to explain the proximity of Turkish Rh frequencies to a broad number of groups—ranging from

Greeks to French, Latvians, and Hungarians—by creating historical narratives of direct gene flow between distinct national entities.[87] This phenomenon attests to the Turkish physical anthropologists' outdated understanding of genetics as a set of simple markers for racial classification, alongside their more familiar anthropometric methods.

Saatçioğlu's emphasis on internal genetic diversity—as well as her oblique acknowledgment of Kurdish and Arab populations residing within Turkey—is remarkable given that she worked for an institution whose *raison d'être* had always been to prove the racial unity of the Turkish Republic. But the work of Saran, Saatçioğlu, and Şentuna appears to have had little immediate impact on the methodological practices of either anthropology or human genetics in Turkey. Firstly, the anthropologists did not maintain their collaborative connections to the blood banks but instead returned to their primary research topics and methods (for Saran, the social problems of immigrants to Istanbul; for Saatçioğlu, anthropometry and skeletal analysis). Secondly, the 1980 coup presaged a major restructuring of Turkish universities. The newly created Higher Education Council (*Yükseköğretim Kurulu*, or YÖK), pressured the anthropology programs at the DTCF to reorganize in 1982.[88] Although no faculty from the anthropology division were directly purged, the experience took its toll. By October 1983, Saatçioğlu had resigned from her position at Ankara University.[89] Her departure effectively suspended the university's physical anthropology program for years. The neoliberal economic paradigm that emerged after the 1980 coup further prolonged the downturn in physical anthropology research, as the discipline's limited marketability and career prospects in the Turkish context attracted few students to rebuild the ranks of anthropologists.[90]

Due to these factors, blood bank directors and medical practitioners like Muzaffer Aksoy continued to dominate the field of human genetics well into the 1990s. Most nationwide ABO and Rh frequency compilations produced by Turkish medical personnel maintain a conservative pattern: data from hospital and blood bank donor records was compared at the national level to the averages of other nations or racial groups, then broken down by region, province, or city to identify genetic variation within Turkey.[91] Very few studies cited scholarly sources for the historical narratives occasionally attached to this data, and none made any reference to Kurds. The only example of a targeted blood survey for anthropological rather than medical purposes was performed in the early 1990s by Mükaddes Gölge, a student of the former Mainz anthropologist Hubert Walter (now posted at the University of Bremen). However, Gölge's serum samples—drawn from patients in Aksoy's Istanbul clinic and shipped frozen to Germany—were still sorted by four broad geographical regions of birth rather than by ethnic or linguistic labels. In their conclusion, Walter and Gölge could only express interest in conducting future ethnic-based studies.[92]

Concurrently, Güher Saruhan, a Turkish immunogeneticist collaborating on HLA research with Julia Bodmer at Oxford, struggled to identify Turkish blood samples in ethnic terms. Saruhan shipped a series of blood samples—donations from the Istanbul national blood bank and an Istanbul University HLA disease association study—to Bodmer's lab. Bodmer sent her a standard set of forms requesting anthropological details about the individuals who had provided the blood. The forms asked if samples came from "a single, well-defined population" and, if so, if this group was "isolated from other populations." They also inquired about the population's ecological environment, living conditions, spoken language, and known history, as well as referrals to persons with "detailed anthropological knowledge" on the population concerned. Saruhan was at a loss as to whether her samples constituted a "single, well-defined population," marking this query "unknown." Though she named the population and its language "Turkish," she described Istanbul as a "multicultural and multi-ethnic" city with an annual influx of hundreds of thousands of migrants from the rest of Turkey.[93] In another fax a few months later, Saruhan confessed, "Unfortunately we do not know much about our population." On the attached forms, she could not provide a reference to any anthropologist or anthropological literature, noting instead that "[f]or political reasons it is very difficult to collect data about the ethnical differences in our country."[94]

NATIONAL BELONGING AND POLITICAL VIOLENCE

In a 2014 interview, Habib Fakhrai recounted the fate of the Arab tribespeople of southwestern Iran whose blood groups he had studied during the 1970s. During the Iran-Iraq War, he explained, the area had been occupied by Iraqi soldiers, who had raped a number of the local women. Alongside the immense human tragedy, he lamented, the unique genetic qualities of this population were now lost, having been contaminated by the violence of war.[95] Fakhrai's narrative accentuates the close relationship between genetic anthropology and the state structures that politically, socially, and economically marginalize certain populations. The field's ongoing quest to study isolated communities does not reckon with the various forms of violence involved in maintaining this isolation. The living conditions and legal status of these research subjects left them particularly vulnerable both to direct military threat and to the discriminatory domestic policies of nationalist regimes. This is not to say that Middle Eastern scientists, as representatives of a politically and socially privileged professional class, were completely insulated from the effects of violence. Had he not escaped Iran, Fereydoun Ala would have faced near-certain death from the notorious Revolutionary Tribunals. Muzaffer Aksoy's own brother Muammer, a politically active Kemalist and law professor, was assassinated in 1990 by a minor Islamist militant group. However, these traumatic incidents were aberrations in their lives and careers. The same cannot be said of the long-term pressures faced by the ethnic and religious minorities living in Iran, Turkey, and other countries whose blood was subjected to scientific scrutiny.

Geneticists' framing and interpretation of ethnic diversity is thus inextricable from the broader questions of how nation-states attempt to control the implications of such diversity and how social and political movements challenge state narratives about ethnicity. In this regard, the emergent genetic anthropology, despite its appeal to "the molecular point of view," did not differ substantially from the patterns of physical anthropology research that emerged with the nation-state. Just as Western colonial and imperial medical infrastructures had laid the foundations for an international scientific community, national-scale networks of professional geneticists owed their existence to state projects aiming to forcibly refashion their diverse citizenry into a uniform mold. Global, regional, and local communities of scientists were therefore deeply embedded in the creation of other communities of belonging—such as nation, ethnicity, and religion—within and beyond their professional work. These various types of social and political communities converged in hierarchical relationships, as researchers and research subjects, to perform the territorial integrity of the nation-state. Internal political dynamics determined whether this performance took the form of a diversity catalog, as in Iran, or a consistent denial of diversity, as in Turkey.

However, just as territorial integrity is a concept grounded in international law, the domestic policies that condemned advocacy for regional autonomy as ethnic separatism were responses to the international scope of Cold War–era violence. The global process of decolonization, coupled with the proxy wars of military superpowers, reinforced existential anxieties about foreign interventions and territorial sovereignty. The inter-state wars between Iran and Iraq, and between Israel and its Arab neighbors, are therefore intimately linked to intra-state suppression of civil strife involving the Kurds and other minority groups. Within this context, geneticists have often behaved opportunistically. Following the Hirszfelds' legacy of blood sampling in war zones and in refugee camps, both Middle Eastern and Western scientists have relied extensively on the blood of soldiers, "pacified" and "resettled" tribes, communities under military occupation, and victims of "population transfers." New regimes of informed consent have not fully resolved the ethics of this ongoing practice. A 2017 study to evaluate the genetic diversity of ethnic groups in Northern Iraq, which collected saliva samples from nearly two hundred Yazidis and Assyrians living in Erbil refugee camps, cites its compliance with a range of informed-consent procedures.[96] Even after Islamic State militants targeted these religious groups for extermination based on their identity, the study unreflexively aimed to articulate their difference from Arabs, Kurds, and Turkmens in genetic terms. Evidently, the presence of interpreters and written consent forms satisfied the institutional ethics committees and camp administrators. After a lengthy discussion on the plausible indigeneity and antiquity of Assyrian and Yazidi lineages, the authors concluded that their data could also have a forensic use for missing persons identification, further underscoring the relationship of genetic anthropology to war and territorial contestation. To conclude this book, I show how the anti-racist, progressive discourses surrounding contemporary human genome projects have so far been unable to overcome the territorial regimes and ethnic concepts produced by a century of conflict.

Conclusion
GENOMES WITHOUT BORDERS?

A YEAR AFTER THE INITIATION OF THE HUMAN GENOME PROJ-
ect in 1990, five eminent geneticists called for an expansion of the massive transnational DNA sequencing effort "to survey human genetic diversity." Led by Stanford's Luigi Luca Cavalli-Sforza, they urged the international Human Genome Organization, with the assistance of the United Nations Educational, Scientific, and Cultural Organization, the World Health Organization, and American funding agencies, to immediately coordinate a worldwide collection of human blood and tissue samples. In particular, they argued, the project must seize the "vanishing opportunity" to sample "isolated" indigenous groups and "record human ethnic and geographic diversity before this possibility is irretrievably lost."[1] As shown in Chapters 1 and 2, the foundational rhetoric of this new Human Genome Diversity Project (HGDP) did not significantly differ from that of interwar salvage anthropology, which anticipated the rapid disappearance of Iranian and Turkish nomadic groups' cultural and genetic distinctiveness. Similarly, the diversity project aimed to draw on an older international infrastructure—the traffic in blood that had defined Cold War genetics (Chapter 3). The main innovation of the HGDP was technological: by directly sequencing the DNA of entire genomes, scientists hoped to reconstruct more-accurate histories of human evolution and migration than those based on inherited conditions like sickle cell disease and favism (Chapters 4 and 5). Furthermore,

Cavalli-Sforza and his colleagues demonstrated a greater professional commitment to antiracism and human rights, which found its way into their HGDP proposal. They noted that the populations they hoped to study, which ranged from "indigenous American populations" to "the Kurds of eastern Turkey," had been "historically vulnerable to exploitation." Therefore, HGDP researchers must consider each "population's needs for medical treatment and other benefits."[2] They hoped that, ultimately, the HGDP would not only contribute to scientific knowledge about human evolution but also marshal genetic evidence to combat racist ideologies.

Despite these progressive motivations, the HGDP soon came under attack by anthropologists and indigenous activists who accused the organizers of racism and "biocolonialism," dubbing the initiative a "Vampire Project" seeking to extract the blood of indigenous peoples whom they expected to "vanish," rather than making it possible for such peoples to survive and flourish. The HGDP foundered amid efforts to reconcile the interests of geneticists and research subjects, with many Native American groups refusing to participate.[3] Yet out of these trials emerged a genuinely new entrepreneurial approach to cataloging human genetic diversity: the Genographic Project (GP). Operating from March 2005 to January 2020 with the support of the National Geographic Society, IBM, and the Waitt Family Foundation, the GP not only sampled coveted indigenous populations but also offered its DNA-sequencing services to the public. Capitalizing on the emerging genetic ancestry industry in North America and Europe through a partnership with Family Tree DNA, the GP marketed sampling kits to consumers to provide them with information about their individual ancestry. This information, represented as proportions of "regional ancestry," claimed to tell consumers where their ancestors "came from" using geographical terminology such as "Southern Europe," "Northern Africa," and "Asia Minor." In this way, the GP satisfied consumers' interests in ethnic and racial ancestry without explicitly committing itself to such categories. The GP's website also portrayed consumers as volunteers contributing their genetic data to a "historic endeavor" to find out "where we originated and how we came to populate the earth." The director and public face of the GP, Spencer Wells, worked with Cavalli-Sforza as a Stanford postdoctoral researcher in the 1990s. Keenly aware of the accusations and misperceptions that felled the HGDP, he spent the years between 2002 and 2005 laying the groundwork for the GP by writing and promoting popular books and documentaries to educate the English-speaking public about anthropological genetics.

Although known within the scientific community mainly for his work on the population genetics of Central Asia, Wells appealed to a widespread fascination with the classical civilizations of the Mediterranean to generate public interest in genetic narratives of history. Specifically, he teamed up with his Lebanese colleague Pierre Zalloua to search for "Phoenician" genetic markers in Y-chromosome DNA in men living near former Phoenician colonies across the Mediterranean. On October 20, 2004, PBS broadcast a documentary called *Quest for the Phoenicians*, which synthesized Wells and Zalloua's work in Lebanon with ongoing archaeological investigations. The documentary's script and cinematography visualized Lebanese fishermen as exotic relics from biblical times, evoking the discourse of the Middle East as genetic crossroads. Wells briefly acknowledged the political sensitivity of the investigation in the Lebanese context, in which Christians had claimed Phoenician heritage to marginalize other religious and ethnic identities. However, Wells enthused, "What the genetic results are telling us definitively is that today's Lebanese people, whether they're Christians or Muslims, are all tied together in this single gene pool, they're all part of one big family. So the genetic results are pulling people together, not ripping them apart." He thus used Lebanon—not long recovered from a fifteen-year civil war, and hence deeply associated with ethnic violence in Western eyes—to make the case for anthropological genetics. Studying genetic diversity could paradoxically show different groups how similar they really were, overcoming their racial and ethnic prejudices to bond over their shared heritage. Through the historically divisive label of "Phoenician" ancestry, Wells polished the core messages that would soon be endlessly repeated in GP press materials: genetic research enables awareness of individual and group relationships to the global human family tree, allowing everyone to celebrate the unity underlying human cultural and biological diversity.[4]

Relatively few Middle Eastern scientists were directly invited to participate in the HGDP or GP, but they have all responded to the ethical challenges and progressive discourses of inclusion associated with these projects. Weighing in on the ethical controversies surrounding the HGDP, Israeli geneticists explicitly declared themselves an example for the rest of the global scientific community to follow. In 1999, Batsheva Bonné-Tamir claimed that her own genetic research never faced accusations of racism or exploitation. Echoing her 1960s critiques of Western collaborators (see Chapter 6), she attributed "the success of the Israeli experience" to the "coupling of the collection of the material for population genetic analysis to the actual provision of especially attentive medical services and genetic counseling" as well as "the frequent inclusion within the research team of dependable members of the studied population, with high competence and authority, such as interpreters, technical aides, or sometimes teachers, nurses, or even doctors."[5] Yet even as Bonné-Tamir performed an enlightened benevolence in advocating for "good relationships and harmony" with research populations in the "Third World," the instrumentalism of her approach surfaced in a comment that "those who pretend to represent these populations sometimes exaggerate artificially" the "exploitation, discrimination, or stigma" faced by these communities in conjunction with the research.[6]

Bonné-Tamir's comments, of course, equally "pretend to represent" the communities studied by Israeli researchers, eliding the power asymmetry between the Ashkenazi technocratic class and the variously marginalized Mizrahim, Samaritans, Armenians, and Palestinians to narrate the "Israeli experience." Meanwhile, her belief in the superior ethical standards of Israeli geneticists stands at odds with the social and cultural effects of their work, which ultimately serves to reify politically manipulated group identities into essentialized biological categories. Psychologists and sociologists have recently taken an interest in analyzing these effects on social group relations. For example, a team of American, European, and Israeli psychologists turned to the Israeli-Palestinian conflict to investigate how genetic discourses might contribute to the resolution or exacerbation of ethnic-nationalist tensions. Following a series of studies conducted mainly on Jewish subjects, the psychologists found that Jewish Israelis who read a simulated news article emphasizing the genetic differences between Jews and Arabs "showed less support for political compromise and [. . .] more support for collective punishment toward Palestinians and more support for the political exclusion of Palestinian citizens of Israel."[7] The psychologists concluded that the rising publicity of research that conflates ethnicity with genetic difference could foreshadow or inflame political violence. Furthermore, this study reaffirmed the co-constitutive roles of Zionist politics and genetic science in the construction of a Jewish biological category and the chronic otherization of Palestinians.

The potential use of genetic research to justify ethnic violence is an explicit concern for Lebanese scientists, including Zalloua, Wells's coinvestigator on the Phoenician DNA project, a geneticist at Lebanese American University, and a principal investigator for the GP. As a Maronite Christian who came of age during the Lebanese Civil War, Zalloua experienced firsthand the violent legacy of Phoenicianism within the sectarian discourses of his fellow Maronites. Yet he fully embraced Wells's idealistic narrative about the peacemaking potential of genetic anthropology. In 2007, he complained to a Reuters reporter, "Whenever I use the word 'Phoenician', people say 'this guy is trying to say we are not Arabs.'"[8] In contrast, he insisted that Phoenician identity could actually bridge Lebanese communalism. Having identified the Y-chromosome haplogroup J2 as a "Phoenician" lineage, Zalloua argued that because J2 haplotypes appeared frequently in each of the Lebanese religious groups, "Phoenician is a heritage for all" rather than for Christians alone.[9] Zalloua strongly aligned his work with the professed values of the GP, believing sincerely in the Project's power "to bring people together by highlighting their shared heritage."[10] So far, there is little evidence that this unifying message has superseded more exclusive notions of Phoenician identity in Lebanon. Zalloua's 2007 appearance on the popular Lebanese talk show *Kalām al-Nās*, which featured DNA tests of the show's crew, inspired hundreds of viewers to volunteer to participate in his research. However, many of these volunteers understood the tests as diagnostic, hoping to learn whether they were "Phoenician or not" based on their Y-DNA haplotype.[11] The geneticized notion of Phoenician ancestry, rather than offering an alternative to the older sectarian discourse, ended up reinforcing the idea that only some Lebanese were true descendants of Phoenician civilization.

These recent cases from Israel and Lebanon, as well as the many historical ones I have presented throughout this book, reveal the significant limitations of Eurocentric visions of a progressive, anti-racist genetic anthropology. Namely, these visions do not account for societies wherein racialized violence is occurring between groups that have been scientifically configured as belonging to the same racial category, as equally "white." The nationalist stakes of such conflicts invoke very different expectations about the relationality (that is, the comparative meaning) of genetic data than those embraced by the leaders of HGDP and GP. Based in Western Europe and North America, these leaders have formulated their anti-racist messaging upon the assumption that racial discrimination occurs predominantly along the lines of physical differences, which can be overcome by exposing the shared evolutionary history hidden away in human genes. Accordingly, they have packaged for the anglophone lay public an academic consensus on human evolution that gradually solidified in the 1990s: all humans living today share a common ancestor who lived on the African continent about 200,000 years ago. The social meaning of this "Out of Africa" model of human origins, as supported first by mitochondrial and then Y-chromosome DNA studies, has been repeatedly communicated with a set of connected tropes. One of these co-opts older protestations that biological races do not exist with the more provocative phrase "We are all African," an attempt to counter anti-Black racism through the argument that the ancestry of any racial group can be genetically traced back to Africa. This abstract notion of racial relatedness is articulated at a personal level by constantly invoking the "extended family." The progressive potential of genetics is alleged to be its ability to show individuals how they fit into the "human family tree," enhancing a sense of kinship between peoples who look and live very differently. These differences, however, must not be devalued—after all, geneticists depend on them for medical and anthropological research. Rather, scientists urge an appreciation for "unity in diversity," framing genetic variation as a desirable patrimony of pluralist and multicultural polities.[12]

As a conclusion to this history of Middle Eastern genetics, I extend my Chapter 7 analysis of Turkey and Iran to briefly investigate how independent genome sequencing projects in those countries have reconciled the nationalist paradigms of earlier decades with emergent genetic discourses designed to promote social tolerance and combat racism. Iranian and Turkish genome projects at the turn of the twenty-first century employed these tropes in a number of ways, including explicit attempts to deconstruct racist ideologies. However, these discourses of kinship ultimately accommodate forms of territorial nationalism rather than break them down. These projects foregrounded nation-state borders to define a biological object, the "national genome," which reanimated academic and public discussions about the problems of defining ethnic identity without providing any substantial resolutions. The century of history I have traced in this book shows that because nationalism and human genetics are so thoroughly co-constituted, neither better technology nor better intentions are sufficient for geneticists to rectify social injustices or even to produce politically neutral data about ancestry—regardless of whether this ancestry is configured as ethnic, racial, or geographic.

TURKISH GENOMICS AS ANTI-RACIST DISCOURSE

In October 2015, the Nobel Prize Committee named Aziz Sancar, a biochemistry professor at the University of North Carolina at Chapel Hill, as one of the year's three winners in chemistry for his work on DNA repair mechanisms. The Turkish media immediately turned Sancar, the first Turkish citizen to receive a Nobel in a scientific field, into a national celebrity. Although many news outlets were quick to make Sancar's award a source of national pride, several also probed him for details of his ethnic origins: because he had been one of eight children born to illiterate farmers in the southeastern province of Mardin, popular speculation abounded as to whether Sancar was "really" Turkish or perhaps of Arab or Kurdish background. His anger at such rumors was widely reprinted, particularly his complaint that a BBC interviewer had "disrespected" him by asking first whether he was "Arab" or "partly Turkish." According to Sancar, he responded, "I don't speak Arabic, I don't speak Kurdish, I am Turkish. . . . I am Turkish and that's it."[13] Sancar's response was hailed as an exemplary rebuke to a Turkish "disease"—namely, an obsession with identity politics that turned Sancar's Nobel Prize into a public discussion not of the scientist's accomplishments but rather of his family's ethnic background.[14] Meanwhile, Sancar's impassioned declarations of gratitude to the Turkish public education system and his resolution to dedicate his Nobel medal to Atatürk cemented his bona fides as a devotee of Turkish civic nationalism.[15]

However, Sancar's public embodiment of an idealized secular Turkish identity glossed over the many processes that molded his own self-representation. Predictably, the media turned to Sancar's relatives to fulfill their quest for a clear ethnic categorization, although the family itself could not agree. On the one hand, Sancar's brother claimed the family descended from a tribe of Oghuz Turks who had immigrated to Anatolia from Central Asia; on the other hand, Mithat Sancar, a cousin serving as a parliament deputy for the *Halkların Demokratik Partisi* (a political party associated with Kurdish interests), claimed the family had Arab and Kurdish ancestry.[16] Reporters also connected Aziz Sancar's insistence on his Turkishness to his brief affiliation with the *Bozkurtlar*, the youth wing of the "ultra-nationalist" political party MHP, during his high school days in the early 1960s.[17] None of the coverage considered how Sancar might have been affected by his professional socialization as a scientist.

At the conclusion of his Nobel lecture in Stockholm, Sancar took the time to express his gratitude to dozens of colleagues and mentors who had contributed to his work. In addition to his American doctoral advisers and collaborators, he mentioned just one of his compatriots by name: "Being a Turk, I wish to acknowledge Muzaffer Aksoy, my internal medicine professor, who encouraged me to go to the United States and do science."[18] As shown in Chapter 4, Aksoy represented a generation of Turkish scientists and physicians devoted to Atatürk's vision of nationalism, and his collaboration with British and American researchers brought the term "Eti-Turk" into international use. Certainly, Aksoy's professional mentorship of Sancar was not innocent of this political orientation. Sancar's emphatic rejection of all ethnic identities other than Turkish strongly resonates with Aksoy's insinuation that any challenge to the population labels he used would represent "extreme nationalism" and "a backward form of thinking."[19] Meanwhile, the Turkish public's interpretation of their scientific careers is constrained by the same essentialized concepts of Turkish identity that their work helped to create.

Although Sancar and Aksoy embody the co-constitution of Kemalist nationalism and Turkish genetic science, a younger generation of Turkish molecular biologists and anthropologists is striving to break down these established tropes. Two initiatives in particular, the Turkey Genome Project and the Anatolian Genetic History Project, are conscious efforts to counteract the Turkish public's conflation of genetics with racial science in the temporal context of renewed racialized violence against Kurds.[20] In 2010, two geneticists at Boğaziçi University in Istanbul, Nesrin Özören and Cemalettin Bekpen, launched the Turkey Genome Project (TGP, *Türkiye Genom Projesi*). Their political sensibilities are evident in the Turkish title of the project, which identifies the genome with the nation-state Turkey (*Türkiye*) rather than the ethnolinguistic designation Turkish (*Türk*), although this distinction is not made within their English-language publications.

Over the next two years, the geneticists conducted a pilot study with the help of Turkish scientists from universities throughout Turkey and North America. They recruited sixteen volunteers from across Turkey to donate blood, ensuring that each individual originated from a different province and claimed at least four generations of residence in a given city. In the English-language publication of their data, the authors explicitly acknowledged what earlier researchers had only implied: "individuals were included in the study irrespective of their mother-tongue/ethnicity; we refer to them collectively as 'Turkish.'"[21] However, whereas their predecessors trumpeted Turkish genetic homogeneity, the TGP anticipated genetic diversity and conveyed their surprise that it was lower than they hypothesized: "We expected that Turkish genomes might exhibit significantly higher nucleotide diversity, given Turkey's location at the crossroads of out-of-Africa migrations, as well as more recent population movements."[22] Within their sixteen samples, they found no evidence of population structure—that is, the genetic data alone did not identify any distinct subpopulations within the group (which might potentially correlate with ethnic, religious, and other social divides). In fact, none of the results of the whole-genome sequencing process contradicted those of earlier Turkish geneticists working with classical markers. The researchers attributed this phenomenon to the major movements of populations across Anatolia in recent centuries and the ensuing high levels of intermarriage and genetic admixture,[23] in line with the hypotheses of previous studies.

In January 2012, the members of the TGP research team convened in Istanbul to discuss the first sequencing results. The Turkish media covered the event in live broadcasts as well as print articles. The news coverage revealed the TGP's uphill battle to explain their work beyond the reductive terms of Turkish nationalism. Ömer Gökçümen, a genetic anthropologist (currently based at the University of Buffalo), repeatedly mentioned the Out of Africa model and emphasized the deep time scale of the project: "Actually we all come from Africa. [...] With this project, for the first time genomes in Turkey and the Middle East are beginning to emerge. Probably we are connected to the first branch coming out of Africa. In fact, perhaps the Europeans are migrants from the agriculturalists who lived in the territories of Turkey ten to fifteen thousand years ago."[24] Meanwhile, Cemalettin Bekpen, when interviewed on the NTV talk show *Bugün Yarın*, attempted to avoid casting the genomic data as evidence for Turkish racial and national distinctiveness, instead describing modern Turks as the products of "cyclic" admixture between ancestral European and Anatolian populations. When the host, Oğuz Haksever, pressed him to resolve the question "Are we Europeans, or are we Orientals?" (*Avrupalı mıyız, doğulu muyuz?*), Bekpen replied, "It is not correct to say [such a thing]" (*Onu söylemek doğru değil*). Trying to redirect the conversation, Bekpen referred back to Gökçümen's presentation of the Out of Africa theory and further emphasized the complicated nature of genetic admixture, ultimately arguing that a reductive assignment of Turks as Europeans or Asians is impossible.

Gökçümen's academic background further illuminates the TGP's approach to wresting Turkish genetics out of its traditionally nationalist framing. Gökçümen is also a leading investigator in the Anatolian Genetic History Project (AGHP; known in Turkish as *Anadolu Popülasyon Tarihi Projesi*), a collaboration between physical anthropologists and geneticists at the University of Pennsylvania and Ankara University that grew out of Gökçumen's doctoral research in anthropology at Penn. In the aftermath of his fieldwork among Central Anatolian villagers between 2005 and 2008, Gökçümen and his Ankara colleague Timur Gültekin made a number of efforts at public outreach. For example, in a 2009 article for the Turkish popular science magazine *Bilim ve Teknik*, they insisted that there was no biological basis for racial classification and cautioned against the "distortion" of genetic findings for racist and discriminatory rhetoric. Specifically, they argued that the DNA-based research of genetic anthropologists disproved the morphology-based racial classifications of earlier physical anthropologists and that there were no diagnostic genes for racial or ethnic origins. At the end of the article, they presented their own AGHP as fulfilling the need for "objective" (*nesnel*) and "non-ideological" studies of Anatolian genetic history.[25] In a similar article aimed at Turkish academics, they expanded on these themes to decry the misunderstanding of genetic research in the Turkish public sphere, especially the tendency of the Turkish media to misinterpret complex genetic findings in terms of ethnic politics and apply modern ethnic labels like "Kurdish," "Turkish," and "Armenian" to populations living thousands of years ago.[26] They exhorted Turkish academics, particularly anthropologists and geneticists, to take greater responsibility for helping the public to "properly understand" the country's genetic diversity and counter a "racist paradigm" (*ırkçı bir paradigm*) of sensationalist media coverage by contributing articles to newspapers, magazines, and websites.[27]

The peer-reviewed publications of the AGHP contained further pointed critiques of Turkish ethnic politics and its influence on academic research, including their own. For instance, out of an ethical concern for the security of their research subjects, the project chose pseudonyms not only for individuals but for whole villages: "Because of the current political sensitivities concerning ethnic-religious identity in Turkey, especially those relating to the Alevis and Kurds, the names of the specific settlements we visited are not identified."[28] In a careful analysis of Y-chromosome and mtDNA data obtained from the villagers, Gökçümen and his colleagues argued that the genetic relationships they observed could best be explained by highly regionalized and localized factors. They charged that the established methods and assumptions of Turkish genetic research, focused on characterizing Turks at the national level and linking genetic data to large-scale historical migrations, did not reflect actual citizens' experiences of migration and family formation. Furthermore, they advocated for researchers to "reflexively present the local voices of those whose identities might be at stake, especially when the echoes of ethnocentrism are louder than any other voices in the nationalistic narrative."[29]

The call of Gökçümen and his colleagues to culturally contextualize genetic research by reducing the use of sweeping national and ethnic labels indicates their interest in transforming research practices outside Turkey, alongside their efforts to deracialize Turkish anthropology. Yet while claiming to take up the cause of marginalized local identities against dominant ethnonational narratives, the AGHP's initial hypothesis disputed the "self-described origins" of the non-Kurdish villagers they studied, who all "claimed an ancestral homeland somewhere outside of the Anatolian peninsula" from which they had emigrated relatively recently. The AGHP researchers attributed these claims to a "rapid nationalization of identity in the last century" and instead expected to find genetic evidence that at least some of these villagers "have been living in Anatolia for centuries and have constructed origin stories to better suit the many changes in regional and national ideological climates."[30] Upon discovering that their genetic data indeed linked these villagers to self-identified homelands, the AGHP scientists downplayed their own hypothesis about the role of contemporary nationalism in the villagers' origin myths.

Furthermore, despite the best efforts of the TGP and AGHP teams to inject some nuance into Turkish discourses of genetic identity, nationalist interpretations of such research proved stubbornly entrenched. This problem is not merely a case of "misunderstanding" on the part of the public or the mass media but also of other Turkish biologists taking a nationalist line and thus opposing the approaches of the TGP and AGHP. For example, a non-TGP-affiliated group of Turkish scientists at Istanbul Bilgi University sequenced the whole genome of a single Turkish individual in 2014 and emphasized distinctive traits in the Turkish genome in comparison to other populations, contradicting the TGP, who had insisted there was nothing particularly unique about the Turkish genome.[31]

In a more extreme case, an assistant professor of molecular biology at Balıkesir University, Osman Çataloluk, published a lengthy book aimed at general audiences titled *The Turk's Genetic History* (*Türk'ün genetik tarihi*) at the end of 2012. Çataloluk linked mtDNA and Y-DNA haplogroups to Turkish migratory history from Central Asia, specifically identifying the Y-DNA haplogroup R1b as "Turkish" because it appears with relatively high frequency in nearly all Turkic-speaking peoples. He insisted that his book was a rebuttal to Turkish and foreign geneticists who had claimed that Turkey's population was only "ten percent Turkish."[32] This is his own representation of a series of findings by Wells, Cavalli-Sforza, and their colleagues suggesting that the *Asian* contribution of genetic material to Turkey's contemporary population ranges from 9 percent to 13 percent. However, he did not contend that Turks are essentially Asian. Instead, he drew on archaeogenetic research to insist on a direct ancestral relationship between living Turks and human remains excavated from Çatal Höyük in the early 1990s. Çataloluk argued that this research proved that Turks had been the "true owners of Anatolia" since Neolithic times but that the European and American scientists involved had willfully ignored this interpretation. While Çataloluk's influence should not be overestimated—his academic connections in Turkey and abroad are not nearly as prestigious as those of the TGP and AGHP— he has marshaled his credentials to gain a platform on several mainstream Turkish talk shows, who invited him to discuss his book throughout 2013.

In other words, the anti-racist ideology of the TGP and AGHP does not reflect a wholesale transformation in the social role of the Turkish scientific community but rather the splintering of their professional class along the lines of a political culture war between hard-line nationalists and those holding more pluralistic values. A similarly politicized academic debate in

China over the evolution of modern Chinese people from African-origin *Homo sapiens* or China-origin *Homo erectus* rests partially on methodological/disciplinary divides between geneticists using DNA and paleoanthropologists using fossils.[33] However, in Turkey, the stakes hinge more specifically on how geneticists interpret the same sets of DNA evidence, with full awareness of the political implications. Some, like Çataloluk, perceive the imperative to universalism and diversity in international genome projects as a hostile Euro-American discourse that undermines Turkey's claims to its national territory. Others, like Bekpen and Gökçümen, perceive the dissemination of this same discourse to the Turkish public as an opportunity to overcome the legacy of racialized discrimination against different ethnic groups in Turkey. However, because the national genome project relies on Turkey's state borders to delineate the boundaries of a putatively biological population, they find both journalists and other scientists recasting their study into a politically laden nationalist framework. Turkish geneticists are burdened with the same fundamental questions about Turkish identity—namely, the country's ethnic position vis-à-vis Europe and Asia—that drove the development of research under Atatürk's rule in the 1930s.

NATIONAL GENOME AS DIVERSITY CATALOG: THE IRANIAN "FAMILY TREE"
Beginning in the late 1990s, Iranian scientists mounted a national-scale response to the Human Genome Diversity Project, variously called the Iranian Human Genome Project (IHGP) or the Human Genome Diversity Project of Iran (HGDPI). A scientific council composed of consulting biologists, including Dariush Farhud, assisted with organizing and coordinating the work of individual research centers across the country. Aiming to study the numerous ethnolinguistic and recognized religious minorities of Iran, the researchers sampled blood from more than 1,900 individuals belonging to 18 different groups during the summer of 2001. Portions of each sample were immediately tested for a range of mitochondrial, autosomal, and Y-chromosomal markers; cell lines were cultivated from the remaining blood for indefinite preservation.[34] A summary of the project published in the *Indian Journal of Human Genetics* emphasized the organizers' attention to and care for the anthropological issues inherent in the research: "Accurate identification of population units for sampling purposes requires extensive knowledge of the social, political and linguistic composition of the region to

be sampled and this stage of the study took a long time, as many as 2 years of collection of all preliminary necessary data."[35] This emphasis demonstrates an awareness of the methodological and ethical disputes between anthropologists and geneticists that had stalled the progress of the HGDP.[36]

The description of the Iranian project's goals and achievements opens with a statement on the diversity of "Iranian ethnicities," which "are very different not only in their origins and languages but also in their cultures, life style and, obviously, their geographical distribution over the country." Identifying and classifying these differences, the author Kambiz Banihashemi explains, is an important matter for national health care in terms of identifying genetic predispositions or resistance to disease. However, the majority of his points expound on the project's national value in ethnohistorical terms. The genome survey had opened "enormous potentials for illuminating our understanding of Iranian ethnicities' history and identity," in addition to linking the work of geneticists "in an unprecedented way with that of anthropologists, archaeologists, biologists, linguists and historians, creating a unique bridge between science and the humanities in Iran for the first time."[37] Furthermore, he declares, the project had contributed a deeper understanding of the specific traits of "Iranian ethnicities" (note that he never designates any of the groups as "minorities," which would expose the power dynamic through which the categories and traits are defined).[38] The project is thus a government-supported extension of the approaches Iranian geneticists adopted in the 1970s (see Chapter 7).

Around the same time as the sampling phase of the IHGP was wrapping up, another group of Iranian geneticists at Tehran's University of Social Welfare and Rehabilitation Sciences established the Iranian Human Mutation Gene Bank (IHMGB). Although its creators imagined that their database would be useful primarily for medical research, a 2003 article in the journal *Human Mutation* aiming to introduce the database to anglophone scientists highlighted ethnic and cultural heterogeneity as an integral feature of the contribution of "the Iranian population" to medical knowledge:

> Iran, a large country with different ethnic groups, tribes, and religions and a population of close to 66 million, offers a highly heterogeneous gene pool to research [...] the purity of many different races in this country has been conserved within geographical borders and by an ancient culture that has always encouraged intrafamilial marriages. All of these factors have created

a population that is remarkably heterogeneous and yet high in consanguinity rate.[39]

The creators of the IHMGB intended to track disease-related mutations, but rather than contributing their data to disease-defined databases, they developed their own database defined by a non-biological criterion, nationality. This database includes many delineations of difference (ethnic groups, tribes, religions, and even "pure races"), yet paradoxically, there remains only "a [single] population"—that of the Iranian nation. Bound to the territory and health-care structure of the nation-state, the Iranian national genome diversity project and national mutation database prevent any amount of genetic differentiation from constituting a Kurd, Armenian, or Zoroastrian as anything other than a variation on an "Iranian" genome. The discourse of "unity in diversity" is therefore not limited to allegedly anti-racist and universalist international endeavors like the Genographic Project but is also instrumentalized to align narratives of ethnic diversity with the nation-state's interests in territorial integrity.

The international sanctions regime against Iran has impeded scientific collaboration and publication, although Iranian scientists themselves disagree over the nature and extent of the damage.[40] Sanctions have certainly made the importation of laboratory equipment and reagents more difficult and costly, which has burdened biological research more than, e.g., theoretical physics or computer science. On the other hand, the sanctions have forced Iranian scientists to become more resourceful and self-sufficient by developing their own equipment and academic programs.[41] Recent visitors to Iran's laboratories, such as correspondents for the journal *Science*, have commented that they are "on par" with those of the West.[42] Regardless, the translational capacity of Iranian gene databases has been hampered by their inability to maintain a stable Internet presence. For example, as of March 2016, none of the published URLs for the IHMGB or the IHGP were functional.

RECONSTITUTING THE CROSSROADS
Despite such political and economic barriers, nationalist approaches to studying the genetics of Iranian populations have been effectively reproduced in the United States through the Stanford-based "Iranian Genome Project" (IGP), a so-called pet project of Iranian-born Mostafa Ronaghi, the

chief technology officer of the American biotechnology company Illumina. Through Ronaghi's coordinating efforts, the bioengineering laboratory of Russ Altman at Stanford University became the project's institutional home in 2010. Although the IGP professed the same basic aims as the research conducted in Iran nearly a decade earlier, its use of next-generation sequencing techniques promised a "one-of-a-kind" genomic study of Iranians residing in the United States. According to the project leader, Stanford medical researcher Roxana Daneshjou, the predominantly Iranian American research team initially hoped to collaborate with geneticists within Iran, but university lawyers warned them that such collaboration could violate U.S. sanctions against Iran. Therefore, the project proceeded completely independently from Iranian geneticists.[43] Nonetheless, the methodological and discursive similarities between the Iranian and Iranian American projects are striking. The IGP mobilized the relatively more multicultural values of the United States and its biomedical regimes of "inclusion" to reproduce a national category that is supposedly coherent enough to be genetically distinct from a category of assimilated whiteness, yet also contains a diversified "family tree" of group identities.[44]

The project was funded primarily by the Parsa Community Foundation, described on its website as "the first Persian community foundation in the U.S. and the leading Persian philanthropic institution." After receiving a $250,000 grant from Parsa at the end of 2010, the project swung into its next phase, recruiting Iranian Americans to donate saliva samples in California. The recruitment efforts of the IGP reflect the scarcity of human, as opposed to material, resources inherent to the study of diaspora communities within the United States. Although the IGP now had plenty of funding and unfettered access to equipment and supplies for genomic sequencing and analysis, the options for tracking down a randomized sample of Iranian American DNA donors were more limited, and the IGP understandably targeted its message for Iranian community organizations. However, the project researchers also hoped to extract data about deep-historical population movements and the biological relationships between different Iranian subpopulations. In other words, their Iranian American sample needed to serve as a "diasporic proxy" standing in for the entire population of Iran, capturing a plurality of Iran's many ethnolinguistic and religious minority groups.[45] To communicate this need in their recruitment appeal, the IGP had to destabilize their own narrative of "Iranian American" as a single ethnic category and acknowledge the internal diversity collapsed by such a category: "We will have the opportunity to draw our family tree and to explore our heritage in a way that has never been done before . . . we will be able to celebrate both the similarities and unique characteristics of the diverse Iranian family."[46]

The IGP's vision of "the diverse Iranian family" substantially aligned with the categories used by the Iranian government's own genomic diversity project. The IGP's participation eligibility survey asked potential recruits to identify their ethnic heritage and spoken language(s), with eighteen options in each category. Respondents were also asked to identify with one of eight religions (two of which—Judaism and Zoroastrianism—are also listed in the ethnic category). The religious identification options, which included Baha'i, Yarsani, and "Sufi" (groups unrecognized by the Islamic Republic who face severe political discrimination), offered the only notable point of difference between the diasporic and Iranian state projects. Otherwise, both adopted the same approach of encapsulating diversity within a population defined a priori as Iranian.

Through its recruitment materials and presentation of preliminary results, the IGP reconfigured Iranian territorial origin and pre-immigration nationality as diagnostic of a biological whole, the "Iranian family." In the process, it reduced a series of ethnic, linguistic, and religious identifications to minority derivations of Iranian status. This reductionism overlooked evidence that many Iranian "subgroups" in the diaspora—such as Kurds, Armenians, Jews, and Baha'is—may not identify themselves primarily as Iranian.[47] Similarly, the participation of Iranian Zoroastrians, Jews, and Baha'is in religious organizations in the United States integrates them with non-Iranian coreligionists. Further sociological research in Los Angeles strongly argued that Iranian "subgroups" formed distinct ethnic communities and "should not be subsumed under an umbrella Iranian category."[48] In contrast, the IGP employs just such an umbrella category—effectively reconstituting the Iranian state's own approach to the "genealogical incorporation" of these groups into a narrative of shared national ancestry.[49] Political sanctions hinder many aspects of Iranian scientific productivity and knowledge exchange. But they have not at all impeded the diasporic reproduction of nationalist ideas about who counts as "genetically" Iranian. A concept of

"Iranian ethnicities" that emerged decades earlier to defend the territorial integrity of the Iranian state has been readily adopted by American institutions and figures at the highest levels of industrial and academic prestige.

Returning to the words of Dariush Farhud with which I opened this book, the "gene drain" from Iran and other countries to the United States has made it possible to reconstitute the Middle Eastern genetic crossroads elsewhere. Diasporic proxies can even enable researchers to completely bypass collaboration with Middle Eastern scientists. But the IGP case confirms that such an approach has little effect on the historical assumptions and population categories embedded in genomic research. Although the capabilities of DNA sequencing far exceed older technologies relying on the cephalic index, blood groups, and inherited disorders, geneticists repeat the process of reifying religious, linguistic, and social differences into biologically detectable ethnic or racial groups. Even when such differences are not substantiated genetically, the very practice of singling out Jews, Kurds, and Zoroastrians for genetic study (sometimes at the behest of the communities themselves) prompts the recitation of historical narratives that situate these groups within or against particular nation-states. Setting out to recover the ancient DNA of Phoenicians as a universal Lebanese, or even human, heritage cannot erase the living memories of anti-Arab violence enacted in the name of racialized Phoenician identity. Despite the benevolent intentions of progressive scientists, contemporary genetic research practices are deeply entangled with, and not always able to overcome, the social divisions produced and exacerbated by twentieth-century nationalism.

This book highlights the careers of Middle Eastern geneticists. But their careers have always been embedded in transregional networks stretching from Mandate-era missionary hospitals to the GP. The use of Middle Eastern nationalist narratives and population labels by both local and foreign scientists shows how nationalism is sustained by particular practices of human genetics research—specifically, the need to describe human populations according to geography and ancestral history, coinciding with the two major constituent elements of the nation-state paradigm. These inherent features of the science underlie a relational notion of "genetic nationalism," a phenomenon not unique to the Middle East but instead formulated on an international scale. Genetic nationalism is not fraudulent or pseudoscientific but has been and continues to be the normative logic of human population genetics as practiced within a global sphere of nation-states. It is the avowed *internationalism* of professional scientific communities—that is, the numerous stages of education, collaboration, and publication that bring Turks, Iranians, Arabs, and Israelis into dialogue with others—that reinforces the use of *nationalist* terminology, even when, or perhaps especially when, this terminology is contested and poorly defined. Accordingly, the persistence of nationalist concepts in Middle Eastern genetic studies does not indicate a rejection of Western models of universal science, nor does it represent the simplistic corruption of objective research through the imposition of retrograde political discourses. Rather, it reflects the thorough integration of Middle Eastern geneticists as collaborators in an international sphere of scientific discourse over the course of the twentieth century.

Notes

INTRODUCTION

Several ideas and paragraphs found in the Introduction were first published in an article called "'Essential Collaborators': Locating Middle Eastern Geneticists in the Global Scientific Infrastructure, 1950s–1970s," in *Comparative Studies of Society and History* 60, no. 1 (2018), 119–49.

1. Saini, *Superior*.
2. On the overall role of science in nationalist identity formation, see, for example, Prakash, *Another Reason*; Mizuno, *Science for the Empire*; Harrison and Johnson, *National Identity*. On the specific relationship between nationalism and human genetics, or "bionationalism," see, for example, Mukharji, "Profiling the Profiloscope"; Hyun, "Blood Purity and Scientific Independence"; Subramaniam, *Holy Science*.
3. Nelkin and Lindee, *The DNA Mystique*, 52.
4. Keller, *The Century of the Gene*.
5. Mourant to Henry Bunjé, January 27, 1965. PP/AEM/D.2, Box 11, Arthur E. Mourant Papers, Wellcome Library, London. Hereafter "Mourant Papers."
6. TallBear, *Native American DNA*, 149–52; Abu El-Haj, *The Genealogical Science*, 27.
7. Gannett and Griesemer, "The ABO Blood Groups," 153.
8. Lipphardt, "The Jewish Community of Rome," 309.
9. Reardon, "Race Without Salvation," 307; Schaffer, *Racial Science and British Society, 1930–62*, 10.

10. See Schreiber, *The Comfort of Kin*; Prager, "'Dangerous Liaisons.'"

11. Existing critiques of methodological nationalism center on the analytical constraints it places on the knowledge produced by the social sciences, including comparative research in science studies; see Wimmer and Glick Schiller, "Methodological Nationalism and Beyond"; Wade et al., *Mestizo Genomics*; Shostak and Beckfield, "Making a Case for Genetics."

12. Abu El-Haj, *The Genealogical Science*, 128.

13. Anderson, *Imagined Communities*.

14. Simpson, "Imagined Genetic Communities," 6. As social scientist Venla Oikkonen notes, the contemporary industry of genetic ancestry testing has made possible new forms of communities in which "the revelation of a genetic connection or similarity is often the only tie that initially connects people." Oikkonen, *Population Genetics and Belonging*, 179.

15. Gottweis and Kim, "Bionationalism, Stem Cells, BSE, and Web 2.0 in South Korea"; Liu, "Making Taiwanese (Stem Cells)"; Kuo, "Techno-Politics of Genomic Nationalism." See also the related concept of "genetic romanticism," which links the practices of contemporary genetic medicine to the folklore studies of nineteenth-century European romantic nationalism: Tupasela, "Genetic Romanticism."

16. Subramaniam, *Holy Science*, 175.

17. Duara, *Rescuing History from the Nation*, 55. Emphasis original.

18. Fan, "The Global Turn in the History of Science," 253.

19. See the historiographical discussion in Shefer-Mossensohn, *Science Among the Ottomans*, 1–10.

20. Schayegh, *Who Is Knowledgeable Is Strong*, 22–24.

21. Lotfalian, "The Iranian Scientific Community and Its Diaspora After the Islamic Revolution."

22. Khosrokhavar, Etemad, and Mehrabi, "Report on Science in Post-Revolutionary Iran—Part II."

23. Anderson and Pols, "Scientific Patriotism," 113.

24. Efron, *Defenders of the Race*; Steinweis, *Studying the Jew*.

25. Kirsh, "Population Genetics in Israel in the 1950s"; Kirsh, "Genetic Studies of Ethnic Communities in Israel"; Falk, *Zionism and the Biology of the Jews*; Abu El-Haj, *The Genealogical Science*.

26. See, for example, Ergin, "Cultural Encounters in the Social Sciences and Humanities."

27. Kurds in Turkey have particularly embraced the language of "internal colonialism" to understand their status vis-à-vis Middle Eastern nation-states. See

Ünlü, "İsmail Beşikçi as a Discomforting Intellectual"; Houston, "An Anti-History of a Non-People." For discussions of internal colonialism concerning Palestinians and Mizrahi Jews in Israel and the Yishuv, see Yiftachel, *Ethnocracy*; Hirsch, "'We Are Here to Bring the West, Not Only to Ourselves.'" For discussions of internal colonialism targeting non-Persian minorities in Iran, see Asgharzadeh, *Iran and the Challenge of Diversity*; Elling, *Minorities in Iran*.

28. See, for example, Salgırlı, "Eugenics for the Doctors"; Hirsch, "'We Are Here to Bring the West, Not Only to Ourselves.'"

29. Jasanoff, "Ordering Knowledge, Ordering Society," 30.

30. TallBear, *Native American DNA*, 26.

31. El Shakry, *The Great Social Laboratory*, 13.

32. Powell, *A Different Shade of Colonialism*, 16–17; Schayegh, "Hygiene, Eugenics, Genetics"; Hirsch, "Zionist Eugenics, Mixed Marriage, and the Creation of a 'New Jewish Type'"; Salgırlı, "Eugenics for the Doctors"; Zia-Ebrahimi, *The Emergence of Iranian Nationalism*.

33. See, for example, Haarmann, "Ideology and History, Identity and Alterity"; Southgate, "The Negative Images of Blacks in Some Medieval Iranian Writings"; Schine, "Conceiving the Pre-Modern Black-Arab Hero"; Sheriff, "The Zanj Rebellion and the Transition from Plantation to Military Slavery."

34. See Selim, "Languages of Civilization"; Fujinami, "'Abd al-Ḥamīd al-Zahrāwī and His Thought Reconsidered;" and Chapter 1 in Çağaptay, *Islam, Secularism, and Nationalism in Modern Turkey*.

35. On the similar process used for Arabic translations of evolutionary biology, see Elshakry, *Reading Darwin in Arabic, 1860–1950*.

36. El Shakry, *The Great Social Laboratory*, 58–61.

37. Anderson, "Racial Conceptions in the Global South," 787.

38. Öktem, "The Legal Notion of Nationality in the Turkish Republic."

39. Zia-Ebrahimi, *The Emergence of Iranian Nationalism*, 46, 165.

40. Gissis, "When Is 'Race' a Race?"; Gribetz, *Defining Neighbors*; Yanow, "From What *Edah* Are You?" See also Tekiner, "Race and the Issue of National Identity in Israel."

41. El Shakry, *The Great Social Laboratory*, 74–86; Falk, *Zionism and the Biology of the Jews*, 15; Kirsh, "Population Genetics in Israel in the 1950s."

42. Maksudyan, *Türklüğü ölçmek*; Maksudyan, "The Turkish Review of Anthropology and the Racist Face of Turkish Nationalism"; Ergin, "Biometrics and Anthropometrics"; Ergin, *Is the Turk a White Man?*

43. El Shakry, *The Great Social Laboratory*, 3.

44. Kowal, Radin, and Reardon, "Indigenous Body Parts, Mutating Temporalities, and the Half-Lives of Postcolonial Technoscience," 471.

45. Houston, "An Anti-History of a Non-People," 32.

46. Schayegh, *Who Is Knowledgeable Is Strong*, 2.

47. Here, I use the idea of the "native informant" not in a conventional anthropological sense but as a specific reference to Gayatri Chakravorty Spivak's critique of postcolonial nationalist intellectuals, in support of applying this critique to the practice of biomedical sciences. Spivak, *A Critique of Postcolonial Reason*.

48. Little, "Human Population Biology in the Second Half of the Twentieth Century," S132.

49. Schaffer et al., *The Brokered World*, xxi–xxx.

50. Schiebinger, *Secret Cures of Slaves*, 13.

51. Mukharji, *Doctoring Traditions*, 24.

52. Bangham, "Blood Groups and Human Groups"; Lindee and Santos, "The Biological Anthropology of Living Human Populations," S7.

53. Mattson, "Nation-State Science," 324.

54. Robertson, "Blood Talks"; Mukharji, "From Serosocial to Sanguinary Identities"; Kent et al., "Building the Genomic Nation."

55. Hilgartner, *Reordering Life*, 82.

56. Hilgartner, 83–89.

57. See Shrum, "Collaborationism."

58. See, for example, Anderson and Pols, "Scientific Patriotism," in juxtaposition with Robinson, "Non-European Foundations of European Imperialism: Sketch for a Theory of Collaboration."

59. See also Bier, *Mapping Israel, Mapping Palestine*, 88–89.

60. See especially McMahon, *National Races*; McMahon, *The Races of Europe*; Mattson, "Nation-State Science"; Mukharji, "Profiling the Profiloscope"; Mukharji, "The Bengali Pharaoh."

61. The literature on these topics is expansive. A representative sampling of scholarship on race and genetics in North America must include: Bliss, *Race Decoded*; Reardon, "The Democratic, Anti-Racist Genome?"; Fullwiley, "The Biologistical Construction of Race"; TallBear, *Native American DNA*; Nelson, *The Social Life of DNA*; Roberts, *Fatal Invention*; Wailoo, Nelson, and Lee, *Genetics and the Unsettled Past*; Koenig, Lee, and Richardson, *Revisiting Race in a Genomic Age*. Scholars who consider both race and nationalism in their analyses of genetics and racial classification include: Bauer, "Virtual Geographies of Belonging"; Sleeboom-Faulkner,

"How to Define a Population"; Sung, "Chinese DNA"; M'charek, *The Human Genome Diversity Project*; Wade et al., *Mestizo Genomics*.

62. Comfort, *The Science of Human Perfection*.
63. TallBear, *Native American DNA*, 146.
64. Cf. Bonnie Effros on the conditions of French colonial research on Berbers: Effros, "Berber Genealogy," 74.
65. Robson, *States of Separation*; Iğsız, *Humanism in Ruins*.
66. Yosmaoğlu, *Blood Ties*, 294.

CHAPTER 1

1. Biddiss, "The Universal Races Congress of 1911," 37.
2. Spiller, *Papers on Inter-Racial Problems*, xiv.
3. Tilley, "Racial Science, Geopolitics, and Empires," 774.
4. Riza Tevfık, "Turkey," 458–59.
5. See McMahon, "Anthropological Race Psychology 1820–1945."
6. Riza Tevfık, "Turkey," 456.
7. Tilley, "Racial Science, Geopolitics, and Empires," 779.
8. *Record of the Proceedings of the First Universal Races Congress*, 78.
9. *Record of the Proceedings of the First Universal Races Congress*, 50.
10. *Record of the Proceedings of the First Universal Races Congress*, 54.
11. *Record of the Proceedings of the First Universal Races Congress*, 67. On Greek race science, see Trubeta, *Physical Anthropology, Race and Eugenics in Greece (1880s–1970s)*; Trubeta, "The 'Strong Nucleus of the Greek Race.'"
12. *Record of the Proceedings of the First Universal Races Congress*, 70.
13. "Races Congress: The concluding meetings; future work." *The Manchester Guardian*, July 31, 1911, p. 9.
14. *Record of the Proceedings of the First Universal Races Congress*, 71.
15. Górny, "Racial Anthropology on the Eastern Front, 1912 to Mid-1920s"; Rhode, "A Matter of Place, Space, and People: Cracow Anthropology, 1870–1920," 127–28.
16. Kennedy and Riga, "Mitteleuropa as Middle America?"; Crampton, "The Cartographic Calculation of Space," 739–42; Turda, "Race, Politics and Nationalist Darwinism in Hungary, 1880–1918," 163; McMahon, *The Races of Europe*, 296.
17. McMahon, *The Races of Europe*, 23–29.
18. See Breisach, *Historiography*, 268–71.
19. See Turda, "From Craniology to Serology"; Kalling and Heapost, "Racial

Identity and Physical Anthropology in Estonia, 1800–1945"; Felder, "'God Forgives—But Nature Never Will.'"

20. Acceptance of the hereditary significance of the cephalic index was widespread but not universal. The most famous critique of the measure came from German-American anthropologist Franz Boas in the 1910s, when he demonstrated that American-born children of immigrants had different cephalic indices than their parents, indicating that head shape was influenced as much (if not more) by environmental conditions as by hereditary factors. Nevertheless, outside the Boasian school, the cephalic index maintained its prominent status in racial anthropology.

21. Bonakdarian, "Negotiating Universal Values and Cultural and National Parameters at the First Universal Races Congress."

22. Hadji Mirza Yahya, "Persia," 145.

23. Hadji Mirza Yahya, 153.

24. Hadji Mirza Yahya, 145, 147.

25. *Record of the Proceedings of the First Universal Races Congress*, 72.

26. Hadji Mirza Yahya, "Persia," 145–46.

27. See, for example, Stocking, "Colonial Situations."

28. Houston, "An Anti-History of a Non-People," 30.

29. Mattson, "Nation-State Science," 332.

30. Mattson, 335.

31. Baum, *The Rise and Fall of the Caucasian Race*, 132–33.

32. See, for example, how French craniometry studies of North Africans "naturalized" French rule in the region. Effros, "Berber Genealogy."

33. Arvidsson, *Aryan Idols*, 47–48.

34. McMahon, *The Races of Europe*.

35. Kaufman, *Reviving Phoenicia*; Elshakry, *Reading Darwin in Arabic, 1860–1950*; Katouzian, "Alborz and Its Teachers."

36. El Shakry, *The Great Social Laboratory*, 58–61; Gualtieri, *Between Arab and White*, 64–66; Marc Gribetz, "'Their Blood Is Eastern.'"

37. See, for example, Chapter 6 in Zia-Ebrahimi, *The Emergence of Iranian Nationalism*; Motadel, "Iran and the Aryan Myth," 130–31.

38. El Shakry, *The Great Social Laboratory*, 18.

39. Reid, *Whose Pharaohs?*, 281–85; Kaufman, *Reviving Phoenicia*.

40. von Luschan, "The Early Inhabitants of Western Asia," 222.

41. Assi, "The Original Arabs."

42. von Luschan, "The Early Inhabitants of Western Asia," 233.

43. Kaufman, *Reviving Phoenicia*.

44. Zurayq, *al- Aʻmāl al-fikriyya al-ʻāmma li-l-Duktūr Qusṭanṭīn Zurayq*, 1:59.

45. One such student, Najla Abu ʻIzz al-Din, took charge of measuring nearly one thousand of her fellow Druze, and her thesis on the "racial origins" of the Druze earned her a doctorate in anthropology from the University of Chicago in 1934. Izzeddin, "The Racial Origins of the Druze."

46. Shanklin, "The Anthropology of Transjordanians," 9.

47. Shanklin and Izzeddin, "Anthropology of the Near East Female," 413; Krischner and Krischner, "The Anthropology of Mesopotamia and Persia A. Armenians, Khaldeans, Suriani (or Aissori) and Christian 'Arabs' from Iraq," 205–17; Kappers, *An Introduction to the Anthropology of the Near East*.

48. Shanklin, "Anthropometry of Syrian Males," 404; Kappers, *An Introduction to the Anthropology of the Near East*, 55.

49. Kappers, *An Introduction to the Anthropology of the Near East*, 24.

50. Kappers, 57. In an earlier draft for the Arabic *al-Kulliyyah*, Kappers states even more strongly that Phoenicians, Palmyrenes, and living Syrian Bedouins "all evolved from one Semitic origin [*nashaʼū jamīʻuhum min aṣl sāmī wāḥid*]." Kappers, "Anthrūbūlūjiyyat al-sharq al-adanī," 358.

51. Huxley, "Preliminary Report of Anthropological Expedition to Syria," 49–50.

52. Huxley, "Samaritans," 675.

53. Huxley, 676.

54. See Efron, *Defenders of the Race*; Morris-Reich, "Jews Between Volk and Rasse."

55. In total, Weissenberg studied sixty-two fellahin from two populations (one coastal and one inland), twenty Samaritans, and fourteen native Jews from Upper Galilee. He also measured sixty-two Jewish settlers from Rishon LeZion and Rosh Pinah, but he did not include them in his tables or analysis. Weissenberg, "Die autochthone Bevölkerung Palästinas in anthropologischer Beziehung," 131.

56. Weissenberg, 137.

57. Szpidbaum, "Die Samaritaner," 140.

58. Gini, "I Samaritani," 145–46.

59. Szpidbaum, "Die Samaritaner," 139.

60. Szpidbaum, 157.

61. Kappers, *An Introduction to the Anthropology of the Near East*, 62–65.

62. Schreiber, *The Comfort of Kin*, 53.

63. Gini, "I Samaritani," 146.

64. Schreiber, *The Comfort of Kin*, 57.

65. Schreiber, 342.

66. "Registration in Robert College for the years 1930–31 and 1931–32," the *Robert College Herald*, October 24, 1931. Box 49, Folder 3, Robert College Records, Rare Book and Manuscript Library, Columbia University.

67. Essay of Nejat Ferit, October 28, 1929. Box 3, Folder 3, Edgar Jacob Fisher Papers, Hoover Institution Archives (hereafter "Fisher Papers").

68. Essays of Mehmed Saki and Ali Feridoun, October 28, 1929. Box 3, Folder 3, Fisher Papers.

69. Essay of Zıhni Haldun, October 14, 1930. Box 3, Folder 4, Fisher Papers.

70. Atakuman, "Cradle or Crucible," 220–25.

71. Kieser, "Türkische Nationalrevolution, anthropologisch gekrönt," 107.

72. Kieser, 112. Note that this is in direct opposition to the Nordic-German theories that would later be endorsed by the Nazis, which held that the "Aryan race" was dolichocephalic.

73. Pittard, "Découverte de la civilisation paléolithique en Asie Mineure"; Pittard, "Contribution à l'étude anthropologique des Turcs d'Asie Mineure," 3, 10, 19.

74. Maksudyan, "The *Turkish Review of Anthropology* and the Racist Face of Turkish Nationalism," 294–95.

75. See the list of studies published in the Turkish Journal of Anthropology between 1930 and 1939 in Maksudyan, *Türklüğü ölçmek*, 196–98.

76. Maarif Vekilliği Neşriyet Müdürlüğü, *Lise Kitapları: Biyoloji II*, 165–66. I thank Murat Gülsaçan for bringing this book to my attention and providing a copy of the chapter.

77. [Kansu], "Anadolu ve Rumeli Türklerinin antropometrik tetkikleri, birinci muhtıra: Boy ve gövde nispetleri," 4.

78. [Kansu], 1.

79. Ergin, *Is the Turk a White Man?*, 196–98. Others who were sent abroad to study physical anthropology in order to staff the Turkish Institute of Anthropology include Seniha Tünakan, who studied with Eugen Fischer at the Kaiser Wilhelm Anthropological Institute in Berlin from 1935 to 1941, and the paleoanthropologist Muzaffer Süleyman Şenyürek, who studied with Earnest A. Hooton at Harvard between 1934 and 1939.

80. See letters from İnan to Atatürk, December 3, 1935, and December 15, 1935, reprinted in İnan, *Prof. Dr. Afet İnan*, 156, 161–63.

81. Letter from İnan to Atatürk, December 15, 1935, reprinted in İnan, 162.

82. Two French editions of her thesis exist: İnan, *L'Anatolie, le pays de la "race" turque* (Genève: Imprimerie Albert Kundig, 1939, and Georg, 1941). A Turkish translation was prepared almost ten years later: İnan, *Türkiye halkının antropolojik karakterleri ve Türkiye tarihi.*

83. Although data forms were collected for 64,000 individuals, İnan explains that about 7 percent of the forms were discarded due to incomplete data, incorrect use (*"mal rédigées"*) or illegibility, leaving a sample size of 59,728. İnan, *L'Anatolie, le pays de la "race" turque*, 1939, 58–59.

84. The nasal index is the ratio of nose breadth to nose height; the skelic index is the ratio of the length of one's leg to the length of one's trunk.

85. İnan, 43.

86. İnan, *Türkiye halkının antropolojik karakterleri ve Türkiye tarihi*, 181.

87. İnan, *L'Anatolie, le pays de la "race" turque*, 1939, 59.

88. İnan, 117.

89. Earlier anthropometric studies conducted by Europeans nearly always characterized the Kurds as dolichocephalic, as in von Luschan, "The Early Inhabitants of Western Asia," 229. Pittard's dissent on this point, discussed below, appears in Pittard, *Race and History*, 366–67.

90. Çağaptay, *Islam, Secularism, and Nationalism in Modern Turkey*, 86.

91. Çağaptay, 108.

92. İnan, *Türkiye halkının antropolojik karakterleri ve Türkiye tarihi*, 181.

93. Iğsız, *Humanism in Ruins*, 6; Robson, *States of Separation*, 72–78.

94. So called in Turkish to distinguish this Arabic-speaking group from the Turkish-speaking Alevis, who belong to different sects.

95. The first academic use of these terms, including an elaborate historical and linguistic justification, appears in Türkmen, *Mufassal Hatay tarihi*. For a full discussion of how the Alawites were rebranded as "Eti-Türks" in line with the Turkish History Thesis, see Pınar, "Türk Tarih Tezi bağlamında Cumhuriyet döneminde Nusayriler."

96. "Büyük devlet haline getirdiğiniz Türkiye ırkdaşları için ve kendi vücudunun bir parçası için haklı isteklerini böyle sergiler: 'Çünku o, Eti imparatorluğu'nun ayrılmaz parçası ve ikinci kültür merkezi idi.' karşılık yanıtını dinlerken itiraf edeyim ki, sabırsızlandım, sinirlendim. Fakat ne de olsa eskiyen bir siyaset karşısında, yeni olan galip gelecek." Letter from İnan to Atatürk, December 10, 1936, reprinted in İnan, *Prof. Dr. Afet İnan*, 174.

97. Tankut, *Nusayriler ve Nusayrilik hakkında*, 10–15. Tankut cites Felix von Luschan, *Völker, Rassen, Sprachen* (Berlin: Welt-Verlag, 1922), 103, although he misattributes the information to a nonexistent page 204.

98. Dilaçar, "Alpin ırk, Türk etnisi, ve Hatay halkı," 16–17. The original source for the cephalic index measurement is not cited, but the value matches Tankut's von Luschan citation. See also von Luschan, "The Early Inhabitants of Western Asia," 231.

99. Shields, *Fezzes in the River*, 167.

100. Watenpaugh, "Creating Phantoms"; Shields, *Fezzes in the River*, 124–36, 182–83.

101. Çağaptay, *Islam, Secularism, and Nationalism in Modern Turkey*, 117–20; Shields, *Fezzes in the River*, 232–39.

102. Vejdani, *Making History in Iran*, 136.

103. Yāsamī, *Kurd va payvastagī-i nizhādī va tārīkhī-i ū*, 100.

104. Yāsamī, 106.

105. Yāsamī, 106.

106. On the 1919 Agreement and the popular unrest it provoked, see Katouzian, "The Campaign Against the Anglo-Iranian Agreement of 1919."

107. Vejdani, *Making History in Iran*, 121–37.

108. Zia-Ebrahimi, *The Emergence of Iranian Nationalism*, 157.

109. Upper-caste Hindu defenders of the Indian caste system similarly incorporated British Orientalist scholarship into their assessments of racial anthropology. Savary, *Evolution, Race and Public Spheres in India*, 93.

110. See Chapters 3 and 4 in Zia-Ebrahimi, *The Emergence of Iranian Nationalism*.

111. Fazeli, *Politics of Culture in Iran*, 53–54.

112. Baḥr al-'Ulūmī, *Kārnāmah-i Anjuman-i Āṣār-i Millī*.

113. Grigor, "Recultivating 'Good Taste,'" 19; Grigor, *Building Iran*, 131–34.

114. Krischner and Krischner, "The Anthropology of Mesopotamia and Persia C. The Anthropology of Persia," 410.

115. Field, *Contributions to the Anthropology of Iran*, 13.

116. Field, 9.

117. Field, 278.

118. Field, 280. Field's medical officer was Walter P. Kennedy, a physician at Baghdad's Royal College of Medicine.

119. Field, 32.

120. Field, 159.

121. Field, 333.
122. Field, 489.
123. Field, 434.
124. For example, Field's 1939 book on the anthropology of Iran was not translated into Persian before 1964, when it was published as Field, *Mardum'shināsī-i īrān, tarjumah-'i duktur-i 'Abdallāh Faryār*.
125. Fazeli, *Politics of Culture in Iran*, 54.
126. Fazeli, 58.
127. "William Haas, 72, Near East Expert, Professor at Columbia Dies," *New York Times*, January 4, 1956, p. 27.
128. See letters exchanged between the German foreign ministry and the German Embassy in Tehran regarding Wilhelm "Willy" Haas, 1934–1935, Files R64067 and R64068, German Reich Files, Political Archive of the German Foreign Ministry (Politisches Archiv des Auswärtigen Amts), Berlin. Hereafter "PAAA."
129. Haas, "Vaẓāyif-i mūzah-i insān-shināsī," 95–96. In the text, he glosses *insān-shināsī-i ijtimā'ī* as "ethnology" and *insān-shināsī-i jismī* as "anthropology."
130. Haqiqi, "'Ilm-i insān-shināsī," 170.
131. Fazeli, *Politics of Culture in Iran*, 55.
132. For proceedings of the conference, see the special issue of *Āmūzish va Parvarish* 8, no. 9 (1317 [1938]).
133. Khuri, *An Invitation to Laughter*, 9–10.
134. On the League's refusal to recognize racial equality in its covenant, see Shimazu, *Japan, Race, and Equality*.
135. Maksudyan, "The *Turkish Review of Anthropology* and the Racist Face of Turkish Nationalism"; Ergin, *Is the Turk a White Man?*; Robson, *States of Separation*, 8.
136. Kay Ustuvān, *Nizhād va zabān*, 2.
137. Kay Ustuvān, 19–21. He glosses *zānīch* as *masqaṭ al-ra's*.
138. In Iran, for example, homegrown publications in the interwar period took anti-British and anti-Russian editorial lines and even served as conduits for German propaganda without any direct support or supervision from the Nazi Party. See Zia-Ebrahimi, *The Emergence of Iranian Nationalism*, 158–59. Meanwhile, members of Lehi, a Zionist Jewish group engaged in terrorist activities against British rule in Mandate Palestine, attempted to negotiate an alliance with Nazi officials up until the end of 1941. See Sofer, *Zionism and the Foundations of Israeli Diplomacy*, 254.
139. See, for example, Gelvin, *The Modern Middle East*, 13; Milani, *The Shah*, 67–68; Abrahamian, *A History of Modern Iran*, 86–87; Katouzian, *The Persians*, 217–18.

140. Motadel, "Iran and the Aryan Myth," 133–34; Khatib-Shahidi, "German Foreign Policy Towards Iran," 254.

141. Mokhtari, *In the Lion's Shadow*. Similarly, European anthropometrists who had worked in the Middle East used their scientific credibility to rescue Jews; Cornelius Kappers used cephalic index measurements to assign Jewish individuals to other European race categories and render them exempt from Nazi racial policies. Zeidman and Cohen, "Walking a Fine Scientific Line."

142. Mehrangiz Dowlatshahi, interviewed by Shirin Sami'i in Paris, May 22, 1984. Foundation for Iranian Studies Oral History Program, transcript pp. 63–64. Taher Ziya'i, interviewed by Mahnaz Afkhami in Washington, D.C., May 12, 1988. Foundation for Iranian Studies Oral History Program, transcript p. 5. Amir Aslan Afshar, interviewed by Mahnaz Afkhami in Nice, September 10, 1988. Foundation for Iranian Studies Oral History Program, transcript p. 1.

143. Foreign Office, Berlin, to all Reich Ministries including the Reich Chancellery and Presidential Chancellery, Nr. 82–35, B 8/4, April 30, 1936. File R 43-II/1498b, Bd. 2 (1936–1943). German Federal Archives (Bundesarchiv), Berlin.

144. "Les turcs promus 'aryens'," *Le Temps*, June 14, 1936.

145. Meeting minutes of July 1, 1936, Foreign Office meeting "on the interpretation of the term 'kindred' [*artverwandt*]." File R99174, Bd. 2 (1936–1940), PAAA.

CHAPTER 2

1. Hirszfeld, *Ludwik Hirszfeld*, 47.

2. Hirschfeld and Hirschfeld, "Serological Differences Between the Blood of Different Races," 677.

3. Hirszfeld, *Ludwik Hirszfeld*, 58.

4. Hirschfeld and Hirschfeld, "Serological Differences Between the Blood of Different Races," 677.

5. Mukharji, "From Serosocial to Sanguinary Identities," 152.

6. Hirschfeld and Hirschfeld, "Serological Differences Between the Blood of Different Races," 677.

7. For example, see the contemporary critique of "lumping vs. splitting" the Indian population category, discussed in Mukharji, "From Serosocial to Sanguinary Identities," 153–57.

8. Gannett and Griesemer, "The ABO Blood Groups," 141.

9. Robson, *States of Separation*.

10. Mazower, *Salonica, City of Ghosts*, 375–88; Naar, "The 'Mother of Israel' or the 'Sephardi Metropolis'?," 106.

11. Schneider, "The History of Research on Blood Group Genetics."

12. Gannett and Griesemer, "The ABO Blood Groups," 131.

13. Marks, "The Legacy of Serological Studies in American Physical Anthropology," 347.

14. Ottenberg, "A Classification of Human Races Based on Geographic Distribution of the Blood Groups."

15. Snyder, "Human Blood Groups: Their Inheritance and Racial Significance," 244–48.

16. Snyder, "The 'Laws' of Serologic Race-Classification Studies in Human Inheritance IV."

17. Shousha, "On the Biochemical Race-Index of the Egyptians," 10. The author's name is commonly written "Aly Tewfik Shousha" in scientific publications and documents related to his work with the World Health Organization, of which he was a founding member. Shusha had trained in medicine and bacteriology at Berlin and Zurich, respectively. Throughout his career in Egypt, he remained a prominent international figure in immunology and public health, and in the 1940s was awarded a British knighthood. World Health Organization, Regional Office for the Eastern Mediterranean, "Biographical note: Sir Aly Tewfik Shousha, Pasha."

18. Altounyan, "Blood Group Percentages for Arabs, Armenians and Jews," 546. Several months earlier, Altounyan's data was also published in the *Lancet* amidst a more general article on blood transfusion at the hospital, with a slightly different explanation of the categories: "In [Syria] the Jews and Armenians represent almost pure racial groups, with little admixture of foreign elements. Under the heading of Arab are included some Turks, Kurds, and a few Greeks, as well as the nomad and town-dwelling Arabs." Altounyan, "A Note on Blood Transfusion in Syria, with an Analysis of 1149 Blood Groupings," 1342.

19. Altounyan, *In Aleppo Once*, 4, 25–26, 33, 67, 97.

20. "Altounyan, Ernest Haik Riddall (1890–1962)."

21. Altounyan, *Chimes from a Wooden Bell*, 66–67, 119.

22. Schneider, "The History of Research on Blood Group Genetics," 300.

23. Mackintosh, "Blood Group Percentages," 732.

24. See Mackintosh, "Evolution of Racial Types of Europe: Its Bearing on the Racial Factor in Disease"; Mackintosh, "The Migratory Factor in Eugenics."

25. See Gannett and Griesemer, "The ABO Blood Groups," 155.
26. Shanklin, "Blood Grouping of Rwala Bedouin," 432.
27. Kennedy and MacFarlane, "Blood Groups in Iraq."
28. "Arabs and American Indians—Same Race: Beirut University Scientist's Discovery."
29. Parr, "Blood Studies on Peoples of Western Asia and North Africa."
30. On the Copts' claims to represent the ancient Egyptians, see Reid, *Whose Pharaohs?*, 281–83; on the Samaritans' claims to represent the ancient Israelites, see Chapter 1.
31. Parr, "Blood Studies on Peoples of Western Asia and North Africa," 20.
32. Parr, 22.
33. Parr, 24.
34. Parr, 22–23.
35. Parr, 25.
36. Kappers, *An Introduction to the Anthropology of the Near East*, 193.
37. Kappers, 27.
38. Kappers, 195.
39. Kossovitch, "Recherches anthropométriques et sérologiques (groupes sanguins) chez les Israélites du Maroc." Kossovitch, a Russian-trained physician, served in the Serbian army during the First World War, stationed in Salonika with the Hirszfelds. In 1924, he secured a position in the laboratory of René Dujarric de la Rivière at the Pasteur Institute, in Paris. Taking advantage of the Institute's network of branches throughout the French colonies, Kossovitch compiled blood group data from around the world for anthropological comparison, leading to a professorship at the School of Anthropology in Paris. He traveled to Morocco in 1931 to conduct research on various populations across the country, including the diverse communities of Jews. See Schneider, "La Recherche sur les Groupes Sanguins avant la Deuxième Guerre Mondiale."
40. Younovitch, "Contribution à l'étude sérologique des Juifs de Yémen," 929.
41. Younovitch, 930–31.
42. Younovitch, "Les caractères sérologiques des juifs asiatiques," 1101.
43. Younovitch, 1102.
44. Younovitch, 1103.
45. Ottenberg, "A Classification of Human Races Based on Geographic Distribution of the Blood Groups"; Snyder, "Human Blood Groups: Their Inheritance and Racial Significance."

46. Shohat, "Sephardim in Israel," 13–15.
47. Sharim, "The Arch of a Sephardic-Mizrahi Ethnic Autonomy in Palestine, 1926 to 1929."
48. Falk, *Zionism and the Biology of the Jews*, 131–32.
49. See Hirsch, "'We Are Here to Bring the West, Not Only to Ourselves'"; Hirsch, "Zionist Eugenics, Mixed Marriage, and the Creation of a 'New Jewish Type.'"
50. Gini, "I Samaritani," 126.
51. Younovitch, "Étude sérologique des juifs samaritains," 971.
52. Younovitch, 970–71.
53. [Kansu], "Kan grupları hakkında," 261.
54. Onur, "Les groupes sanguins chez les Turcs," 521.
55. That is, "reine Türken und nicht Mischlinge oder Angehörige anderer Völker." Babacan, "Über die Blutgruppen bei Türken," 1.
56. He reported his results in Turkish at a meeting of the Turkish Microbiology Society on December 24, 1936, and in the medical journal *Pratik Doktor*. Onur, "Türklerde kan grupları."
57. Onur, "Les groupes sanguins chez les Turcs," 522.
58. Onur, "Türklerde kan grupları," 6.
59. Öztuncay and Ertem, *Ottoman Arcadia*.
60. Ergin, "Biometrics and Anthropometrics," 285.
61. Irmak, "Anadolu yürüklerinin kan gruplarına dair araştırmalar," 3645. The paper was also published in German several months later: Irmak, "Die ersten Blutgruppenuntersuchungen bei Nomaden in Kleinasien."
62. Irmak, "Anadolu yürüklerinin kan gruplarına dair araştırmalar," 3648.
63. Irmak, "Türk ırkının biyolojisine dair araştırmalar (kan gruplar ve parmak izleri)," 844.
64. Onur, "Kan grupları bakımından Türk ırkının menşei hakkında bir etüd," 850.
65. "Bu neticeler, tarih buluşları ile de karşılaştırılacak olursa, pek haksız olmıyarak, bizi şu kanaate sevk eder: *Türk ırki, A tipini Avrupaya getiren ana köktür.*" Onur, 850–51. Emphasis original.
66. "Lakin, medeni alemin yapısında en eski fakat en üstad mimar olduğuna inandığımız Türkün bu hak ve şerefini *kendi asil kanında*, aramak, bulmak ve dünyaya tanıtmak, yine Türk oğlunun en başlı ve en mukaddes borcudur." Onur, *İnsan ve hayvanlarda kan grupları*, vi. Emphasis original.
67. Wyman and Boyd, "Human Blood Groups and Anthropology," 197–98.

68. Letters from Lyle Boyd to her parents, December 24, 1935, and January 3, 1936. Papers of Lyle Gifford Boyd, 1921–1982, Schlesinger Library, Harvard University. Hereafter "Lyle Boyd papers."

69. Boyd and Boyd, "New Data on Blood Groups and Other Inherited Factors in Europe and Egypt," 61–62.

70. On the Village Welfare Service and its programs for rural development, see Dodd, "The Village Welfare Service in Lebanon, Syria and Palestine."

71. Lyle Boyd to her parents, September 4, 1937; typescript accounts of "Rualla expeditions," August 1937 and October 1937. Lyle Boyd papers.

72. Lyle Boyd to her parents, November 25, 1937. Lyle Boyd papers.

73. Boyd and Boyd, "The Blood Groups of the Rwala Bedouin," 445.

74. Boyd and Boyd, "Blood Groups and Inbreeding in Syria," 325.

75. Kayssi, Boyd, and Boyd, "Blood Groups of the Bedouin near Baghdad," 298.

76. Boyd and Boyd, "Blood Groups and Types in Baghdad and Vicinity," 399.

77. Lyle Boyd to her parents, November 7 and November 15, 1937. Lyle Boyd papers.

78. Previous compilations include Steffan, *Handbuch der Blutgruppenkunde*; Dujarric de la Rivière and Kossovitch, *Les groupes sanguins*.

79. Boyd, "Blood Groups."

80. See letters from Haldane to Boyd, June 26, 1939 (HALDANE/5/1/2/2/5); Boyd to Haldane, August 14, 1939 (HALDANE/5/1/2/2/10); Haldane to Boyd, August 31, 1939 (HALDANE/5/1/2/2/11). All housed in Special Collections, University College London. Hereafter "UCL."

81. Haldane to Boyd, September 1, 1939. HALDANE/5/1/2/2/12, UCL.

82. Boyd to Haldane, March 28, 1940. HALDANE/5/1/2/2/19, UCL.

83. Boyd, "Critique of Methods of Classifying Mankind." Boyd extensively cited Haldane's manuscript, expecting it to appear before his own paper, but ultimately, Boyd's work made it to print in advance of Haldane. Boyd to Haldane, April 19, 1940. HALDANE/5/1/2/2/21, UCL.

84. Letters from Boyd to Haldane, January 13, 1941 (HALDANE/5/1/2/2/26); March 19, 1941 (HALDANE/5/1/2/2/28); and November 23, 1941 (HALDANE/5/1/2/2/30, UCL).

85. Haldane, "The Blood-Group Frequencies of European Peoples, and Racial Origins," 464.

86. Haldane, 463.

87. Boyd to Haldane, August 14, 1939. HALDANE/5/1/2/2/10, UCL.

88. See Boyd's note in Haldane, "The Blood-Group Frequencies of European Peoples, and Racial Origins," 478.

89. Haldane, 461–62.

90. Haldane, 477.

91. Marks, "The Legacy of Serological Studies in American Physical Anthropology," 347.

92. Ammar, "Physical Measurements and Serology of the People of Sharqiya (Egypt)," 148–51.

93. El Shakry, *The Great Social Laboratory*, 69–83.

94. See Braun and Öktem, "Türklerde kan grupları tevziatı"; Derman; Özek, "Kan grupları üzerinde araştırmalar."

95. Aygen, "Türklerin Kan Grupları ve Kan Gruplarının Antropolojik Karakterlerle İlgisi Üzerine Bir Araştırma," 26.

96. Aygen, *Türklerin kan grupları ve kan gruplarının antropolojik karakterlerle ilgisi üzerine bir araştırma*; Myres, "Shorter Notes: Anthropological Studies in Turkey."

97. See Barkan, *Retreat of Scientific Racism*.

98. See Barkan, "Mobilizing Scientists Against Nazi Racism, 1933–1939."

99. "The Problem of Race."

100. Reardon, *Race to the Finish*, 23–31.

CHAPTER 3

Several ideas and paragraphs found in Chapters 3, 5, and 6 were first published in an article called "'Essential Collaborators': Locating Middle Eastern Geneticists in the Global Scientific Infrastructure, 1950s–1970s," in *Comparative Studies of Society and History* 60, no. 1 (2018), 119–49.

1. Botting, *Island of the Dragon's Blood*, 224.

2. Botting, 226.

3. Lord Rennell et al., "The Oxford University Expedition to Socotra: Discussion," 208.

4. Lister et al., "The Blood Groups and Haemoglobin of the Bedouin of Socotra," 85.

5. Mourant, *The Distribution of the Human Blood Groups*, 76.

6. On the other hand, Jews and the "culturally Turkish" were singled out as invalid racial groupings: "National, religious, geographical, linguistic and cultural groups do not necessarily coincide with racial groups. . . . Muslims and Jews are no

more races than are Roman Catholics and Protestants; nor are people who live in Iceland or Britain or India, or who speak English or any other language, or who are culturally Turkish or Chinese and the like, thereby describable as races." *The Race Concept: Results of an Inquiry*, 12.

7. Boyd, *Genetics and the Races of Man*; Boyd, "Genetics and the Human Race."
8. On "centers of calculation," see Latour, *Science in Action*.
9. Bangham, "Blood Groups and Human Groups," 6.
10. Radin, "Latent Life," 6.
11. Wiener, Belkin, and Sonn, "Distribution of the A1-A2-B-O, M-N, and Rh Blood Factors Among Negroes in New York City," 193–94.
12. Hemoglobin is the protein within red blood cells (erythrocytes) that binds to and transports oxygen molecules throughout the body. Haptoglobin is the protein that binds to free-floating hemoglobin in blood plasma that has been released from dying erythrocytes, enabling it to be safely removed from the bloodstream. Transferrin is the protein that binds to and transports iron ions through blood plasma. Immunoglobulins, also commonly called antibodies, are immune system proteins that identify and mark pathogens (such as viruses and bacteria) for destruction.
13. Radin, "Unfolding Epidemiological Stories," 70.
14. Radin, 68; Radin, "Latent Life," 16.
15. Mourant, *Blood and Stones*, 60–62.
16. See Chapter 1 in Redhead, "Histories of Sickle Cell Anaemia in Postcolonial Britain, 1948–1997."
17. Mourant, *Blood and Stones*, 64.
18. Bangham, "Blood Groups and Human Groups," 74.
19. See Anderson, "Objectivity and Its Discontents," 565–67, as well as Chapter 7 in Bangham, *Blood Relations*.
20. See Hilgartner, *Reordering Life*, 84–85, 87–89.
21. Burton, "'Essential Collaborators.'"
22. Kayssi, "The Rh Blood Groups of the Population of Baghdad."
23. Donegani et al., "The Blood Groups of the People of Egypt," 379.
24. Donegani et al., 379.
25. Donegani et al., 382.
26. Mourant, "The Blood Groups of the Peoples of the Mediterranean Area," 225–26.
27. Abdoosh and El-Dewi, "The Blood Groups of Egyptians."

28. El-Dewi, "The Rh Groups Among Egyptians."

29. Abdoosh and El-Dewi, "The Blood Groups of Egyptians," 719.

30. Abdoosh and El-Dewi, 717.

31. Gohar, "A Preliminary Note on the Distribution of ABO Blood Groups and Incidence of Rh Negatives in Egyptians"; El-Dewi, "The Rh Groups Among Egyptians."

32. Mourant, *The Distribution of the Human Blood Groups*, 76.

33. Mourant to Gelpi, November 1, 1971. PP/AEM/K.41, Box 30, Mourant Papers. Compare this to his attitude toward Israeli genetic anthropologist Batsheva Bonné-Tamir during the same time period; see Chapter 6.

34. Interviews with Richard Daggy and Armand P. Gelpi, oral histories conducted in 1996 by Carole Hicke in "Health and Disease in Saudi Arabia: The Aramco Experience, 1940s–1990s," Regional Oral History Office, Bancroft Library, University of California, Berkeley, 1998.

35. Jones, *Desert Kingdom*, 102–10.

36. Lord Rennell et al., "The Oxford University Expedition to Socotra: Discussion," 208.

37. Maranjian et al., "The Blood Groups and Haemoglobins of the Saudi Arabians," 394.

38. Ikin, "The Incidence of the Blood Antigens in Different Populations," 74.

39. Ikin, 75.

40. Maranjian et al., "The Blood Groups and Haemoglobins of the Saudi Arabians," 400.

41. Maranjian et al., 399; Lehmann, Maranjian, and Mourant, "Distribution of Sickle-Cell Hæmoglobin in Saudi Arabia," 493.

42. "Biographical particulars," document concerning Gurevitch's appointment to a clinical bacteriology lectureship at the Hadassah-Hebrew University Medical School, 1947. Joseph Gurevitch file, Central Archives, Hebrew University of Jerusalem; Yerahmi'el Roz'anski, "Prof. Yosef Gurevits' z.l., 17.11.1898-20.10.1960," *Harefuah* 62, no. 7 (1962): 280. See also Kirsh, "Genetic Studies of Ethnic Communities in Israel," 182.

43. DellaPergola, "Demography in Israel at the Dawn of the Twenty-First Century," 24. For detailed statistics by country of origin, see also Yahil, "Israel's Immigration Policy," 446.

44. Yahil, "Israel's Immigration Policy," 447.

45. See, for example, Shohat, "Sephardim in Israel," 18–20; Chapter 6 in Segev, *1949*.

46. Koenig, "East Meets West in Israel," 169–70.

47. Eisenstadt, "The Oriental Jews in Israel," 200.

48. The most notorious example is an oft-cited 1949 editorial by a *Haaretz* reporter, Aryeh Gelblum, which railed against immigrants from North Africa: "This is the immigration of a race (*geza'*) unlike any we have known before. [...] The primitiveness of these people is unsurpassable..." Gelblum, *Haaretz*, April 22, 1949. Reproduced in the original Hebrew in Segev, *1949: ha-Yiśre'elim ha-rishonim*, 158–59. An English translation of the full article is also available in Segev, *1949*, 159–60.

49. See Picard, "Yaḥaso shel 'iton ha-Arets la-'aliyatam shel yehude tsefon afriḳah."

50. Gurevitch, Brzezinski, and Polishuk, "Rh Factor and Haemolytic Disease of the Newborn in Jerusalem Jews," 944.

51. Gurevitch, Hermoni, and Polishuk, "Rh Blood Types in Jerusalem Jews."

52. Kirsh, "Population Genetics in Israel in the 1950s," 643; Kirsh, "Genetic Studies of Ethnic Communities in Israel," 186; Abu El-Haj, *The Genealogical Science*, 87–99.

53. Stark and Henner, "Bediḳot hemaṭologyot etsel torme dam ba-tel-aviv ye-ḥaluḳat ha-tormim li-ḳevutsot dam A, B, O"; Zilberstein and Goldstein, "Suge ha-dam etsel nashim yehudiyot mi-'edot ha-mizraḥ."

54. Brzezinski et al., "Blood Groups in Jews from the Yemen," 335. For the poor conditions prevailing in Rosh Ha'ayin and other *ma'abarot* until 1951, see Chapters 4–7 in Hakohen, *Immigrants in Turmoil*.

55. Gurevitch et al., "Blood Groups in Jews from Tripolitania," 260. See also Gurevitch et al., "Blood Groups in Jews from Cochin, India"; Gurevitch, Hasson, and Margolis, "Blood Groups in Persian Jews"; Margolis, Gurevitch, and Hasson, "Blood Groups in Jews from Morocco and Tunisia."

56. Mourant, *The Distribution of the Human Blood Groups*, 73–74; Gurevitch, Hermoni, and Margolis, "Blood Groups in Kurdistani Jews"; Gurevitch and Margolis, "Blood Groups in Jews from Iraq."

57. See, e.g., Lipphardt, "The Jewish Community of Rome."

58. Kirsh, "Population Genetics in Israel in the 1950s," 637, 642, 649; Abu El-Haj, *The Genealogical Science*, 276n27.

59. Margolis, Gurevitch, and Hermoni, "Blood Groups in Ashkenazi Jews," 201;

Gurevitch, Margolis, and Hermoni, "Blood Groups in Sephardic Jews"; Margolis, Gurevitch, and Hermoni, "Blood Groups in Sephardic Jews."

60. Gurevitch, Margolis, and Hermoni, "Blood Groups in Sephardic Jews," 276; Margolis, Gurevitch, and Hermoni, "Blood Groups in Sephardic Jews," 198.

61. Mourant, "The Blood Groups of the Jews," 164–65.

62. Mourant, 169.

63. Mourant, *The Distribution of the Human Blood Groups*, 102.

64. Bidar, "The MN Blood Groups of the Population of Azarbaijan, Iran"; Afkari, "Groupes sanguins ABO et Rh (D) dans la province du Khorassan."

65. Ansari, *Modern Iran*, 107–19; Oberling, *The Qashqā'i Nomads of Fārs*; Shaw, "'Strong, United and Independent'"; Elling, "War of Clubs."

66. Podliachouk et al., "Serum Factor Gma Among the Iranians," 718.

67. Podliachouk et al., "Les facteurs sériques Gm(a), Gm(b), Gm(x) et Gm-like chez les Iraniens," 498.

68. See Boué and Boué, "Étude sur la répartition des groupes sanguins en Iran"; Boué and Boué, "Étude sur la répartition des groupes sanguins en Iran II." I analyze the Boués' work in detail in the chapter of a forthcoming book: "'They Say They Are Kurds': Informants and Identity Work at the Iranian Pasteur Institute," in *Invisible Labour: Power and Politics in Scientific Knowledge Production* (eds. Jenny Bangham, Xan Chacko, and Judith Kaplan), Rowman & Littlefield International.

69. Beckett, "ABO Blood Groups in Kerman, South Persia," 141.

70. See Chapter 1 of Elling, *Minorities in Iran*, 15–44.

71. Elling, 27.

72. Nijenhuis, "Blood Group Frequencies in Iran," 726.

73. Nijenhuis, 723.

74. Nijenhuis, 735.

75. Nijenhuis, 735.

76. See the discussion of Anthony C. Allison, Mourant, and the Oxford University Exploration Club in Chapter 7 of Bangham, *Blood Relations*.

77. Ikin, Mourant, and Lehmann, "The Blood Groups and Haemoglobin of the Assyrians of Iraq"; Ikin, "The Incidence of the Blood Antigens in Different Populations."

78. Ikin, Mourant, and Lehmann, "The Blood Groups and Haemoglobin of the Assyrians of Iraq," 110.

79. Ikin, Mourant, and Lehmann, "The Blood Groups and Haemoglobin of the Assyrians of Iraq," 111.

80. See Robson, *States of Separation*.

81. Mourant, *The Distribution of the Human Blood Groups and Other Polymorphisms*.

CHAPTER 4

A version of this chapter appeared as "Red Crescents: Race, Genetics, and Sickle Cell Disease in the Middle East," in *Isis* 110, no. 2 (2019), 250–69.

1. Madmoni-Gerber, *Israeli Media and the Framing of Internal Conflict*.

2. Hava Ulman, interview with Dr. George Mendel, February 1, 1993, "Sub-Topic: Doctors and nurses in transit camps" [in Hebrew], Hebrew University Institute for Contemporary Jewry Oral Documentation Center and Yad Yitzhak Ben-Zvi Institute for the Study of the Land of Israel and Its Settlements. Transcript copied for "Cohen-Kedmi Commission on the Yemenite Children – Witness Invited by the Commission – Prof. Mendel George – 241/96," pp. 31–32. Accessed March 10, 2020. Available http://www.archives.gov.il/archives/#/Archive/0b07170684ee7d96/File/0b07170680ae905b.

3. Newman, "When Israeli Doctors Allegedly Tested Yemenites for 'Negro Blood'"; Altman, "Ha-Nisuyim Be-Yalde Teman Neḥṣafim: 'Mukraḥ Li-Heyot Lahem Dam Kushi.'"

4. Pauling et al., "Sickle Cell Anemia, a Molecular Disease."

5. Strasser, "Linus Pauling's 'Molecular Diseases'"; Wailoo and Pemberton, *The Troubled Dream of Genetic Medicine*, 122–24.

6. Tapper, *In the Blood*; Wailoo, *Dying in the City of the Blues*; Wailoo and Pemberton, *The Troubled Dream of Genetic Medicine*. A key theme of this literature is racial health disparities and the role of sickle cell disease in catalyzing African American health-care activism; see also Nelson, *Body and Soul*.

7. Herrick, "Peculiar Elongated and Sickle-Shaped Red Blood Corpuscles in a Case of Severe Anemia. 1910."

8. Tapper, *In the Blood*, 58–61; Fullwiley, *The Encultured Gene*, 161–63.

9. Dacie, "Hermann Lehmann."

10. Lehmann and Raper, "Distribution of the Sickle-Cell Trait in Uganda, and Its Ethnological Significance."

11. Choremis et al., "Sickle-Cell Anaemia in Greece."

12. Lehmann, "Sickle-Cell Anaemia in Greece."

13. Dreyfuss and Benyesch, "Sickle-Cell Trait in Yemenite Jews."

14. For a critical account of Coon's North African fieldwork, see Anderson, "The Anomalous Blonds of the Maghreb."

15. Dreyfuss, Mundel, and Benyesch, "Sickling in Oriental Jews," 197.
16. Lehmann, "Distribution of the Sickle Cell Gene," 106.
17. Lehmann and Cutbush, "Sickle-Cell Trait in Southern India," 404.
18. Lehmann, "The Sickle-Cell Trait: Not an Essentially Negroid Feature."
19. Dreyfuss et al., "An Investigation of Blood-Groups and a Search for Sickle-Cell Trait in Yemenite Jews."
20. Lehmann, "Distribution of the Sickle Cell Gene." Neither Lehmann nor Ikin describes their sampling procedures in their publications, but it is most likely they tested blood collected from hospital patients and staff. According to Adel Aulaqi, who worked as a clerk in Aden's Keith-Falconer Mission Hospital during his adolescence in the 1950s, the Civil Hospital as well as the Mission Hospital treated patients of all social strata. Although the senior doctors at the Civil Hospital were predominantly British and Indian, English-speaking Adeni Arabs worked as technical assistants, and *akhdām* were employed to handle cleaning and waste removal. Interview with Dr. Adel Aulaqi by the author in London, May 9, 2018; Aulaqi, *From Barefoot Doctors to Professors of Medicine*.
21. Allison, "The Distribution of the Sickle-Cell Trait in East Africa and Elsewhere."
22. Playfair, *A History of Arabia Felix or Yemen*, 15–16; Arnaud and Vayssière, "Les Akhdam de l'Yémen," 379; de Maltzan, "Notes de Voyage Sur Les Régions Du Sud de l'Arabie," 145–46.
23. Walters, "Perceptions of Social Inequality in the Yemen Arab Republic," 94; Meneley, *Tournaments of Value*, 14. In recent decades, in concert with a rise in collective political activism, members of the community have adopted the term *al-muhamashīn*, "the marginalized"; however, for historical consistency I use the term "*akhdām*" throughout this chapter.
24. Hunter, *An Account of the British Settlement of Aden in Arabia*, 32; Hamilton, "The Social Organization of the Tribes of the Aden Protectorate," 150. Interestingly, francophone writers were less inclined to accept this narrative, with some even claiming that the *akhdām* were descendants of the ancient Himyarites: Arnaud and Vayssière, "Les Akhdam de l'Yémen," 380–82; de Maltzan, "Notes de Voyage Sur Les Régions Du Sud de l'Arabie," 146.
25. Coon, *Measuring Ethiopia and Flight into Arabia*, 301–4.
26. Lehmann, "Distribution of the Sickle Cell Gene," 113.
27. Chadarevian, "Following Molecules," 182; Silverman, "The Blood Group 'Fad' in Post-War Racial Anthropology."

28. Lehmann, "Distribution of the Sickle Cell Gene," 113.

29. Lehmann, "Distribution of the Sickle Cell Gene," 120–21.

30. Aksoy, "Brief Note: Hemoglobin S in Eti-Turks and the Allewits in Lebanon," 657.

31. "Muzaffer Aksoy sözlü tarih [oral history]." Undated interview of Muzaffer Aksoy in Istanbul, probably early 1990s. DVD held by the National Library, Ankara, Turkey, call number FL CD 2010 DK 201. Hereafter "Aksoy oral history."

32. Winter, *A History of the 'Alawis*, 265.

33. Winter, 266; Akyol, "Memory and Strategies of Identity Formation in Hatay"; Shields, *Fezzes in the River*, 179; Procházka-Eisl, *The Plain of Saints and Prophets*, 67, 74–75.

34. Procházka-Eisl, *The Plain of Saints and Prophets*, 23, 70–72.

35. Aksoy, "Sickle-Cell Trait in South Turkey," 589.

36. Lehmann, "Sickle-Cell Trait in South Turkey."

37. "iki ünlü İngiliz bilim adamı, Doktor Hermann Lehmann ve Doktor Allison, bu konuda benimle çalışmak istedikleri belirdeler. Ben Doktor Lehmann'a yeğledim." Aksoy oral history.

38. Aksoy, "Sickle-Cell Anemia in South Turkey," 460.

39. Aksoy et al., "Hemoglobin E in Asia," 57P; Aksoy and Lehmann, "The First Observation of Sickle-Cell Haemoglobin E Disease," 1248; Aksoy and Lehmann, "Sickle-Cell-Thalassaemia Disease in South Turkey," 734.

40. Jonxis and Delafresnaye, *Abnormal Haemoglobins*, ix.

41. Lehmann, "Distribution of Variations in Human Haemoglobin," 206–8.

42. Aksoy, "Abnormal Haemoglobins in Turkey," 216. Emphasis added.

43. Aksoy et al., "Blood Groups, Haemoglobins, and Thalassemia in Turks in Southern Turkey and Eti-Turks," 938.

44. Ikin, "Blood Group Distribution in the Near East."

45. Based on Aksoy's acknowledgments, he was assisted in Tripoli by Dr. Hanna Ghantus, an AUB alum, with the coordination facilitated by Musa Khalil Ghantus, the associate dean of AUB.

46. Aksoy, "Brief Note: Hemoglobin S in Eti-Turks and the Allewits in Lebanon," 657.

47. Aksoy, 658.

48. James V. Neel, referee comments for M. Aksoy's "Hemoglobin S in Eti-Turks and the Allewits in Lebanon," November 11, 1960. Neel also disagreed with Aksoy's initial rejection of the African admixture hypothesis, insisting that it had been neither proved nor disproved by Aksoy's data. William Dameshek file, Folder 3, Box 17,

James V. Neel papers, American Philosophical Society, Philadelphia. Hereafter "Neel papers, APS." The second reviewer, Amoz I. Chernoff, was even harsher, writing that Aksoy's data was statistically insignificant and the entire discussion section was "speculative, unsupported by critical data and essentially meaningless." Chernoff, referee comments for M. Aksoy's "Hemoglobin S in Eti-Turks and the Allewits in Lebanon," November 22, 1960. Dameshek file, Folder 3, Box 17, Neel papers, APS.

49. Duruel, Altay, and Ulutin, *Bilime adanmış bir ömür*, 38.

50. See the list of Aksoy's lifetime publications, ibid., 151–167.

51. See Chapters 7 and 8 in Pelt, *Military Intervention and a Crisis of Democracy in Turkey*.

52. Allison, "The Sickle-Cell Trait in the Mediterranean Area," 24.

53. Shahid and Abu Haydar, "Sickle Cell Disease in Syria and Lebanon," 268.

54. Dabbous and Firzli, "Sickle Cell Anemia in Lebanon—Its Predominance in the Mohammedans," 227.

55. Dabbous and Firzli, 228.

56. Bakir and Al-Qaysi, "Sickle Cell Anaemia in Iraq: First Case Report."

57. Kamel and Awny, "Origin of the Sickling Gene"; Ibrahim et al., "Hereditary Blood Factors and Anthropometry of the Inhabitants of the Egyptian Siwa Oasis"; Selim et al., "Genetic Markers and Anthropometry in the Populations of the Egyptian Oases of El-Kharga and El-Dakhla"; Kamel, "Heterogeneity of Sickle Cell Anaemia in Arabs"; Kamel and Moafy, "Some Aspects of Thalassemia and the World Common Hemoglobinopathies in the Middle East."

58. Lehmann, Maranjian, and Mourant, "Distribution of Sickle-Cell Hæmoglobin in Saudi Arabia," 493.

59. Interview with Armand P. Gelpi, an oral history conducted in 1996 by Carole Hicke in "Health and Disease in Saudi Arabia: The Aramco Experience, 1940s–1990s," Regional Oral History Office, Bancroft Library, University of California, Berkeley, 1998. Hereafter "Gelpi oral history."

60. Gelpi, "Glucose-6-Phosphate Dehydrogenase Deficiency, the Sickling Trait, and Malaria in Saudi Arab Children"; Gelpi, "Sickle Cell Disease in Saudi Arabs."

61. See Chapter 4 in Nelson, *Body and Soul*.

62. Gelpi, "Migrant Populations and the Diffusion of the Sickle-Cell Gene," 263.

63. Kulozik et al., "Geographical Survey of Bs-Globin Gene Haplotypes"; Allison, "The Discovery of Resistance to Malaria of Sickle-Cell Heterozygotes."

64. Gelpi oral history. In contrast, the anthropologist Frank Livingstone argued that the haplotype data could support an ultimate single origin of HbS in Arabia,

which subsequently mutated at different times and places in Africa. Livingstone, "Who Gave Whom Hemoglobin S."

65. Gelpi, "Migrant Populations and the Diffusion of the Sickle-Cell Gene," 261.

66. Gelpi, "Sickle Cell Disease in Saudi Arabs," 95.

67. Hardy and Ragbeer, "Homozygous HbE and HbSE Disease in a Saudi Family," 50–51.

68. Aluoch et al., "Sickle Cell Anaemia Among Eti-Turks," 52–53.

69. Lehmann, Maranjian, and Mourant, "Distribution of Sickle-Cell Hæmoglobin in Saudi Arabia."

70. Ikin, "The Incidence of the Blood Antigens in Different Populations," 78–79.

71. Aksoy and Tanrıkulu, "The Hemoglobin E Syndromes. I. Hemoglobin E in Eti-Turks."

72. Aksoy and Erdem, "Abnormal Hemoglobins and Thalassemia in Eti-Turks Living in Antakya."

73. Lehmann, "Sickle Cell Anemia 35 Years Ago," 74.

74. Lehmann and Huntsman, *Man's Haemoglobins*, 306; Jackson, "'In Ways Unacademical'"; Sachs Collopy, "Race Relationships."

75. "Sonradan bu isim Batı ülkelerinin 'orak hücre ile ilgili yayınlarında' aynen kullanıldı." Duruel, Altay, and Ulutin, eds., *Bilime adanmış bir ömür*, 38.

76. Bier, *Mapping Israel, Mapping Palestine*, 129.

CHAPTER 5

1. Lederer, "A New Form of Acute Haemolytic Anaemia." Biographical information on Lederer supplemented by: Maria Hull, letter to the editor, *AJR Journal* 11, no. 8 (August 2011).

2. Nkhoma et al., "The Global Prevalence of Glucose-6-Phosphate Dehydrogenase Deficiency."

3. For a history of these projects and their ethical problems, see Comfort, "The Prisoner as Model Organism."

4. Carson et al., "Enzymatic Deficiency in Primaquine-Sensitive Erythrocytes."

5. Motulsky, "Metabolic Polymorphisms and the Role of Infectious Diseases in Human Evolution."

6. Sheba, "Jewish Migration in Its Historical Perspective," 1333–34.

7. Sheba, 1334.

8. See, for example, Rejwan, *The Last Jews in Baghdad*; Madmoni-Gerber, *Israeli Media and the Framing of Internal Conflict*, 9, 100; Shohat, "The Invention of the Mizrahim."

9. Packard, *The Making of a Tropical Disease*; Bashkin, *Impossible Exodus*, 28–30.

10. Davidovich and Shvarts, "Health and Hegemony"; Davidovitch and Margalit, "Public Health, Racial Tensions, and Body Politic"; Seidelman, Troen, and Shvarts, "'Healing' the Bodies and Souls of Immigrant Children."

11. Schieber, "Target-Cell Anaemia: Two Cases in Bucharan Jews."

12. Szeinberg, Sheba, and Adam, "Selective Occurrence of Glutathione Instability in Red Blood Corpuscles of the Various Jewish Tribes," 1045; Szeinberg et al., "Studies on Erythrocytes in Cases with Past History of Favism and Drug-Induced Acute Hemolytic Anemia"; Szeinberg, Asher, and Sheba, "Studies on Glutathione Stability in Erythrocytes of Cases with Past History of Favism or Sulfa-Drug-Induced Hemolysis"; Szeinberg, Sheba, and Adam, "Enzymatic Abnormality in Erythrocytes of a Population Sensitive to Vicia Faba or Haemolytic Anaemia Induced by Drugs."

13. Szeinberg and Sheba, "Hemolytic Trait in Oriental Jews Connected with an Hereditary Enzymatic Abnormality of Erythrocytes," 166.

14. Marks and Gross, "Erythrocyte Glucose-6-Phosphate Dehydrogenase Deficiency."

15. "Excerpt from Daily News Bulletin: Szold Award made to two Tel Aviv medical researchers," May 4, 1959. NY AR195564/4/33/17/385, "Israel: Medical, Tel Hashomer Hospital" Files, Records of the New York Headquarters of the American Jewish Joint Distribution Committee, 1955–1964, JDC Archives. Hereafter "JDC Archives."

16. See Bondy, *Shiba*, 153–55; Shvarts et al., "Medical Selection and the Debate over Mass Immigration in the New State of Israel (1948–1951)," 24–26.

17. Sheba, "Nisayon le-shiḥzur nedidat bene-yiśra'el be-'ezrat tavḥinim biyokhimim," 39; Goldschmidt, *The Genetics of Migrant and Isolate Populations*, 106; Sheba, "Reconstructing Jewish Migration with the Aid of Biochemical Tests," 95; Sheba, "Jewish Migration in Its Historical Perspective," 1338.

18. Kirsh, "Population Genetics in Israel in the 1950s"; Abu El-Haj, *The Genealogical Science*.

19. Abu El-Haj, *The Genealogical Science*, 99.

20. Moses Leavitt to Charles Jordan, May 18, 1959. NY AR195564/4/33/17/385, "Israel: Medical, Tel Hashomer Hospital" Files, JDC Archives.

21. Jordan to Leavitt, May 6, 1959. NY AR195564/4/33/17/385, "Israel: Medical, Tel Hashomer Hospital" Files, JDC Archives.

22. See letters: Leavitt to Abram Abeloff, May 12, 1959. NY AR195564/4/33/17/385; Leavitt to Jordan, May 18, 1959. NY AR195564/4/33/17/385 "Israel: Medical, Tel

Hashomer Hospital" Files. Abraham Gonik to Leavitt, October 22, 1964. NY AR195564/4/8/1/41, "Ethiopia: Administration, General" Files, JDC Archives.

23. Typed draft manuscript of "A Survey of Some Genetical Characters in Ethiopian Tribes," probably written by Chaim Sheba. This draft is undated but must have been produced between January 1960 and May 1962. Available at the National Library of Israel, reference number 003538559.

24. Bondy, *Shiba*, 309.

25. Summerfield, *From Falashas to Ethiopian Jews*, 128.

26. Roshwald, "Marginal Jewish Sects in Israel II," 347–51.

27. Bishku, "Israel and Ethiopia," 46. See also Mooreville, "Eyeing Africa."

28. Najjar, Prince, and Fontaine, "The 1958 Malaria Epidemic in Ethiopia."

29. See Sheba, Adam, and Bat-Miriam, "A Survey of Some Genetical Characters in Ethiopian Tribes: Introduction." The following thirteen articles in the same issue catalog the various traits tested, including ABO, Rh, and MNS blood groups; haptoglobin and transferrin types; the presence of G6PD deficiency, abnormal hemoglobins, color vision defects, and mid-digital hair; fingerprint patterns and morphological traits such as hair type and eye color; and the ability to taste PTC and roll one's tongue.

30. Bondy, *Shiba*, 309.

31. See the discrepancy between Sheba's concluding discussion in the draft manuscript (National Library of Israel), which avoids commenting on the implications of the findings for the Falasha's Jewishness, and that of the final publication, which emphasizes the stark genetic differences between the Falasha and all other tested Jewish communities: Barnicot, "A Survey of Some Genetical Characters in Ethiopian Tribes: Concluding Discussion." See also Mourant, *The Distribution of the Human Blood Groups, and Other Polymorphisms*, 81; Mourant, *The Genetics of the Jews*, 39.

32. Goldschmidt, *The Genetics of Migrant and Isolate Populations*.

33. Goldschmidt, 2.

34. Goldschmidt, 8.

35. Goldschmidt, 100–106.

36. Szeinberg and Sheba, "Hemolytic Trait in Oriental Jews Connected with an Hereditary Enzymatic Abnormality of Erythrocytes"; Sheba, "Nisayon le-shiḥzur nedidat bene-yiśra'el be-'ezrat tavḥinim biyokhimim"; Sheba et al., "Epidemiologic Surveys of Deleterious Genes in Different Population Groups in Israel"; Goldschmidt, *The Genetics of Migrant and Isolate Populations*, 100–106; Sheba et al.,

"Distribution of Glucose-6-Phosphate-Dehydrogenase Deficiency Among Various Communities in Israel"; Sheba, "Geneṭiḵah shel shevaṭe-yiśra'el"; Sheba, "Reconstructing Jewish Migration with the Aid of Biochemical Tests"; Sheba, "Jewish Migration in Its Historical Perspective."

37. Sheba, "Jewish Migration in Its Historical Perspective."

38. Sheba et al., "Epidemiologic Surveys of Deleterious Genes in Different Population Groups in Israel."

39. A fact of which Sheba was fully aware, as evidenced by his own comments that he had been unable to convince his colleagues on various points; see Goldschmidt, *The Genetics of Migrant and Isolate Populations*, 104; Sheba, "Jewish Migration in Its Historical Perspective," 1340; Bonné, "Chaim Sheba (1908–1971)," 312.

40. Bonné, "Are There Hebrews Left?"

41. Walker and Bowman, "Glutathione Stability of the Erythrocytes of Iranians."

42. Bowman and Walker, "Virtual Absence of Glutathione Instability of the Erythrocytes Among Armenians in Iran."

43. Bowman and Walker, 222.

44. Bowman and Walker, 222.

45. Sanasarian, *Religious Minorities in Iran*, 35–40.

46. Bowman, "Haptoglobin and Transferrin Differences in Some Iranian Populations."

47. Bowman and Walker, "The Origin of Glucose-6-Phosphate Dehydrogenase Deficiency in Iran," 586. This account reflects shifting community attitudes after the 1930s, when Zoroastrian communal organizations began taking firmer stances against intermarriage with non-Zoroastrians and eliding historical cases where outsiders had occasionally been able to convert to Zoroastrianism and marry into the community. See Amighi, *The Zoroastrians of Iran*, 190.

48. Oral History Interview with James Bowman, June 26–28, 2006, Oral History of Human Genetics Collection (Ms. Coll. no. 316), History & Special Collections Division, Louise M. Darling Biomedical Library, UCLA. Hereafter "Bowman oral history, 2006."

49. Bowman and Walker, "The Origin of Glucose-6-Phosphate Dehydrogenase Deficiency in Iran," 584.

50. Bowman and Walker, 584.

51. Bowman oral history, 2006.

52. Bowman oral history, 2006.

53. Bowman, "Comments on Abnormal Erythrocytes and Malaria," 160.

54. Bowman, 160.

55. Bowman and Frischer, "Malaria, G-6-P.D. Deficiency, and Sickle-Cell Trait."

56. Interview with Armand P. Gelpi, an oral history conducted in 1996 by Carole Hicke in "Health and Disease in Saudi Arabia: The Aramco Experience, 1940s–1990s," Regional Oral History Office, Bancroft Library, University of California, Berkeley, 1998.

57. Gelpi, "Glucose-6-Phosphate Dehydrogenase Deficiency in Saudi Arabia: A Survey."

58. Shaker, Onsi, and Aziz, "The Frequency of Glucose-6-Phosphate Dehydrogenase Deficiency in the Newborns and Adults in Kuwait," 611.

59. Bowman wrote a letter to the journal's editor chastising the Kuwaiti researchers for these deliberate omissions. Bowman, "Glucose-6-Phosphate Dehydrogenase Deficiency in Kuwait."

60. Ragab, El-Alfi, and Abboud, "Incidence of Glucose-6-Phosphate Dehydrogenase Deficiency in Egypt," 23; Taleb et al., "Sur La Déficience En Glucose-6-Phosphate-Déshydrogénase Dans Les Populations Autochtones Du Liban."

61. CENTO, *CENTO Conference on Combating Malnutrition in Preschool Children*.

62. *First Seminar of Favism in Iran, July 6, 1965*, 1.

63. Donoso, Hedayat, and Khayatian, "Favism, with Special Reference to Iran."

64. *First Seminar of Favism in Iran, July 6, 1965*, 22, 57–58.

65. Hedayat, Amirshahy, and Khademy, "Frequency of G-6-PD Deficiency Among Some Iranian Ethnic Groups."

66. Beaconsfield et al., "Glucose 6 Phosphate Dehydrogenase Deficiency in Iran and Its Relation to Physiopathological Processes."

67. Beaconsfield, Mahboubi, and Rainsbury, "Epidemiologie Des Glukose-6-Phosphat-Dehydrogenase-Mangels."

68. Beaconsfield, "Malaria, Glucose-6-Phosphate Dehydrogenase Deficiency, and Hb S."

69. Lehmann et al., "The Hereditary Blood Factors of the Kurds of Iran," 196.

70. The blood samples were analyzed for ten blood group antigen systems, six plasma protein systems, nine erythrocyte enzyme systems, and abnormal hemoglobins. Lehmann et al., "The Hereditary Blood Factors of the Kurds of Iran"; Lightman, Carr-Locke, and Pickles, "The Frequency of PTC Tasters and Males Defective in Colour Vision in a Kurdish Population in Iran."

71. Lehmann et al., "The Hereditary Blood Factors of the Kurds of Iran," 205.

72. Godber et al., "The Hereditary Blood Factors of the Yemenite and Kurdish Jews," 182.

73. Suárez-Díaz, "Indigenous Populations in Mexico."

74. Suárez-Díaz, "Blood Diseases in the Backyard."

75. See Bowman et al., "Hemoglobin and Red Cell Enzyme Variation in Some Populations of the Republic of Vietnam with Comments on the Malaria Hypothesis."

76. Szeinberg, "Investigation of Genetic Polymorphic Traits in Jews," 1173–74.

77. Zaidman et al., "Red Cell Glucose-6-Phosphate Dehydrogenase Deficiency in Ethnic Groups in Israel."

78. Luzzatto and Battistuzzi, "Glucose-6-Phosphate Dehydrogenase," 295–300.

79. Kurdi-Haidar et al., "Origin and Spread of the Glucose-6-Phosphate Dehydrogenase Variant (G6PD-Mediterranean) in the Middle East," 1017.

80. Kurdi-Haidar et al., 1017.

81. Oppenheim et al., "G6PD Mediterranean Accounts for the High Prevalence of G6PD Deficiency in Kurdish Jews," 294.

CHAPTER 6

1. Pallister to Mourant, August 1, 1967. PP/AEM/K/116, box 33, Mourant Papers.

2. Bier, *Mapping Israel, Mapping Palestine*, 171.

3. Bier, 202–3.

4. Little, "Human Population Biology in the Second Half of the Twentieth Century," S132.

5. Burton, "Essential Collaborators," 141.

6. Mourant to Boyd, June 11, 1965. PP/AEM/K.13, Box 28, Mourant Papers.

7. Springfield College Undergraduate Catalog, 1964–1965, p. 114. Archived catalogs of the American University of Beirut show Smith's original appointment as an assistant professor of biology in 1955–1956 (p. 21) and his promotion to an associate professor of biology in 1956–1957 (p. 24). Curiously, in academic years 1957–1958 (p. 135) and 1958–1959 (p. 137), he is identified as an "associate professor of engineering sciences."

8. Mourant, *Blood and Stones*, 77.

9. Smith to Speakman, October 30, 1961. Folder 1, Harry Madison Smith Papers, Faculty Records, Springfield College Archives and Special Collections, Springfield, Massachusetts. Hereafter "Smith Papers."

10. Speakman to Smith, December 28, 1961; Smith to Speakman, February 13, 1962. Folder 1, Smith Papers.

11. Smith to Speakman, March 2, 1963. Folder 1, Smith Papers.

12. Smith et al., "Geographic Variation and Human Population Genetics Among the Indigenous Peoples of Lebanon."

13. "Summary Progress Report" for Smith's NIH grant covering the period September 1, 1961, to May 31, 1963. Folder 1, Smith Papers.

14. Smith to Britton C. McCabe, February 9, 1963. Folder 1, Smith Papers.

15. Her father, Alfred Abraham Bonné, was a prominent economist who worked for the Jewish Agency and the Hebrew University of Jerusalem. After the Peel Commission Report was issued in 1937, he served on the Jewish Agency's Population Transfer Committee, wherein he devised a plan for the "complete and forcible evacuation" of Palestinians from the proposed Jewish state. Masalha, *Expulsion of the Palestinians*, 102–6.

16. Bonné-Tamir, *Ḥayim 'im ha-genim*, 22–24.

17. Bonné, "The Samaritans: A Demographic Study."

18. Bonné-Tamir, *Ḥayim 'im ha-genim*, 43–50.

19. Mourant to Boyd, November 22, 1962. PP/AEM/K.13, Box 28, Mourant Papers.

20. Bonné-Tamir, *Ḥayim 'im ha-genim*, 50.

21. Mourant to Boyd, December 10, 1962. PP/AEM/K.13, Box 28, Mourant Papers.

22. Bonné to Mourant, December 17, 1962. PP/AEM/K.8, Box 28, Mourant Papers.

23. Bonné-Tamir, *Ḥayim 'im ha-genim*, 51–53. The monetary amounts are in U.S. dollars, worth about $4 and $25, respectively, in 2020.

24. Bonné to Mourant, February 19, 1963. PP/AEM/K.8, Box 28, Mourant Papers. Perplexed by the Samaritans' behavior, Mourant wrote to an American colleague, "Unfortunately we came up against social and psychological factors which we did not understand, and work came to a full stop." Mourant to L. C. Dunn, September 27, 1963. PP/AEM/K.21, Box 29, Mourant Papers.

25. Bonné-Tamir, *Ḥayim 'im ha-genim*, 54.

26. Mourant to Bonné, April 25, 1963. PP/AEM/K.8, Box 28, Mourant Papers.

27. Bonné to Mourant, December 3, 1963. PP/AEM/K.8, Box 28, Mourant Papers. Bonné-Tamir, *Ḥayim 'im ha-genim*, 55.

28. Bonné to Mourant, October 10, 1963. PP/AEM/K.8, Box 28, Mourant Papers.

29. Bonné, "Genes and Phenotypes in the Samaritan Isolate," 17.

30. Woodd-Walker, Smith, and Clarke, "The Blood Groups of the Timuri and Related Tribes in Afghanistan"; Sunderland and Smith, "The Blood Groups of the Shi'a in Yazd, Central Iran."

31. Clarke to Mourant, August 23, 1966. PP/AEM/K.157, Box 34, Mourant Papers.
32. Mourant to Clarke, July 8, 1966. PP/AEM/K.156, Box 34, Mourant Papers.
33. Mourant to Shahid, September 6, 1966. PP/AEM/K.157, Box 34, Mourant Papers.
34. Ruffié and Taleb, *Étude hémotypologique des ethnies libanaises*, 14.
35. Taleb and Ruffié, "Hémotypologie des populations jordaniennes."
36. Collins and Weiner, *Human Adaptability*, 118.
37. Ruffié and Taleb, *Étude hémotypologique des ethnies libanaises*, 98.
38. Taleb and Ruffié, "Hémotypologie des populations jordaniennes," 281–82.
39. Bonné to Mourant, February 15, 1966. PP/AEM/K.9, Box 28, Mourant Papers.
40. Bonné to Mourant, July 20, 1966. PP/AEM/K.9, Box 28, Mourant Papers.
41. Bonné to Mourant, October 23, 1966. PP/AEM/K.9, Box 28, Mourant Papers.
42. Mourant to Bonné, July 8 and July 15, 1966. PP/AEM/K.9, Box 28, Mourant Papers.
43. Mourant to E. B. Worthington, "Proposal for the Recognition of the Serological Population Genetics Laboratory as an IBP Centre," October 3, 1966. PP/AEM/K.396, Box 43, Mourant Papers.
44. Alain Marengo-Rowe, "Haemoglobins and Blood Groups in Southern Arabia," January 16, 1967. PP/AEM/K.116, Box 33, Mourant Papers.
45. Marengo-Rowe to Mourant, January 27, 1967. PP/AEM/K.116, Box 33, Mourant Papers.
46. Bonné to Mourant, January 12, 1967. PP/AEM/K.10, Box 28, Mourant Papers.
47. Bonné to Donald Tills, December 19, 1966. PP/AEM/K.9, Box 28, Mourant Papers.
48. Marengo-Rowe to Mourant, February 23, March 9, and March 16, 1967. PP/AEM/K.116, Box 33, Mourant Papers.
49. Tills to Bonné, April 1967. PP/AEM/K.10, Box 28, Mourant Papers.
50. Marengo-Rowe to Mourant, May 2, May 7, and May 18, 1967. PP/AEM/K.116, Box 33, Mourant Papers.
51. Bonné to Mourant, June 9, 1967. PP/AEM/K.10, Box 28, Mourant Papers.
52. Marengo-Rowe to Mourant, June 22, 1967. PP/AEM/K.116, Box 33, Mourant Papers.
53. Bonné-Tamir, *Ḥayim 'im ha-genim*, 64.
54. Mourant to Bonné, June 26, 1967. PP/AEM/K.10, Box 28, Mourant Papers.
55. Mourant to Bonné, August 11, 1967. PP/AEM/K.10, Box 28, Mourant Papers.
56. Mourant to Bonné, June 9, 1967. PP/AEM/K.10, Box 28, Mourant Papers.
57. Bonné to Tills, June 23, 1967. PP/AEM/K.10, Box 28, Mourant Papers.

58. Bonné to Mourant, January 29, 1968. PP/AEM/K.11, Box 28, Mourant Papers.

59. Bonné to Mourant, November 1, 1967, and December 2, 1968. PP/AEM/K.10–11, Box 28, Mourant Papers; Bonné-Tamir, *Ḥayim 'im ha-genim*, 98–99.

60. Bonné-Tamir, 101.

61. Bonné to Mourant, April 9, 1968. PP/AEM/K.11, Box 28, Mourant Papers.

62. Bonné to Mourant, April 17, 1969. PP/AEM/K.11, Box 28, Mourant Papers.

63. Glassner, "The Bedouin of Southern Sinai Under Israeli Administration"; Lavie, *The Poetics of Military Occupation*, 70–72.

64. Mourant to Bonné, December 1, 1966, and Bonné to Mourant, December 11, 1966. PP/AEM/K.9, Box 28; Tills to Bonné, October 10, 1968. PP/AEM/K.11, Box 28, Mourant Papers. See Mourant and Tills, "Phosphoglucomutase Frequencies in Habbanite Jews and Icelanders."

65. Bonné to Tills, March 2, 1968. PP/AEM/K.11, Box 28, Mourant Papers.

66. Bonné to Mourant, December 2, 1968. PP/AEM/K.11, Box 28, Mourant Papers.

67. Bonné, "The Beduins of South-Sinai."

68. Bonné to Mourant, April 17, 1969. PP/AEM/K.11, Box 28, Mourant Papers. Emphasis original.

69. Mourant sought the assistance of his colleague Hermann Lehmann, on his way to visit relatives in Israel, as a mediator: "I am particularly anxious to maintain good relations with Sheba himself and all those colleagues such as Adam, Szeinberg etc with whom we have worked in harmony before. I also want to be on good terms with Batsheva but she is, as you know, a very difficult personality although a brilliant and charming woman." Mourant to Lehmann, August 4, 1969. PP/AEM/K.336, Box 41, Mourant Papers.

70. Bonné to M. Hauge, April 28, 1970. PP/AEM/K.12, Box 28, Mourant Papers.

71. Bonné to Mourant, April 29, 1970. PP/AEM/K.12, Box 28, Mourant Papers.

72. Tills to Bonné, May 5, 1970. PP/AEM/K.12, Box 28, Mourant Papers. Emphasis added.

73. Tills to Bonné, May 5, 1970. PP/AEM/K.12, Box 28, Mourant Papers.

74. See Bonné et al., "The Habbanite Isolate I. Genetic Markers in the Blood"; Bonné et al., "South-Sinai Beduin: A Preliminary Report on Their Inherited Blood Factors."

75. Bonné to Mourant, May 26, 1970. PP/AEM/K.12, Box 28, Mourant Papers. Emphasis original.

76. On methods of resistance to knowledge-control regimes, see Hilgartner, *Reordering Life*, 17–18.

77. Bonné-Tamir, *Ḥayim 'im ha-genim*, 102.

78. Draft letter from Batsheva Bonné-Tamir to Ephraim Gazit, undated [summer 1976]. MS Bodmer 94, fol. 1, Bodleian Library, Oxford University. Hereafter "MS Bodmer (box, folder), Bodleian Library."

79. Bonné to Tills, May 12, 1967. PP/AEM/K.10, Box 28, Mourant Papers.

80. Bodmer et al., "Study of the HL-A System in a Yemenite Jewish Population in Israel."

81. Nevo to Bodmer, June 1, 1971. MS Bodmer 28, fol. 3, Bodleian Library.

82. Bodmer et al., "Study of the HL-A System in a Moslem Arab Population in Israel."

83. Bonné-Tamir et al., "HLA Polymorphism in Israel 8. The Armenian Community in Jerusalem."

84. Bonné to Walter Bodmer, March 14, 1974. MS Bodmer 94, fol. 1; Nevo to Bodmer, June 2, 1975. MS Bodmer 28, fol. 3, Bodleian Library.

85. Pappé, *The Forgotten Palestinians*, 96.

86. Firro, *The Druzes in the Jewish State*, 205–6.

87. Pappé, *The Forgotten Palestinians*, 131–32.

88. Nevo to Walter and Julia Bodmer, December 17, 1976, and April 28, 1977. MS Bodmer 28, fol. 3, Bodleian Library.

89. Batsheva Bonné's surname after her marriage to Zvi Tamir in May 1975; Bonné-Tamir to Walter and Julia Bodmer, June 6, 1975. MS Bodmer 94, fol. 1, Bodleian Library.

90. Nevo to Bonné-Tamir, June 7, 1976. MS Bodmer 28, fol. 3, Bodleian Library.

91. Nevo to Walter and Julia Bodmer, August 17 and August 30, 1976. MS Bodmer 28, fol. 3, Bodleian Library.

92. See Nevo to Walter Bodmer, June 13, 1978. MS Bodmer 28, fol. 2, Bodleian Library; Cleve et al., "Genetic Studies on the Gc Subtypes." Nevo's subsequent publications ultimately emphasized the genetic similarity between Muslim and Druze Arabs. Nevo and Cleve, "Gc Subtypes in the Middle East"; Nevo, "Genetic Blood Markers in Arab Druze of Israel"; Nevo et al., "Serum Protein Polymorphisms in Arab Moslems and Druze of Israel."

93. Walter Bodmer to Nevo, June 13, 1978. MS Bodmer 28, fol. 2, Bodleian Library.

94. Nevo to Julia Bodmer, May 22, 1980. MS Bodmer 61, fol. 1, Bodleian Library.

95. Bonné-Tamir, *Ḥayim 'im ha-genim*, 110.

96. See Der Matossian, "The Armenians of Palestine 1918–48."

97. Bonné-Tamir et al., "HLA Polymorphism in Israel 8. The Armenian Community in Jerusalem," 231.

98. Bonné-Tamir et al., 230–31.

99. Bonné-Tamir to Walter and Julia Bodmer, October 10, 1976, and Walter Bodmer to Bonné-Tamir, June 8, 1977. MS Bodmer 1895, fol. 1, Bodleian Library.

100. The problem of unequal "collaborations" between researchers from the Global North and South persists to this day. See the accounts of French scientists' conception of Senegalese collaborators as "blood senders" during the 1980s and 1990s in Fullwiley, *The Enculturated Gene*, 188–96.

101. Bonné-Tamir to Mourant, May 7, 1975. PP/AEM/K.12, Box 28, Mourant Papers.

102. Bonné to Tills, November 22, 1968. PP/AEM/K.11, Box 28, Mourant Papers.

103. Selim et al., "Genetic Markers and Anthropometry in the Populations of the Egyptian Oases of El-Kharga and El-Dakhla"; Ibrahim et al., "Hereditary Blood Factors and Anthropometry of the Inhabitants of the Egyptian Siwa Oasis"; Azim et al., "Genetic Blood Markers and Anthropometry of the Populations in Aswan Governorate, Egypt."

104. Hamza et al., "Hereditary Blood Factors of North Sinai Inhabitants."

CHAPTER 7

1. "Der Fall Bajatzadeh," extra page issued by *Gutenbergblatt*, 1968. Stadtarchiv Mainz (Mainz City Archives), ZGS / N 6, 8. Another copy available in Hamid Shawkat collection, Box 14, Folder 20, Hoover Institution Archives. A week later, during the same state visit, West German student Benno Ohnesorg was killed by a police officer while demonstrating against the Shah in Berlin; this event galvanized the growth of leftist student groups across Germany, including the militant anarchist 2 June Movement.

2. See Walter's obituary by Hossfeld, "Nachruf Prof. Dr. Rer. Nat. Habil. Dr. Med. h. c. Hubert Walter, 14 April 1930–6 Dezember 2008."

3. Bajatzadeh, "Untersuchungen Zur Populationsgenetik Persiens."

4. Memo from Iranian Student Union board members to the Johannes Gutenberg University rector's office, December 20, 1966. Hamid Shawkat collection, Box 14, Folder 7, Hoover Institution Archives.

5. Letter from Iranian Student Union board members to Günter König, March 15, 1967. Hamid Shawkat collection, Box 14, Folder 7, Hoover Institution Archives.

6. Bajatzadeh and Walter, "Serumprotein Polymorphisms in Iran," 41; Bajatzadeh and Walter, "Investigations on the Distribution of Blood and Serum Groups in Iran," 402, 409.

7. Bajatzadeh and Walter, "Investigations on the Distribution of Blood and Serum Groups in Iran," 412.

8. Bajatzadeh and Walter, 414.

9. Walter and Bajatzadeh, "Studies on the Distribution of the Human Red Cell Acid Phosphatase Polymorphism in Iranians and Other Populations"; Bajatzadeh and Walter, "Studies on the Population Genetics of the Ceruloplasmin Polymorphism"; Bajatzadeh and Walter, "Investigations on the Distribution of Blood and Serum Groups in Iran"; Bajatzadeh and Bernhard, "Untersuchungen Zur Verteilung Der Haupttypen Der Fingerbeermuster in Iran (Persien)."

10. Marks, "The Origins of Anthropological Genetics," S165.

11. Elling, *Minorities in Iran*, 41.

12. See Chapter 2 in Elling, *Minorities in Iran*; Samii, "The Nation and Its Minorities."

13. Fazeli, *Politics of Culture in Iran*, 107.

14. Ünlü, "İsmail Beşikçi as a Discomforting Intellectual," 13.

15. Bozarslan, "Kurds and the Turkish State," 349.

16. Landau, "Arab and Turkish Universities," 5.

17. Kaplan, "Din-u Devlet All Over Again?," 119.

18. Yavuz, "Political Islam and the Welfare (Refah) Party in Turkey," 67–68.

19. A useful illustration is the 1974 abnormal hemoglobins symposium, held in Istanbul, seventeen years after the 1957 conference on the same topic discussed in Chapter 4. This time around, the symposium was organized and funded exclusively by local sources, the Istanbul University Medical Faculty and the Turkish Society of Hematology, and presided over by Muzaffer Aksoy. In 1957, Aksoy had been the only Turkish presenter out of twenty-five researchers, who hailed mostly from Europe and North America. At the 1974 symposium, out of forty-five presentations, eighteen involved Turkish scientists. Of those, thirteen were the products of all-Turkish research teams, and five of international collaborations between Turkish and foreign scientists. Aksoy, *International Istanbul Symposium*.

20. Kariminezhad, Khavari Khorasani, and Omidi, "Tārīkhchih-i zhinitīk dar īrān va jahān, bakhsh-i dahhum."

21. Habibollah Fakhrai, interview by the author in San Diego, California, December 22, 2014. Hereafter "Fakhrai interview, 2014."

22. Ameri, "Blood Transfusion Services in Iran."

23. Recommendation letter for Victor Alan Clarke by Fereydoun Ala, December 10, 1981. PP/AEM/K.226, Box 37, Mourant Papers.

24. Fakhrai et al., "Comparison of ABO Blood System in Persian and Arabic Speaking Khuzistanies."

25. Fakhrai interview, 2014.

26. Farhud, interviewed by Ehsan Amini in Tehran, November 8, 2015. Hereafter "Farhud interview, 2015."

27. See, for example, Farhud, Amirshahi, and Hedayat, "Ta'yīn-i anvā'-i hāptūglūbīn dar ustān-i sāḥilī (Bandar-i 'Abbās)"; Mirdamadi, Fotuhi, and Sonbolestan, "Nisbat-i dar ṣad gurūhhā-yi khūnī-i ABO va RH dar 9753 nafar zan az marāji'īn bih markaz-i pizishkī-i Riẓa Shāh-i Kabīr, Dānishgāh-i Iṣfahān."

28. Farhud and Walter, "Hp Subtypes in Iranians"; Farhud et al., "Electrophoretic Investigation of Some Red Cell Enzymes in Iran"; Farhud and Walter, "Polymorphism of C'3 in German, Bulgarian, Iranian and Angola Population."

29. Farhud interview, 2015.

30. Mahluji and Livingston, "Muṭala'ah-i ba'ẓ ī az khuṣūṣiyāt-i mardum-shināsī dar panj dihkadah-i fārs."

31. The anonymous article took an anti-racist position, clarifying that no clear boundaries between races existed and that racial discrimination had no scientific basis. "Mabnā-yi nizhādhā-yi insānī."

32. Aleh-Azhar, Ikhvan, and Shahidi, "Natāyij-i muqaddamātī-i bar'rasī-i zhinitīkī-i 'ashāyir (īl-i qashqā'ī)"; Motabar, Reiss-Sadat, and Tabatabai, "Prevalence of High Blood Pressure in Qashqai Tribe, Southern Iran, 1973."

33. Montazemi, "Bar'rasī-i mīzān-i farāvānī-i gurūhhā-yi khūnī dar īrān."

34. Bowman oral history, 2006.

35. Beck, "Nomads and Urbanites, Involuntary Hosts and Uninvited Guests," 432.

36. For the role of Kurush-i Kabir Hospital (later Sapir Hospital) during the 1979 revolution, see Sternfeld, "The Revolution's Forgotten Sons and Daughters."

37. Mohagheghpour, Tabatabai, and Mohammad, "Distribution of HLA Antigens in Zoroastrians," 258.

38. Mohagheghpour, Tabatabai, and Mohammad, 259.

39. Tabatabai, Mohammad, and Mohagheghpour, "HLA Antigens in Two Iranian Populations," 309.

40. The authors further identified Armenians as racially "Mongolo-Aryan," citing Henry Field's 1939 book. Tabatabai, Mohammad, and Mohagheghpour, 310. For further analysis of this research in light of concurrent Israeli studies of Armenians, see Burton, "Essential Collaborators."

41. Tabatabai, "Bar'rasī-i muqāyasah'ī-i zhinitīk-i īrāniyān-i armanī va yahūdī."

42. Bill, "The Politics of Student Alienation."

43. Fakhrai interview, 2014.

44. Milani, *Eminent Persians*, 1056–57.

45. Kariminezhad, Khavari Khorasani, and Omidi, "Tārīkhchih-i zhinitīk dar īrān va jahān, bakhsh-i dahhum."

46. For example, his colleague at the School of Public Health, Kanwarjit Singh Sawhney, helped foster connections between Farhud's Tehran unit and Eric Sunderland, Sawhney's former adviser at Durham University. See Sawhney, Sunderland, and Farhud, "Study of Red Cell Enzyme Systems in Tehran and Isfahan Iranians." Sunderland, in turn, introduced Farhud's work to his British/Indian colleagues Derek F. Roberts and Surinder Singh Papiha, who had built up the human genetics department at the nearby Newcastle University.

47. Clarke to Mourant, June 12, 1982, and August 30, 1983. PP/AEM/K.226, Box 37, Mourant Papers.

48. Amirshahi, "A Serological-Genetic Study of Iranian and Neighbouring Populations"; Amirshahi et al., "A Genetic Study of Iranian Populations: Serum Proteins."

49. Akbari et al., "Serogenetic Investigations of Two Populations of Iran," 371–72; Akbari et al., "Genetic Differentiation Among Iranian Christian Communities"; Akbari et al., "Population Genetics of the Persians and Other Peoples in Iran."

50. Elling, *Minorities in Iran*, 47–50.

51. See Chapter 2 in Sanasarian, *Religious Minorities in Iran*.

52. Elling, *Minorities in Iran*, 47–50.

53. Seyedna, "A Genetic and Demographic Investigation of the Zoroastrians of Iran," 91, 118–19. Not all of the genetic markers Seyedna tested supported his conclusions; see Papiha, Seyedna, and Sunderland, "Phosphoglucomutase (PGM) and Group-Specific Component Isoelectric Focusing Among Zoroastrians of Iran."

54. Amirshahi, "A Serological-Genetic Study of Iranian and Neighbouring Populations," 4.

55. Amirshahi, 10; Amirshahi et al., "Serum Proteins and Erythrocyte Enzymes of Populations in Iran," 75.

56. Amirshahi et al., "Serum Proteins and Erythrocyte Enzymes of Populations in Iran," 79.

57. Amirshahi et al., "Population Genetics of the Peoples of Iran I.," 8.

58. Akbari et al., "Population Genetics of the Persians and Other Peoples in Iran," 198–99.

59. Akbari et al., "Serogenetic Investigations of Two Populations of Iran," 371–72. See also Papiha et al., "Isoelectric Focusing on Vitamin D Binding Protein (Gc): Genetic Diversity in the Populations of Iran."

60. Akbari et al., "Population Genetics of the Persians and Other Peoples in Iran," 214–15.

61. Akbari et al., "Genetic Differentiation Among Iranian Christian Communities," 84.

62. Akbari et al., "Population Genetics of the Persians and Other Peoples in Iran," 215.

63. Walter et al., "Investigations on the Ethnic Variability of the ABO Blood Group Polymorphism in Iran," 302.

64. Walter, Azimian, and Farhud, "Red Cell Enzyme Polymorphisms in Three Iranian Population Samples: Tabriz, Yazd, Mashhad," 125.

65. See Chapter 4 in Elling, *Minorities in Iran*.

66. Farhud, *Īrān-i farhangī*, 426.

67. Asgharzadeh, *Iran and the Challenge of Diversity*, 124–26.

68. Farzanfar, "Ethnic Groups and the State: Azaris, Kurds and Baluch of Iran," 45.

69. Farhud, *Īrān-i farhangī*, 357.

70. See representative works like Büyükyüksel, *Türkiyede Kan Grupları Dağılımı*; Binyıldız and Büyükyüksel, "Türkiye'de kan grupları dağılımı"; Büyükyüksel, "Groupes sanguins ABO et Rh (D) dans la population turque."

71. Aksoy and Erdem, "Haemoglobin H Disease Study of an Eti-Turk Family"; Aksoy and Erdem, "Abnormal Hemoglobins and Thalassemia in Eti-Turks Living in Antakya"; Altay et al., "Hemoglobin S and Some Other Hemoglobinopathies in Eti-Turks"; Aksoy, Erdem, and Dinçol, "A Survey for Hemoglobin Variants, Thalassemia, Glucose-6-Phosphate Dehydrogenase Deficiency and Haptoglobin Types in Greek Population Living in a Turkish Island, İmroz"; Aksoy, Erdem, and Dinçol, "Three Turkish Families with Delta-Beta-Thalassemia"; Dinçol, Aksoy, and Erdem, "Beta-Thalassaemia with Increased Haemoglobin A2 in Turkey."

72. World Health Organization et al., *The Long Road to Malaria Elimination in Turkey*, 40–41.

73. Akoğlu et al., "Glucose-6-Phosphate Dehydrogenase Deficiency in Çukurova Province, Turkey," 27.

74. Saran, "Kan grupları ve Türkiye'deki dağılımı."

75. Saatçioğlu, *ABO genleri yönünden Türkiye'nin yeri ve bu ülkedeki gensel çeşitlilik üzerine biyometrik bir inceleme*, 22–24.

76. Saatçioğlu, "An Analysis of the ABO Gene Frequencies in Turkey," *Journal of Human Evolution* 8 (1979): 367; see also Saatçioğlu, *ABO genleri yönünden Türkiye'nin yeri ve bu ülkedeki gensel çeşitlilik üzerine biyometrik bir inceleme*, 3–4.

77. Magnarella, "The Development of Turkish Social Anthropology."

78. Unat and Erol, "Doç. Dr. Armağan Saatçioğlu," 279.

79. Saatçioğlu, *ABO genleri yönünden Türkiye'nin yeri ve bu ülkedeki gensel çeşitlilik üzerine biyometrik bir inceleme*, iii.

80. Kansu, "Rassengeschichte der Türkei."

81. Saatçioğlu, *ABO genleri yönünden Türkiye'nin yeri ve bu ülkedeki gensel çeşitlilik üzerine biyometrik bir inceleme*, 129–30.

82. Saatçioğlu, 137.

83. Saatçioğlu, "An Analysis of the ABO Gene Frequencies in Turkey," 370.

84. Saatçioğlu, 372.

85. Diamond et al., "The Case of Dr. Besikci."

86. Saatçioğlu, *ABO genleri yönünden Türkiye'nin yeri ve bu ülkedeki gensel çeşitlilik üzerine biyometrik bir inceleme*, 160; Saatçioğlu, "An Analysis of the ABO Gene Frequencies in Turkey," 372.

87. Şentuna, "Rh Gen Frekansları Yönünden Türkiye'nin Yeri," 173–75.

88. Erdentuğ, "The Pioneering Anthropologists of Turkey," 40.

89. Unat and Erol, "Doç. Dr. Armağan Saatçioğlu," 280. Nermin Erdentuğ, the division chair, retired two years later.

90. Erdentuğ, "The Pioneering Anthropologists of Turkey," 42.

91. Ergün and Aksoy, "Yurt içi ABO ve Rh kan grupları araştırması"; Akbay et al., "Türkiye'de kan gruplarının coğrafi bölgelere göre dağılımı ve bunun kan depolanmasındaki önemi"; Ergün and Yardımcı, "Türkiye genelinde ABO kan grupları ve Rh faktörünün dağılımı"; Akbaş, Aydın, and Cenani, "ABO Blood Subgroup Allele Frequencies in the Turkish Population"; Önde and Kence, "Distribution of ABO and Rh Gene Frequencies over 67 Provinces of Turkey."

92. Walter et al., "Genetic Serum Protein Markers (HP, GC, TF, PI) in Four Turkish Population Samples."

93. Forms faxed from Güher Saruhan to Julia Bodmer, November 30, 1993. MS Bodmer 1111, fol. 1, Bodleian Library.

94. Fax from Saruhan to Julia Bodmer, March 2, 1995. MS Bodmer 1111, fol. 1, Bodleian Library.

95. Fakhrai interview, 2014.

96. Dogan et al., "A Glimpse at the Intricate Mosaic of Ethnicities from Mesopotamia."

CONCLUSION

Sections of the Conclusion were originally developed in "Narrating Ethnicity and Diversity in Middle Eastern National Genome Projects," *Social Studies of Science* 48, no. 5 (2018), 762–86.

1. Cavalli-Sforza et al., "Call for a Worldwide Survey of Human Genetic Diversity," 490.
2. Cavalli-Sforza et al., 490.
3. Reardon, *Race to the Finish*.
4. See Chapter 3 in Nash, *Genetic Geographies*; and Chapters 14 and 15 in Sommer, *History Within*.
5. Bonné-Tamir, "The HGDP Can Be and Should Be Pursued in Harmony with Its Subjects: An Account of the Israeli Experience," 292.
6. Bonné-Tamir, 292.
7. Kimel et al., "Living in a Genetic World," 695.
8. Perry, "In Lebanon DNA May Yet Heal Rifts."
9. Maroon, "A Geneticist with a Unifying Message"; Zalloua et al., "Y-Chromosomal Diversity in Lebanon Is Structured by Recent Historical Events"; Zalloua et al., "Identifying Genetic Traces of Historical Expansions."
10. Maroon, "A Geneticist with a Unifying Message."
11. Marc Haber, interviewed by the author on October 8, 2018, at the Wellcome Sanger Institute, Hinxton, UK.
12. For further analyses and critiques of these tropes, see TallBear, "Genomic Articulations of Indigeneity"; Nash, *Genetic Geographies*; Reardon, "The Democratic, Anti-Racist Genome?"; Sommer, *History Within*.
13. "Sancar: Ben Türk'üm o Kadar"; "Aziz Sancar'a BBC'den Şoke Eden Soru!"
14. Dündar, "Aziz Sancar Türk Mü, Kürt Mü?"; Kongar, "Aziz Sancar'ın Düşündürdükleri . . ."
15. Canikligil, "Bu Ödülü 19 Mayıs'ta Anıtkabir'de Atatürk'e Sunacağım."
16. "Nobel Ödüllü Sancar'ı Ailesi Anlattı."
17. Tanış, "Nobel Panteri Aziz Sancar."
18. Aziz Sancar, Nobel Lecture, Stockholm University, Sweden, December 8, 2015.
19. Duruel, Altay, and Ulutin, *Bilime adanmış bir ömür*, 38.
20. Ergin, "The Racialization of Kurdish Identity in Turkey."
21. Alkan et al., "Whole Genome Sequencing of Turkish Genomes Reveals Functional Private Alleles and Impact of Genetic Interactions with Europe, Asia and Africa," 965.

22. Alkan et al., 965.
23. Alkan et al., 966.
24. Erşan, "Türklere özel bir genom bulunamadı."
25. Gültekin and Gökçümen, "Genetik Bilgi ve Antropoloji," 55.
26. Gökçümen and Gültekin, "Genetik ve Kamusal Alan," 25.
27. Gökçümen and Gültekin, 27–28.
28. Gökçümen et al., "Biological Ancestries, Kinship Connections, and Projected Identities in Four Central Anatolian Settlements," 120.
29. Gökçümen et al., 128.
30. Gökçümen et al., 119–20.
31. Cumhuriyet, "Türklerin genetik şifresi çözüldü."
32. Çataloluk, *Türk'ün genetik tarihi*, 9.
33. Cheng, "Is Peking Man Still Our Ancestor?"
34. Farhud et al., "Progress of Education, Research and Services in Medical Genetics, in Some Institutions of Iran"; Banihashemi, "Iranian Human Genome Project."
35. Banihashemi, "Iranian Human Genome Project."
36. Reardon, *Race to the Finish*.
37. Banihashemi, "Iranian Human Genome Project."
38. Banihashemi did not directly reference Aryanism in his description of the IHGP, but there is no question that it played a major role in how the project's leading scientists understood the structure of ethnic diversity in Iran. These scientists have coauthored publications describing the Iranian population as consisting "of several ethnic groups most of whom are of 'Aryan origin'" (Rezaee et al., "Beta-Thalassemia in Iran," 1) and defining Aryans as "the largest ethnicity in Iran," comprising the ethnic Persian ("Pars") majority and "non-Persian Aryan groups such as the Gilak and the Kurds." Banoei et al., "Diversity and Relationship Between Iranian Ethnic Groups," 821–22.
39. Najmabadi et al., "The Iranian Human Mutation Gene Bank," 147.
40. "Sanctions on Scientific Publication"; Habibi, Rashidi, and Feldman, "Emerging Concerns About Iran's Scientific and Medical Future"; Habibzadeh and Vessal, "Scientific Research in Iran"; Marshall, "Scientific Journals Adapt to New U.S. Trade Sanctions on Iran"; Rezaee-Zavareh, Karimi-Sari, and Alavian, "Iran, Sanctions, and Research Collaborations"; Afshari and Bhopal, "Iran, Sanctions, and Collaborations."
41. Lotfalian, "The Iranian Scientific Community and Its Diaspora After the Islamic Revolution."

42. Stone, "Unsanctioned Science."
43. Roxana Daneshjou, personal communication, March 21, 2016.
44. Epstein, *Inclusion*.
45. Benjamin, "A Lab of Their Own."
46. Home page of irangenes.com.
47. Kelly, "Transnational Diasporic Identities"; Graham and Khosravi, "Home Is Where You Make It."
48. Bozorgmehr, "Internal Ethnicity," 400.
49. Nash, "The Politics of Genealogical Incorporation."

Bibliography

ARCHIVAL COLLECTIONS AND UNPUBLISHED WORKS

Muzaffer Aksoy Oral History (*sözlü tarih*). DVD call number: FL CD 2010 DK 201. National Library (Milli Kütüphane), Ankara, Turkey.

American University of Beirut Catalogues, 1955–1962.

Anthropology Institute, uncataloged papers. Johannes Gutenberg University Archives, Mainz, Germany.

Sir Walter and Lady Julia Bodmer Papers (MS Bodmer). Bodleian Library, Oxford University.

James E. Bowman Oral History. Oral History of Human Genetics Collection (Ms. Coll. no. 316), History & Special Collections Division, Louise M. Darling Biomedical Library, University of California, Los Angeles.

Lyle Gifford Boyd Papers, 1921–1982. Schlesinger Library, Harvard University, Cambridge, Massachusetts.

Cohen-Kedmi Commission on the Yemenite Children: Witness Invited by the Commission: Prof. Mendel George, 241/96. Israel State Archives, Jerusalem. Available via http://www.archives.gov.il.

Edgar Jacob Fisher Papers. Hoover Institution Archives, Stanford University, California.

Armand P. Gelpi Oral History. Interview by Carole Hicke in "Health and Disease in Saudi Arabia: The Aramco Experience, 1940s–1990s," Regional Oral History Office, Bancroft Library, University of California, Berkeley.

German Federal Archives (Bundesarchiv), Lichterfelde, Berlin, Germany.

German Reich Files, Political Archive of the German Foreign Ministry (Politisches Archiv des Auswärtigen Amts), Berlin, Germany.

Joseph Gurevitch Personnel File. Central Archives, Hebrew University of Jerusalem, Israel.

J. B. S. Haldane Papers. Special Collections, University College London.

"Israel: Medical, Tel Hashomer Hospital" Files and "Ethiopia: Administration, General" Files. Records of the New York Headquarters of the American Jewish Joint Distribution Committee, 1955–1964. JDC Archives, New York. Available via www.archives.jdc.org.

Modern and Contemporary History Collection (ZGS). Mainz City Archives (Stadtarchiv), Germany.

Arthur E. Mourant Papers. Wellcome Library, London.

James V. Neel Papers. American Philosophical Society, Philadelphia, Pennsylvania.

Oral History Collection. Foundation for Iranian Studies, Bethesda, Maryland.

Robert College Records. Rare Book and Manuscript Library, Columbia University, New York.

Hamid Shawkat Collection. Hoover Institution Archives, Stanford University, California.

Harry Madison Smith Papers. Faculty Records, Springfield College Archives and Special Collections, Springfield, Massachusetts.

Springfield College Undergraduate Catalog, 1964–1965.

"A Survey of Some Genetical Characters in Ethiopian Tribes." Undated typescript, reference number 003538559, National Library of Israel, Jerusalem.

PUBLISHED PRIMARY AND SECONDARY SOURCES

Abdoosh, Y. B., and Salah El-Dewi. "The Blood Groups of Egyptians." *Journal of the Royal Egyptian Medical Association* 32, no. 10 (1949): 715–26.

Abrahamian, Ervand. *A History of Modern Iran*. Cambridge; New York: Cambridge University Press, 2008.

Abu El-Haj, Nadia. *The Genealogical Science: The Search for Jewish Origins and the Politics of Epistemology*. Chicago; London: University of Chicago Press, 2012.

Afkari, Abdolhossein. "Groupes sanguins ABO et Rh (D) dans la province du Khorassan." *Transfusion* 10, no. 4 (1967): 407–8.

Afshari, Reza, and Raj S. Bhopal. "Iran, Sanctions, and Collaborations." The *Lancet* 387, no. 10023 (March 2016): 1055–56. https://doi.org/10.1016/S0140-6736(16)00628-0.

Akbari, Mohammad Taghi, Surinder Singh Papiha, Derek F. Roberts, and Dariush Daneshvar Farhud. "Genetic Differentiation Among Iranian Christian Communities." *American Journal of Human Genetics* 38 (1986): 84–98.

Akbari, Mohammad Taghi, Surinder Singh Papiha, Derek F. Roberts, and Dariush Daneshvar Farhud. "Population Genetics of the Persians and Other Peoples in Iran." *Zeitschrift Für Morphologie Und Anthropologie* 76, no. 2 (1986): 197–217.

Akbari, Mohammad Taghi, Surinder Singh Papiha, Derek F. Roberts, and Dariush Daneshvar Farhud. "Serogenetic Investigations of Two Populations of Iran." *Human Heredity* 34 (1984): 371–77.

Akbaş, Fahri, Müge Aydın, and Asim Cenani. "ABO Blood Subgroup Allele Frequencies in the Turkish Population." *Anthropologischer Anzeiger* 61, no. 3 (2003): 257–60.

Akbay, Tahir, Çakır Güney, Pekcan Demiröz, Ali Şengül, and Fikri Kocabalkan. "Türkiye'de kan gruplarının coğrafi bölgelere göre dağılımı ve bunun kan depolanmasındaki önemi." *Gülhane Askeri Tıp Akademisi Bülteni* 31 (1989): 391–402.

Akoğlu, Tevfık, Faruk L. Özer, S Çiğ, M Kümi, A Erdoğan, and H Anil. "Glucose-6-Phosphate Dehydrogenase Deficiency in Çukurova Province, Turkey." *International Journal of Epidemiology* 10, no. 1 (1981): 27–29.

Aksoy, Muzaffer. "Abnormal Haemoglobins in Turkey." In *Abnormal Haemoglobins: A Symposium Organized by the Council for International Organizations of Medical Sciences, Established Under the Joint Auspices of UNESCO and WHO*, edited by J. H. P. Jonxis and J. F. Delafresnaye, 216–35. Oxford: Blackwell Scientific, 1959.

Aksoy, Muzaffer. "Brief Note: Hemoglobin S in Eti-Turks and the Allewits in Lebanon." *Blood* 17, no. 5 (1961): 657–59.

Aksoy, Muzaffer. "Sickle-Cell Anemia in South Turkey. A Study of Fifteen Cases in Twelve White Families." *Blood* 11, no. 5 (1956): 460–72.

Aksoy, Muzaffer. "Sickle-Cell Trait in South Turkey." *The Lancet* 265, no. 6864 (1955): 589–90.

Aksoy, Muzaffer, and Hermann Lehmann. "Sickle-Cell-Thalassaemia Disease in South Turkey." *British Medical Journal* 1, no. 5021 (March 30, 1957): 734–39.

Aksoy, Muzaffer, and Hermann Lehmann. "The First Observation of Sickle-Cell Haemoglobin E Disease." *Nature* 179, no. 4572 (1957): 1248–49.

Aksoy, Muzaffer, and Kamil Tanrıkulu. "The Hemoglobin E Syndromes. I. Hemoglobin E in Eti-Turks." *Blood* 15, no. 5 (1960): 606–9.

Aksoy, Muzaffer, and Şakir Erdem. "Abnormal Hemoglobins and Thalassemia in Eti-Turks Living in Antakya." *Medical Bulletin of Istanbul* 1 (1968): 296–301.

Aksoy, Muzaffer, and Şakir Erdem. "Haemoglobin H Disease Study of an Eti-Turk Family." *Human Heredity* 18, no. 1 (1968): 12–22.

Aksoy, Muzaffer, ed. *International Istanbul Symposium on Abnormal Hemoglobins and Thalassemia: August 24–27, 1974, Istanbul, Turkey.* Ankara: TÜBİTAK, 1975.

Aksoy, Muzaffer, Elizabeth W. Ikin, Arthur E. Mourant, and Hermann Lehmann. "Blood Groups, Haemoglobins, and Thalassemia in Turks in Southern Turkey and Eti-Turks." *British Medical Journal* 2, no. 5102 (October 18, 1958): 937–39.

Aksoy, Muzaffer, G. W. G. Bird, Hermann Lehmann, Arthur E. Mourant, H. Thein, and R. L. Wickremasinghe. "Hemoglobin E in Asia." *Journal of Physiology* 130 (Supplemental) (1955): 56P–57P.

Aksoy, Muzaffer, Şakir Erdem, and Günçağ Dinçol. "A Survey for Hemoglobin Variants, Thalassemia, Glucose-6-Phosphate Dehydrogenase Deficiency and Haptoglobin Types in Greek Population Living in a Turkish Island, İmroz." In *International Istanbul Symposium on Abnormal Hemoglobins and Thalassemia: August 24–27, 1974, Istanbul, Turkey*, edited by Muzaffer Aksoy, 191–96. Ankara: TÜBİTAK, 1975.

Aksoy, Muzaffer, Şakir Erdem, and Günçağ Dinçol. "Three Turkish Families with Delta-Beta-Thalassemia." In *International Istanbul Symposium on Abnormal Hemoglobins and Thalassemia: August 24–27, 1974, Istanbul, Turkey*, edited by Muzaffer Aksoy, 365–69. Ankara: TÜBİTAK, 1975.

Akyol, Esra Demirci. "Memory and Strategies of Identity Formation in Hatay." *International Journal of Turkish Studies* 21, no. 1/2 (2015): 117–36.

Aleh-Azhar, Hishmat, Mehdi Ikhvan, and Hossein Khatib Shahidi, eds. "Natāyij-i muqaddamātī-i bar'rasī-i zhinitīkī-i 'ashāyir (īl-i qashqā'ī)." In *Maqālāt arā'ih shudah dar panjumīn Kungrih-'i zhinitīk-i īrān: Dānishgāh-i Tabrīz, sivvum va chahārum va panjum-i mihr māh 1352*, 189–205. Tabrīz: Dānishgāh-i Tabrīz, 1973.

Alkan, Can, Pinar Kavak, Mehmet Somel, Omer Gokcumen, Serkan Ugurlu, Ceren Saygi, Elif Dal, et al. "Whole Genome Sequencing of Turkish Genomes Reveals Functional Private Alleles and Impact of Genetic Interactions with Europe, Asia and Africa." *BMC Genomics* 15, no. 1 (2014): 963. https://doi.org/10.1186/1471-2164-15-963.

Allison, A. C. "The Discovery of Resistance to Malaria of Sickle-Cell Heterozygotes." *Biochemistry and Molecular Biology Education* 30, no. 5 (2002): 279–87.

Allison, A. C. "The Distribution of the Sickle-Cell Trait in East Africa and Elsewhere, and Its Apparent Relationship to the Incidence of Subtertian Malaria." *Transactions of the Royal Society of Tropical Medicine and Hygiene* 48, no. 4 (July 1954): 312–18. https://doi.org/10.1016/0035-9203(54)90101-7.

Allison, A. C. "The Sickle-Cell Trait in the Mediterranean Area." *Man* 53 (February 1953): 23–24.

Altay, Çiğdem, Sevgi Yetgin, Şinası Özsoylu, and A. Kutsal. "Hemoglobin S and Some Other Hemoglobinopathies in Eti-Turks." *Human Heredity* 28, no. 1 (1978): 56–61.

Altman, Yair. "Ha-Nisuyim Be-Yalde Teman Neḥśafim: 'Mukraḥ Le-Hiyot La-Hem Dam Kushi.'" *Yisra'el Ha-Yom*, June 14, 2017. http://www.israelhayom.co.il/article/483661.

"Altounyan, Ernest Haik Riddall (1890–1962)." In *Plarr's Lives of the Fellows Online*. Royal College of Surgeons of England, December 20, 2013. http://livesonline.rcseng.ac.uk/biogs/E004837b.htm.

Altounyan, Ernest H. R. "A Note on Blood Transfusion in Syria, with an Analysis of 1149 Blood Groupings." The *Lancet* 210, no. 5443 (1927): 1342–43. https://doi.org/10.1016/S0140-6736(00)55911-X.

Altounyan, Ernest H. R. "Blood Group Percentages for Arabs, Armenians and Jews." *British Medical Journal* 1, no. 3508 (March 31, 1928): 546.

Altounyan, Taqui. *Chimes from a Wooden Bell: A Hundred Years in the Life of a Euro-Armenian Family*. London; New York: I. B. Tauris, 1990.

Altounyan, Taqui. *In Aleppo Once*. London: John Murray, 1969.

Aluoch, J. R., Y Kilinç, M. Aksoy, G. T. Yüregir, I. Bakioğlu, A. Kutlar, F. Kutlar, and T. H. J. Huisman. "Sickle Cell Anaemia Among Eti-Turks: Haematological, Clinical and Genetic Observations." *British Journal of Haematology* 64, no. 1 (1986): 45–55.

Ameri, Ali. "Blood Transfusion Services in Iran." In *Encyclopædia Iranica*. New York: Encyclopaedia Iranica Foundation, July 29, 2013. http://www.iranicaonline.org/articles/blood-transfusion.

Amighi, Janet Kestenberg. *The Zoroastrians of Iran: Conversion, Assimilation, or Persistence*. New York: AMS Press, 1990.

Amirshahi, Parivash. "A Serological-Genetic Study of Iranian and Neighbouring Populations." PhD diss., Durham University, 1983.

Amirshahi, Parivash, Dariush Daneshvar Farhud, Eric Sunderland, S. H. Tavakoli, and P. Daneshmand. "A Genetic Study of Iranian Populations: Serum Proteins." *Iranian Journal of Public Health* 17, no. 1–4 (1988): 75–80.

Amirshahi, Parivash, Eric Sunderland, Dariush Daneshvar Farhud, S. H. Tavakoli, P. Daneshmand, and Surinder Singh Papiha. "Population Genetics of the Peoples of Iran I. Genetic Polymorphisms of Blood Groups, Serum Proteins and Red Cell Enzymes." *International Journal of Anthropology* 7, no. 3 (July 1, 1992): 1–10. https://doi.org/10.1007/BF02447604.

Amirshahi, Parivash, Eric Sunderland, Dariush Daneshvar Farhud, S.H. Tavakoli, P. Daneshmand, and Surinder Singh Papiha. "Serum Proteins and Erythrocyte Enzymes of Populations in Iran." *Human Heredity* 39 (1989): 75–80.

Ammar, Abbas. "Physical Measurements and Serology of the People of Sharqiya (Egypt)." *The Journal of the Royal Anthropological Institute of Great Britain and Ireland* 70, no. 2 (1940): 147–170. https://doi.org/10.2307/2844368.

Anderson, Benedict. *Imagined Communities: Reflections on the Origin and Spread of Nationalism.* London; New York: Verso, 2006.

Anderson, Warwick. "Objectivity and Its Discontents." *Social Studies of Science* 43, no. 4 (August 1, 2013): 557–76.

Anderson, Warwick. "Racial Conceptions in the Global South." *Isis* 105, no. 4 (December 2014): 782–92.

Anderson, Warwick. "The Anomalous Blonds of the Maghreb: Carleton Coon Invents the African Nordics." In *Expeditionary Anthropology: Teamwork, Travel and the "Science of Man,"* edited by Martin Thomas and Amanda Harris, 150–74. New York: Berghahn, 2018.

Anderson, Warwick, and Hans Pols. "Scientific Patriotism: Medical Science and National Self-Fashioning in Southeast Asia." *Comparative Studies in Society and History* 54, no. 01 (January 2012): 93–113. https://doi.org/10.1017/S0010417511000600.

Ansari, Ali M. *Modern Iran: The Pahlavis and After.* 2nd ed. London; New York: Routledge, 2014.

"Arabs and American Indians—Same Race: Beirut University Scientist's Discovery." The *Palestine Post*, December 25, 1938.

Arnaud, T. H., and A. Vayssière. "Les Akhdam de l'Yémen, Leur Origine Probable, Leurs Mœurs." *Journal Asiatique* 15 (1850): 376–87.

Arvidsson, Stefan. *Aryan Idols: Indo-European Mythology as Ideology and Science.* Translated by Sonia Wichmann. Chicago; London: University of Chicago Press, 2006.

Asgharzadeh, Alireza. *Iran and the Challenge of Diversity: Islamic Fundamentalism, Aryanist Racism, and Democratic Struggles.* New York: Palgrave Macmillan, 2007.

Assi, Seraj. "The Original Arabs: The Invention of the 'Bedouin Race' in Ottoman Palestine." *International Journal of Middle East Studies* 50, no. 2 (May 2018): 213–32. https://doi.org/10.1017/S002074381800003X.

Atakuman, Çiğdem. "Cradle or Crucible: Anatolia and Archaeology in the Early Years of the Turkish Republic (1923–1938)." *Journal of Social Archaeology* 8, no. 2 (2008): 214–35. https://doi.org/10.1177/1469605308089965.

Aulaqi, Adel. *From Barefoot Doctors to Professors of Medicine in 75 Years: A Brief History of Modern Medicine in Yemen 1940–2015.* Chesham, UK: Eidat Shams Medical History Publishing, 2017.

Aygen, Nermin. "Türklerin Kan Grupları ve Kan Gruplarının Antropolojik Karakterlerle İlgisi Üzerine Bir Araştırma." *Ankara Üniversitesi Dil ve Tarih-Coğrafya Fakültesi Dergisi* 11, no. 1 (1943): 17–27.

Aygen, Nermin. *Türklerin kan grupları ve kan gruplarının antropolojik karakterlerle ilgisi üzerine bir araştırma* (A Research on the Blood Groups of Turks and on the Correlation of These Groups with Anthropological Characters). Ankara: Ideal Basımevi, 1946.

Azim, Aida A., Karim Kamel, M.F. Gaballah, Fadila H. Sabry, W. Ibrahim, Osaima Selim, and Nadia Moafy. "Genetic Blood Markers and Anthropometry of the Populations in Aswan Governorate, Egypt." *Human Heredity* 24, no. 1 (1974): 12–23. https://doi.org/10.1159/000152632.

"Aziz Sancar'a BBC'den Şoke Eden Soru!" *Sabah*, October 8, 2015. http://www.sabah.com.tr/gundem/2015/10/08/aziz-sancara-bbcden-soke-eden-soru.

Babacan, Ethem. "Über die Blutgruppen bei Türken." *Zeitschrift für Immunitätsforschung* 90, no. 1 (1937): 1–4.

Baḥr al-'Ulūmī, Ḥusayn. *Kārnāmah-i Anjuman-i Āṣār-i Millī az āghāz tā 2535 Shāhanshāhī/1301–1355 Hijrī-i Khvurshīdī*. Tehran: Chāpkhānah-i Mīhan, 2535.

Bajatzadeh, Mansur. "Untersuchungen Zur Populationsgenetik Persiens." PhD diss., Johannes Gutenberg-Universität, Mainz, 1967.

Bajatzadeh, Mansur, and Hubert Walter. "Investigations on the Distribution of Blood and Serum Groups in Iran." *Human Biology* 41, no. 3 (1969): 401–15.

Bajatzadeh, Mansur, and Hubert Walter. "Serumprotein Polymorphisms in Iran." *Humangenetik* 6 (1968): 40–54.

Bajatzadeh, Mansur, and Hubert Walter. "Studies on the Population Genetics of the Ceruloplasmin Polymorphism." *Humangenetik* 8 (1969): 134–36.

Bajatzadeh, Mansur, and Wolfram Bernhard. "Untersuchungen Zur Verteilung Der Haupttypen Der Fingerbeermuster in Iran (Persien)." *Zeitschrift Für Morphologie Und Anthropologie* 61, no. 1 (1969): 72–80.

Bakir, Farhan, and Majid Al-Qaysi. "Sickle Cell Anaemia in Iraq: First Case Report." *Journal of the Faculty of Medicine Baghdad* 6, no. 1 (January 1964): 26–31.

Bangham, Jenny. "Blood Groups and Human Groups: Collecting and Calibrating Genetic Data After World War Two." *Studies in History and Philosophy of Science Part C: Studies in History and Philosophy of Biological and Biomedical Sciences* 47 (2014): 74–86.

Bangham, Jenny. *Blood Relations: Transfusion and the Making of Human Genetics*. Chicago: University of Chicago Press, 2020.

Banihashemi, Kambiz. "Iranian Human Genome Project: Overview of a Research

Process Among Iranian Ethnicities." *Indian Journal of Human Genetics* 15, no. 3 (2009): 88. https://doi.org/10.4103/0971-6866.60182.

Banoei, Mohammad Mehdi, Morteza Hashemzadeh Chaleshtori, Mohammah Hossein Sanati, Mehdi Shafa Sharait Panahi, Tayebeh Majidizadeh, Maryam Rostami, Massoumeh Dehghan Manshadi, and Masoud Golalipour. "Diversity and Relationship Between Iranian Ethnic Groups: Human Dopamine Transporter Gene *(DATi)* VNTR Genotyping." *American Journal of Human Biology* 19, no. 6 (November 2007): 821–26. https://doi.org/10.1002/ajhb.20647.

Barkan, Elazar. "Mobilizing Scientists Against Nazi Racism, 1933–1939." In *Bones, Bodies, Behavior: Essays on Biological Anthropology*, edited by George W. Stocking, 180–205. Madison, WI: University of Wisconsin Press, 1988.

Barkan, Elazar. *Retreat of Scientific Racism: Changing Concepts of Race in Britain and the United States Between the World Wars*. Cambridge; New York: Cambridge University Press, 1992.

Barnicot, Nigel. "A Survey of Some Genetical Characters in Ethiopian Tribes: Concluding Discussion." *American Journal of Physical Anthropology* 20, no. 2 (1962): 208–208B.

Bashkin, Orit. *Impossible Exodus: Iraqi Jews in Israel*. Stanford: Stanford University Press, 2017.

Bauer, Susanne. "Virtual Geographies of Belonging: The Case of Soviet and Post-Soviet Human Genetic Diversity Research." *Science, Technology, & Human Values* 39, no. 4 (July 2014): 511–37. https://doi.org/10.1177/0162243914528739.

Baum, Bruce David. *The Rise and Fall of the Caucasian Race: A Political History of Racial Identity*. New York: New York University Press, 2006.

Beaconsfield, Peter. "Malaria, Glucose-6-Phosphate Dehydrogenase Deficiency, and Hb S." *British Medical Journal* 2, no. 5545 (April 15, 1967): 174–75. https://doi.org/10.1136/bmj.2.5545.174-b.

Beaconsfield, Peter, Ezzatollah Mahboubi, and Rebecca Rainsbury. "Epidemiologie Des Glukose-6-Phosphat-Dehydrogenase-Mangels." *Münchener Medizinische Wochenschrift* 109 (1967): 1950–52.

Beaconsfield, Peter, Ezzatollah Mahboubi, Badri Khademi, Rebecca Rainsbury, Esther Aghai, and Chamseddine Mofidi. "Glucose 6 Phosphate Dehydrogenase Deficiency in Iran and Its Relation to Physiopathological Processes." *Acta Medica Iranica* 9, no. 1–2 (1966): 35–42.

Beck, Lois. "Nomads and Urbanites, Involuntary Hosts and Uninvited Guests." *Middle Eastern Studies* 18, no. 4 (October 1982): 426–44. https://doi.org/10.1080/00263208208700524.

Beckett, P. H. T. "ABO Blood Groups in Kerman, South Persia." *Man* 56 (1956): 141.

Benjamin, Ruha. "A Lab of Their Own: Genomic Sovereignty as Postcolonial Science Policy." *Policy and Society* 28, no. 4 (December 2009): 341–55. https://doi.org/10.1016/j.polsoc.2009.09.007.

Beutler, E. "Glucose-6-Phosphate Dehydrogenase Deficiency: A Historical Perspective." *Blood* 111, no. 1 (January 1, 2008): 16–24. https://doi.org/10.1182/blood-2007-04-077412.

Bidar, Abdoollah. "The MN Blood Groups of the Population of Azarbaijan, Iran." *Man* 57 (1957): 103.

Biddiss, Michael D. "The Universal Races Congress of 1911." *Race* 13, no. 1 (July 1971): 37–46. https://doi.org/10.1177/030639687101300103.

Bier, Jess. *Mapping Israel, Mapping Palestine: How Occupied Landscapes Shape Scientific Knowledge.* Cambridge, MA: MIT Press, 2017.

Bill, James A. "The Politics of Student Alienation: The Case of Iran." *Iranian Studies* 2, no. 1 (March 1969): 8–26. https://doi.org/10.1080/00210866908701371.

Binyıldız, Pervin, and Cemal Büyükyüksel. "Türkiye'de kan grupları dağılımı." *İstanbul Üniversitesi Tıp Fakültesi Dergisi* 42 (1979): 166–75.

Bishku, Michael B. "Israel and Ethiopia: From a Special to a Pragmatic Relationship." *Journal of Conflict Studies* 14, no. 2 (1994): 39–62.

Bliss, Catherine. *Race Decoded: The Genomic Fight for Social Justice.* Stanford: Stanford University Press, 2012.

Bodmer, Julia, Batsheva Bonné, Walter F. Bodmer, Susan Black, Ammon Ben-David, and Sara Ashbel. "Study of the HL-A System in a Yemenite Jewish Population in Israel." In *Histocompatibility Testing 1972: Report of an International Workshop & Conference: Colloque de l'Institut National de La Santé et de La Recherche Médicale, Held at Evian, France 23–27 May 1972*, edited by Jean Dausset and Jacques Colombiani, 125–32. Copenhagen: Munksgaard, 1973.

Bodmer, Julia, Sarah Nevo, Walter F. Bodmer, and Anne Coukell. "Study of the HL-A System in a Moslem Arab Population in Israel." In *Histocompatibility Testing 1972: Report of an International Workshop & Conference: Colloque de l'Institut National de La Santé et de La Recherche Médicale, Held at Evian, France 23–27 May 1972*, edited by Jean Dausset and Jacques Colombiani, 117–23. Copenhagen: Munksgaard, 1973.

Bonakdarian, Mansour. "Negotiating Universal Values and Cultural and National Parameters at the First Universal Races Congress." *Radical History Review* 2005, no. 92 (April 1, 2005): 118–32. https://doi.org/10.1215/01636545-2005-92-118.

Bondy, Ruth. *Shiba: rofe le-khol adam.* Tel Aviv: Zemorah Bitan Modan, 1981.

Bonné, Batsheva. "Are There Hebrews Left?" *American Journal of Physical Anthropology* 24, no. 2 (1966): 135–45.

Bonné, Batsheva. "Chaim Sheba (1908–1971)." *American Journal of Physical Anthropology* 36, no. 3 (1972): 308–13.

Bonné, Batsheva. "Genes and Phenotypes in the Samaritan Isolate." *American Journal of Physical Anthropology* 24, no. 1 (1966): 1–19.

Bonné, Batsheva. "The Beduins of South-Sinai." *Proceedings of the Tel-Hashomer Hospital* 7, no. 3/4 (1968): 67–73.

Bonné, Batsheva. "The Samaritans: A Demographic Study." *Human Biology* 35, no. 1 (1963): 61–89.

Bonné, Batsheva, Marilyn Godber, Sarah Ashbel, Arthur E. Mourant, and Donald Tills. "South-Sinai Beduin: A Preliminary Report on Their Inherited Blood Factors." *American Journal of Physical Anthropology* 34, no. 3 (1971): 397–408.

Bonné, Batsheva, Sara Ashbel, M. Modai, M. J. Godber, Arthur E. Mourant, D. Tills, and B. G. Woodhead. "The Habbanite Isolate I. Genetic Markers in the Blood." *Human Heredity* 20, no. 6 (1970): 609–22.

Bonné-Tamir, Batsheva. Ḥayim 'im ha-genim: ḥamishim shenot meḥḳar ba-geneṭiḳah shel 'edot Yiśra'el. Yerushalayim: Karmel, 2010.

Bonné-Tamir, Batsheva. "The HGDP Can Be and Should Be Pursued in Harmony with Its Subjects: An Account of the Israeli Experience." *Politics and the Life Sciences* 18, no. 2 (1999): 291–93.

Bonné-Tamir, Batsheva, Julia G. Bodmer, D. Young, and Sara Ashbel. "HLA Polymorphism in Israel 8. The Armenian Community in Jerusalem." *Tissue Antigens* 11, no. 3 (March 1978): 230–34. https://doi.org/10.1111/j.1399-0039.1978.tb01254.x.

Botting, Douglas. *Island of the Dragon's Blood*. London: Hodder and Stoughton, 1958.

Boué, André, and Joelle Boué. "Étude sur la répartition des groupes sanguins en Iran." *Le Sang* 26 (1955): 708–13.

Boué, André, and Joelle Boué. "Étude sur la répartition des groupes sanguins en Iran II." *Annales de l'Institut Pasteur* 91, no. 6 (1956): 898–911.

Bowman, James E. "Comments on Abnormal Erythrocytes and Malaria." *The American Journal of Tropical Medicine and Hygiene* 13, no. 1 (January 1, 1964): 159–61. https://doi.org/10.4269/ajtmh.1964.13.159.

Bowman, James E. "Haptoglobin and Transferrin Differences in Some Iranian Populations." *Nature* 201 (1964): 88.

Bowman, James E. "Glucose-6-Phosphate Dehydrogenase Deficiency in Kuwait." *American Journal of Human Genetics* 19, no. 3 (1967): 460.

Bowman, James E., and Deryck G. Walker. "The Origin of Glucose-6-Phosphate Dehydrogenase Deficiency in Iran: Theoretical Considerations." In *Proceedings of the Second International Congress of Human Genetics (Rome, September 6–12, 1961)*, edited by Luigi Gedda, 1:583–86. Rome: Instituto G. Mendel, 1963.

Bowman, James E., and Deryck G. Walker. "Virtual Absence of Glutathione Instability of the Erythrocytes Among Armenians in Iran." *Nature* 191, no. 4785 (1961): 221–22.

Bowman, James E., and Henri Frischer. "Malaria, G-6-P.D. Deficiency, and Sickle-Cell Trait." *British Medical Journal* 1, no. 5394 (1964): 1378.

Bowman, James E., Paul E. Carson, Henri Frischer, Robin D. Powell, Edward J. Colwell, Llewellyn J. Legters, Andrew J. Cottingham, Stephen C. Boone, and Wesley W. Hiser. "Hemoglobin and Red Cell Enzyme Variation in Some Populations of the Republic of Vietnam with Comments on the Malaria Hypothesis." *American Journal of Physical Anthropology* 34, no. 3 (May 1971): 313–24. https://doi.org/10.1002/ajpa.1330340302.

Boyd, William C. "Blood Groups." *Tabulae Biologicae* 17, no. 2 (1939): 113–240.

Boyd, William C. "Critique of Methods of Classifying Mankind." *American Journal of Physical Anthropology* 27, no. 3 (1940): 333–64.

Boyd, William C. "Genetics and the Human Race." *Science* 140, no. 3571 (1963): 1057–64.

Boyd, William C. *Genetics and the Races of Man: An Introduction to Modern Physical Anthropology*. Boston: Little, Brown & Co., 1950.

Boyd, William C., and Lyle G. Boyd. "Blood Groups and Inbreeding in Syria." *American Journal of Physical Anthropology* 28, no. 3 (1941): 319–30.

Boyd, William C., and Lyle G. Boyd. "Blood Groups and Types in Baghdad and Vicinity." *Human Biology* 13, no. 3 (1941): 398–404.

Boyd, William C., and Lyle G. Boyd. "New Data on Blood Groups and Other Inherited Factors in Europe and Egypt." *American Journal of Physical Anthropology* 23, no. 1 (July 1937): 49–70. https://doi.org/10.1002/ajpa.1330230106.

Boyd, William C., and Lyle G. Boyd. "The Blood Groups of the Rwala Bedouin." *Journal of Immunology* 34 (1938): 441–46.

Bozarslan, Hamit. "Kurds and the Turkish State." In *The Cambridge History of Turkey*, edited by Reşat Kasaba, 4:333–56, 2008.

Bozorgmehr, Mehdi. "Internal Ethnicity: Iranians in Los Angeles." *Sociological Perspectives* 40, no. 3 (January 1997): 387–408. https://doi.org/10.2307/1389449.

Braun, Hugo, and Ziya Öktem. "Türklerde kan grupları tevziatı." *İstanbul Seririyatı* 3 (March 8, 1938).

Breisach, Ernst. *Historiography: Ancient, Medieval, and Modern.* 3rd ed. Chicago: University of Chicago Press, 2007.

Brzezinski, Aron, Joseph Gurevitch, David Hermoni, and G. Mundel. "Blood Groups in Jews from the Yemen." *Annals of Eugenics* 16, no. 1 (1951): 335–37.

Burton, Elise K. "'Essential Collaborators': Locating Middle Eastern Geneticists in the Global Scientific Infrastructure, 1950s–1970s." *Comparative Studies in Society and History* 60, no. 1 (2018): 119–49.

Büyükyüksel, Cemal. "Groupes sanguins ABO et Rh (D) dans la population turque." *Revue Française de Transfusion* 16, no. 4 (1973): 403–10.

Büyükyüksel, Cemal. *Türkiyede Kan Grupları Dağılımı.* Ankara: Türkiye Kızılay Derneği Genel Merkezi Sağlık Yayınları, 1969.

Çağaptay, Soner. *Islam, Secularism, and Nationalism in Modern Turkey: Who Is a Turk?* London; New York: Routledge, 2006.

Canikligil, Razi. "Bu Ödülü 19 Mayıs'ta Anıtkabir'de Atatürk'e Sunacağım." *Hürriyet*, December 11, 2015. http://www.hurriyet.com.tr/bu-odulu19-mayista-anitka birde-ataturke-sunacagim-40025570.

Carson, Paul E., C. Larkin Flanagan, C. E. Ickes, and Alf S. Alving. "Enzymatic Deficiency in Primaquine-Sensitive Erythrocytes." *Science* 124, no. 3220 (1956): 484–85.

Çataloluk, Osman. *Türk'ün genetik tarihi.* İstanbul: Togan Yayıncılık, 2012.

Cavalli-Sforza, Luigi Luca, Allan C. Wilson, Charles R. Cantor, Robert M. Cook-Deegan, and Mary-Claire King. "Call for a Worldwide Survey of Human Genetic Diversity: A Vanishing Opportunity for the Human Genome Project." *Genomics* 11, no. 2 (October 1991): 490–91. https://doi.org/10.1016/0888-7543(91)90169-F.

CENTO. *CENTO Conference on Combating Malnutrition in Preschool Children.* Islamabad: Office of the U.S. Economic Coordinator for CENTO Affairs, 1968.

Chadarevian, Soraya de. "Following Molecules: Hemoglobin Between the Clinic and the Laboratory." In *Molecularizing Biology and Medicine: New Practices and Alliances, 1910s–1970s*, edited by Soraya de Chadarevian and Harmke Kamminga, 171–201. Amsterdam: Harwood Academic Publishers, 1998.

Cheng, Yinghong. "'Is Peking Man Still Our Ancestor?'—Genetics, Anthropology, and the Politics of Racial Nationalism in China." *The Journal of Asian Studies* 76, no. 03 (August 2017): 575–602. https://doi.org/10.1017/S0021911817000493.

Choremis, C., N. Zervos, V. Constantinides, and L. Zannos. "Sickle-Cell Anaemia in Greece." The *Lancet* 257, no. 6665 (May 1951): 1147–49. https://doi.org/10.1016/S0140-6736(51)92657-8.

Cleve, H., W. Patutschnick, S. Nevo, and G. G. Wendt. "Genetic Studies on the Gc Subtypes." *Human Genetics* 44, no. 2 (1978): 117–22. https://doi.org/10.1007/BF00295404.

Collins, K. J., and J. S. Weiner. *Human Adaptability: A History and Compendium of Research in the International Biological Programme*. London: Taylor & Francis, 1977.

Collopy, Peter Sachs. "Race Relationships: Collegiality and Demarcation in Physical Anthropology." *Journal of the History of the Behavioral Sciences* 51, no. 3 (July 2015): 237–60. https://doi.org/10.1002/jhbs.21728.

Comfort, Nathaniel. "The Prisoner as Model Organism: Malaria Research at Stateville Penitentiary." *Studies in History and Philosophy of Science Part C: Studies in History and Philosophy of Biological and Biomedical Sciences* 40, no. 3 (September 2009): 190–203. https://doi.org/10.1016/j.shpsc.2009.06.007.

Comfort, Nathaniel C. *The Science of Human Perfection: How Genes Became the Heart of American Medicine*. New Haven, CT: Yale University Press, 2012.

Coon, Carleton S. *Measuring Ethiopia and Flight into Arabia*. London: Jonathan Cape, 1936.

Crampton, Jeremy W. "The Cartographic Calculation of Space: Race Mapping and the Balkans at the Paris Peace Conference of 1919." *Social & Cultural Geography* 7, no. 5 (October 2006): 731–52. https://doi.org/10.1080/14649360600974733.

Dabbous, Ibrahim A., and Salim S. Firzli. "Sickle Cell Anemia in Lebanon—Its Predominance in the Mohammedans." *Zeitschrift Für Morphologie Und Anthropologie* 58, no. 3 (1968): 225–31.

Dacie, John. "Hermann Lehmann, 8 July 1910–13 July 1985." *Biographical Memoirs of the Fellows of the Royal Society* 34 (1988): 406–49.

Davidovitch, Nadav, and Avital Margalit. "Public Health, Racial Tensions, and Body Politic: Mass Ringworm Irradiation in Israel, 1949–1960." *The Journal of Law, Medicine & Ethics* 36, no. 3 (September 2008): 522–29. https://doi.org/10.1111/j.1748-720X.2008.300.x.

Davidovich, Nadav, and Shifra Shvarts. "Health and Hegemony: Preventive Medicine, Immigrants and the Israeli Melting Pot." *Israel Studies* 9, no. 2 (2004): 150–79.

DellaPergola, Sergio. "Demography in Israel at the Dawn of the Twenty-First Century." In *Jews in Israel: Contemporary Social and Cultural Patterns*, edited by Uzi Rebhun and Chaim I. Waxman, 20–44. Lebanon, NH: Brandeis University Press, 2004.

Der Matossian, Bedross. "The Armenians of Palestine 1918–48." *Journal of Palestine Studies* 41, no. 1 (2011): 24–44.

Derman, Halil. *Tıb Dünyası* 15 (1942): 4831.

Diamond, Stanley, Richard A. Falk, Robert L. Heilbroner, and Edward J. Nell. "The Case of Dr. Besikci." *New York Review of Books*, August 12, 1982.

Dilaçar, Agop. "Alpin ırk, Türk etnisi, ve Hatay halkı." In *CHP Konferanslar Serisi Kitap 19*. Ankara: Recep Ulusoğlu Basımevi, 1940.

Dinçol, Günçağ, Muzaffer Aksoy, and Şakir Erdem. "Beta-Thalassaemia with Increased Haemoglobin A2 in Turkey." *Human Heredity* 29, no. 5 (1979): 272–78.

Dodd, Stuart C. "The Village Welfare Service in Lebanon, Syria and Palestine." *Journal of the Royal Central Asian Society* 32, no. 1 (January 1945): 87–90. https://doi.org/10.1080/03068374508731159.

Dogan, Serkan, Cemal Gurkan, Mustafa Dogan, Hasan Emin Balkaya, Ramazan Tunc, Damla Kanliada Demirdov, Nihad Ahmed Ameen, and Damir Marjanovic. "A Glimpse at the Intricate Mosaic of Ethnicities from Mesopotamia: Paternal Lineages of the Northern Iraqi Arabs, Kurds, Syriacs, Turkmens and Yazidis." Edited by Chuan-Chao Wang. *PLOS ONE* 12, no. 11 (November 3, 2017): e0187408. https://doi.org/10.1371/journal.pone.0187408.

Donegani, Joyce A., Karima A. Ibrahim, Elizabeth W. Ikin, and Arthur E. Mourant. "The Blood Groups of the People of Egypt." *Heredity* 4, no. 3 (December 1950): 377–82. https://doi.org/10.1038/hdy.1950.30.

Donoso, Gonzalo, Habibollah Hedayat, and H. Khayatian. "Favism, with Special Reference to Iran." *Bulletin of the World Health Organization* 40 (1969): 513–19.

Dreyfuss, F., G. Mundel, and M. Benyesch. "Sickling in Oriental Jews." *Acta Haematologica* 9, no. 3 (1953): 193–99. https://doi.org/10.1159/000204284.

Dreyfuss, Fritz, and M. Benyesch. "Sickle-Cell Trait in Yemenite Jews." *Nature* 167 (1951): 950.

Dreyfuss, Fritz, Elizabeth W. Ikin, Hermann Lehmann, and Arthur E. Mourant. "An Investigation of Blood-Groups and a Search for Sickle-Cell Trait in Yemenite Jews." The *Lancet* 260, no. 6743 (1952): 1010–12.

Duara, Prasenjit. *Rescuing History from the Nation: Questioning Narratives of Modern China*. Chicago: University of Chicago Press, 1996.

Dujarric de la Rivière, René, and Nicolas Kossovitch. *Les groupes sanguins*. Paris: J. B. Baillière, 1936.

Dündar, Can. "Aziz Sancar Türk Mü, Kürt Mü?" *Cumhuriyet*, October 9, 2015. http://

www.cumhuriyet.com.tr/koseyazisi/384070/Aziz_Sancar_Turk_mu__Kurt _mu_.html.

Duruel, Nursel, Çiğdem Altay, and Orhan N. Ulutin, eds. *Bilime adanmış bir ömür: Muzaffer Aksoy*. Ankara: Türkiye Bilimler Akademisi, 2005.

Effros, Bonnie. "Berber Genealogy and the Politics of Prehistoric Archaeology and Craniology in French Algeria (1860s–1880s)." *The British Journal for the History of Science* 50, no. 1 (March 2017): 61–81. https://doi.org/10.1017/S0007087417000024.

Efron, John M. *Defenders of the Race: Jewish Doctors and Race Science in Fin-de-Siècle Europe*. New Haven, CT: Yale University Press, 1994.

Eisenstadt, S. N. "The Oriental Jews in Israel: A Report on a Preliminary Study in Culture-Contacts." *Jewish Social Studies* 12, no. 3 (1950): 199–222.

El Shakry, Omnia S. *The Great Social Laboratory: Subjects of Knowledge in Colonial and Postcolonial Egypt*. Stanford: Stanford University Press, 2007.

El-Dewi, Salah. "The Rh Groups Among Egyptians." *Journal of the Royal Egyptian Medical Association* 32, no. 11/12 (1949): 883–94.

Elling, Rasmus Christian. "War of Clubs: Struggle for Space in Abadan and the 1946 Oil Strike." In *Violence and the City in the Modern Middle East*, edited by Nelida Fuccaro, 189–210. Stanford: Stanford University Press, 2016.

Elling, Rasmus Christian. *Minorities in Iran: Nationalism and Ethnicity After Khomeini*. New York: Palgrave Macmillan, 2013.

Elshakry, Marwa. *Reading Darwin in Arabic, 1860–1950*. Chicago: University of Chicago Press, 2013.

Epstein, Steven. *Inclusion: The Politics of Difference in Medical Research*. Chicago: University of Chicago Press, 2007.

Erdentuğ, Nermin Aygen. "The Pioneering Anthropologists of Turkey: Personal Profiles in Socio-Political Context." *Turkish Studies Association Bulletin* 22, no. 2 (1998): 13–44.

Ergin, Murat. "Biometrics and Anthropometrics: The Twins of Turkish Modernity." *Patterns of Prejudice* 42, no. 3 (2008): 281–304. https://doi.org/10.1080/00313220802204038.

Ergin, Murat. "Cultural Encounters in the Social Sciences and Humanities: Western Émigré Scholars in Turkey." *History of the Human Sciences* 22, no. 1 (2009): 105–30. https://doi.org/10.1177/0952695108099137.

Ergin, Murat. "The Racialization of Kurdish Identity in Turkey." *Ethnic and Racial Studies* 37, no. 2 (January 28, 2014): 322–41. https://doi.org/10.1080/01419870.2012.729672.

Ergin, Murat. *Is the Turk a White Man?: Race and Modernity in the Making of Turkish Identity*. Leiden: Brill, 2017.

Ergün, Ahmet, and Nevzat Aksoy. "Yurt içi ABO ve Rh kan grupları araştırması." *Ankara Tıp Mecmuası* 40 (1987): 319–28.

Ergün, Ahmet, and Serdar Yardımcı. "Türkiye genelinde ABO kan grupları ve Rh faktörünün dağılımı." *Ankara Tıp Mecmuası* 46 (1993): 527–33.

Erşan, Mesude. "Türklere özel bir genom bulunamadı." *Hürriyet*, January 21, 2012. http://www.hurriyet.com.tr/turklere-ozel-bir-genom-bulunamadi-19737173.

Fakhrai, Habibollah, Fereydun Karimi Boushehri, Aghdas Rasouli-Nia, Mansour Marzban, Nahid Tayebi, Shahbaz Yosefi, and Kazem Kazempour. "Comparison of ABO Blood System in Persian and Arabic Speaking Khuzistanies." In *The Program and Abstracts of the Seventh National Genetics Congress at Jundi Shapur University, Ahvaz, Iran, March 14–17, 1978*, 43. Ahvaz: Jundi Shapur University, 1978.

Falk, Raphael. *Zionism and the Biology of the Jews*. New York: Springer International Publishing, 2017.

Fan, Fa-ti. "The Global Turn in the History of Science." *East Asian Science, Technology and Society* 6, no. 2 (January 1, 2012): 249–58. https://doi.org/10.1215/18752160-1626191.

Farhud, D. D., R. Ananthakrishnan, Hubert Walter, and J. Loser. "Electrophoretic Investigation of Some Red Cell Enzymes in Iran." *Human Heredity* 23 (1973): 263–66.

Farhud, Dariush Daneshvar. *Īrān-i farhangī*. Tehran: Ṣabā, 2013.

Farhud, Dariush Daneshvar, A. S. Lotfi, M. Hashemzadeh Chaleshtori, M. Akhondi, and H. Sadighi. "Progress of Education, Research and Services in Medical Genetics, in Some Institutions of Iran." *Iranian Journal of Public Health* 38, no. 1 (2009): 115–18.

Farhud, Dariush Daneshvar, and Hubert Walter. "Hp Subtypes in Iranians." *Human Heredity* 22 (1972): 184–89.

Farhud, Dariush Daneshvar, and Hubert Walter. "Polymorphism of C'3 in German, Bulgarian, Iranian and Angola Population." *Humangenetik* 17 (1973): 161–64.

Farhud, Dariush Daneshvar, Parivash Amirshahi, and Shua' al-din Hedayat. "Taʿyīn-i anvāʿ-i hāptūglūbīn dar ustān-i sāḥilī (Bandar-i ʿAbbās)." *Iranian Journal of Public Health* 7, no. 4 (1978): 181–93.

Farzanfar, Ramesh. "Ethnic Groups and the State: Azaris, Kurds and Baluch of Iran." PhD diss., Massachusetts Institute of Technology, 1992.

Fazeli, Nematollah. *Politics of Culture in Iran: Anthropology, Politics and Society in the Twentieth Century*. London; New York: Routledge, 2006.

Felder, Björn M. "'God Forgives—But Nature Never Will': Racial Identity, Racial Anthropology, and Eugenics in Latvia 1918–1940." In *Baltic Eugenics: Bio-Politics, Race and Nation in Interwar Estonia, Latvia and Lithuania, 1918–1940*, edited by Björn M. Felder and Paul J. Weindling, 115–46. Leiden: Brill, 2013. https://brill.com/view/book/edcoll/9789401209762/B9789401209762-s007.xml.

Field, Henry. *Contributions to the Anthropology of Iran*. Publications of the Field Museum of Natural History. Anthropological Series, 29.1. Chicago: Field Museum of Natural History, 1939.

Firro, Kais. *The Druzes in the Jewish State: A Brief History*. Leiden: Brill, 1999.

First Seminar of Favism in Iran, July 6, 1965. Tehran: Ministry of Health, Food & Nutrition Institute of Iran, 1965.

Fujinami, Nobuyoshi. "'Abd al-Ḥamīd al-Zahrāwī and His Thought Reconsidered: An Intellectual Portrait of the Arab Nationalist as an Ottoman Politician." *Journal of Ottoman Studies* 51 (2018): 239–63.

Fullwiley, Duana. "The Biologistical Construction of Race: 'Admixture' Technology and the New Genetic Medicine." *Social Studies of Science* 38, no. 5 (2008): 695–735. https://doi.org/10.1177/0306312708090796.

Fullwiley, Duana. *The Enculturated Gene: Sickle Cell Health Politics and Biological Difference in West Africa*. Princeton: Princeton University Press, 2011.

Gannett, Lisa, and James R. Griesemer. "The ABO Blood Groups: Mapping the History and Geography of Genes in Homo Sapiens." In *Classical Genetics Research and Its Legacy: The Mapping Cultures of Twentieth Century Genetics*, edited by Hans-Jörg Rheinberger and Jean-Paul Gaudillière, 119–72. Oxon, UK: Routledge, 2004.

Gelpi, Armand P. "Glucose-6-Phosphate Dehydrogenase Deficiency in Saudi Arabia: A Survey." *Blood* 25, no. 4 (1965): 486–93.

Gelpi, Armand P. "Glucose-6-Phosphate Dehydrogenase Deficiency, the Sickling Trait, and Malaria in Saudi Arab Children." *Journal of Pediatrics* 71, no. 1 (July 1967): 138–46. https://doi.org/10.1016/S0022-3476(67)80246-4.

Gelpi, Armand P. "Migrant Populations and the Diffusion of the Sickle-Cell Gene." *Annals of Internal Medicine* 79 (1973): 258–64.

Gelpi, Armand P. "Sickle Cell Disease in Saudi Arabs." *Acta Haematologica* 43 (1970): 89–99.

Gelvin, James L. *The Modern Middle East: A History*. 3rd ed. New York: Oxford University Press, 2011.

Gini, Corrado. "I Samaritani." *Genus* 1, no. 1/2 (1934): 117–46.

Gissis, Snait B. "When Is 'Race' a Race? 1946–2003." *Studies in History and Philosophy of Science Part C: Studies in History and Philosophy of Biological and Biomedical Sciences* 39, no. 4 (December 2008): 437–50. https://doi.org/10.1016/j.shpsc.2008.09.006.

Glassner, Martin Ira. "The Bedouin of Southern Sinai Under Israeli Administration." *Geographical Review* 64, no. 1 (1974): 31–60.

Godber, M. J., A. C. Kopec, Arthur E. Mourant, D. Tills, and Eliahu E. Lehmann. "The Hereditary Blood Factors of the Yemenite and Kurdish Jews." *Philosophical Transactions of the Royal Society of London* 266 (1973): 169–84.

Gohar, M. A. "A Preliminary Note on the Distribution of ABO Blood Groups and Incidence of Rh Negatives in Egyptians." *Journal of the Royal Egyptian Medical Association* 32, no. 10 (1949): 727–28.

Gökçümen, Ömer, and Timur Gültekin. "Genetik ve Kamusal Alan." *Ankara Üniversitesi Dil ve Tarih-Coğrafya Fakültesi Dergisi* 49, no. 1 (2009): 19–31.

Gökçümen, Ömer, Timur Gültekin, Yeşim Doğan Alakoç, Ayşim Tuğ, Erksin Güleç, and Theodore G. Schurr. "Biological Ancestries, Kinship Connections, and Projected Identities in Four Central Anatolian Settlements: Insights from Culturally Contextualized Genetic Anthropology." *American Anthropologist* 113, no. 1 (March 2011): 116–31. https://doi.org/10.1111/j.1548-1433.2010.01310.x.

Goldschmidt, Elisabeth, ed. *The Genetics of Migrant and Isolate Populations: Proceedings of a Conference on Human Population Genetics in Israel Held at the Hebrew University, Jerusalem, 1961*. Baltimore: Williams & Wilkins, 1963.

Górny, Maciej. "Racial Anthropology on the Eastern Front, 1912 to Mid-1920s." In *National Races: Transnational Power Struggles in the Sciences and Politics of Human Diversity, 1840–1945*, edited by Richard McMahon, 271–94. Lincoln: University of Nebraska Press, 2019.

Gottweis, Herbert, and Byoungsoo Kim. "Bionationalism, Stem Cells, BSE, and Web 2.0 in South Korea: Toward the Reconfiguration of Biopolitics." *New Genetics and Society* 28, no. 3 (September 2009): 223–39. https://doi.org/10.1080/14636770903162437.

Graham, M., and S. Khosravi. "Home Is Where You Make It: Repatriation and Diaspora Culture Among Iranians in Sweden." *Journal of Refugee Studies* 10, no. 2 (January 1, 1997): 115–33. https://doi.org/10.1093/jrs/10.2.115.

Gribetz, Jonathan Marc. "'Their Blood Is Eastern': Shahin Makaryus and *Fin de Siècle* Arab Pride in the Jewish 'Race.'" *Middle Eastern Studies* 49, no. 2 (March 2013): 143–61. https://doi.org/10.1080/00263206.2012.759107.

Gribetz, Jonathan Marc. *Defining Neighbors: Religion, Race, and the Early Zionist-Arab Encounter.* Jews, Christians, and Muslims from the Ancient to the Modern World. Princeton and Oxford: Princeton University Press, 2014.

Grigor, Talinn. "Recultivating 'Good Taste': The Early Pahlavi Modernists and Their Society for National Heritage." *Iranian Studies* 37, no. 1 (March 2004): 17–45. https://doi.org/10.1080/0021086042000232929.

Grigor, Talinn. *Building Iran: Modernism, Architecture, and National Heritage Under the Pahlavi Monarchs.* New York: Periscope Publishing, 2009.

Gualtieri, Sarah M. A. *Between Arab and White: Race and Ethnicity in the Early Syrian American Diaspora.* Berkeley: University of California Press, 2009.

Gültekin, Timur, and Ömer Gökçümen. "Genetik Bilgi ve Antropoloji." *Bilim ve Teknik*, January 2009: 50–55.

Gurevitch, Joseph, and Emmanuel Margolis. "Blood Groups in Jews from Iraq." *Annals of Human Genetics* 19 (1955): 257–59.

Gurevitch, Joseph, Aron Brzezinski, and Zeev Polishuk. "Rh Factor and Haemolytic Disease of the Newborn in Jerusalem Jews." *The Lancet* 250, no. 6487 (1947): 943–44.

Gurevitch, Joseph, David Hermoni, and Emmanuel Margolis. "Blood Groups in Kurdistani Jews." *Annals of Eugenics* 17, no. 1 (1952): 94–95.

Gurevitch, Joseph, David Hermoni, and Zeev Polishuk. "Rh Blood Types in Jerusalem Jews." *Annals of Eugenics* 16, no. 1 (1951): 129–30.

Gurevitch, Joseph, Elijah Hasson, and Emmanuel Margolis. "Blood Groups in Persian Jews." *Annals of Human Genetics* 21 (1956): 135–38.

Gurevitch, Joseph, Elijah Hasson, Emmanuel Margolis, and C. Poliakoff. "Blood Groups in Jews from Cochin, India." *Annals of Human Genetics* 19, no. 4 (1955): 254–56.

Gurevitch, Joseph, Elijah Hasson, Emmanuel Margolis, and C. Poliakoff. "Blood Groups in Jews from Tripolitania." *Annals of Human Genetics* 19, no. 4 (1955): 260–61.

Gurevitch, Joseph, Emmanuel Margolis, and David Hermoni. "Blood Groups in Sephardic Jews." In *Proceedings of the Seventh Congress of the International Society of Blood Transfusion, Rome, September 3–6, 1958*, edited by L. Holländer, 274–76. Basel: S. Karger, 1959.

Haarmann, Ulrich W. "Ideology and History, Identity and Alterity: The Arab Image of the Turk from the Abbasids to Modern Egypt." *International Journal of Middle East Studies* 20, no. 2 (1988): 175–96.

Haas, Professor [Wilhelm]. "Vaẓāyif-i mūzah-i insān-shināsī." *Taʿlīm va Tarbiyat* 6, no. 12 (1315 [1936]): 94–97.

Habibi, Gholamreza, Armin Rashidi, and Marc D. Feldman. "Emerging Concerns About Iran's Scientific and Medical Future." The *Lancet* 368, no. 9540 (September 2006): 985. https://doi.org/10.1016/S0140-6736(06)69403-8.

Habibzadeh, Farrokh, and Karim Vessal. "Scientific Research in Iran: Forgotten Factors." The *Lancet* 368, no. 9546 (October 2006): 1494. https://doi.org/10.1016/S0140-6736(06)69635-9.

Hadji Mirza Yahya. "Persia." In *Papers on Inter-Racial Problems Communicated to the First Universal Races Congress, Held at the University of London, July 26–29, 1911*, edited by Gustav Spiller, 143–54. London; Boston: P. S. King & Son; World's Peace Foundation, 1911.

Hakohen, Devorah. *Immigrants in Turmoil: Mass Immigration to Israel and Its Repercussions in the 1950s and After*. Modern Jewish History. Syracuse, NY: Syracuse University Press, 2003.

Haldane, J. B. S. "The Blood-Group Frequencies of European Peoples, and Racial Origins." *Human Biology* 12, no. 4 (1940): 457–80.

Hamilton, R. A. B. "The Social Organization of the Tribes of the Aden Protectorate." *Journal of The Royal Central Asian Society* 30, no. 2 (May 1943): 142–57. https://doi.org/10.1080/03068374308731086.

Hamza, Sawsan, Karim Kamel, Nadia Moafy, W. N. Ibrahim, Osaima Selim, Aida Azim, Fadila Sabry, and Kirk C. Hoerman. "Hereditary Blood Factors of North Sinai Inhabitants." *Human Biology* 48, no. 1 (February 1976): 193–202.

Haqiqi, Fazlallah. "'Ilm-i insān-shināsī." *Taʻlīm va Tarbiyat* 7, no. 3 (1316 [1937]): 165–70.

Hardy, M. J., and M. S. Ragbeer. "Homozygous HbE and HbSE Disease in a Saudi Family." *Hemoglobin* 9, no. 1 (January 1985): 47–52. https://doi.org/10.3109/03630268508996981.

Harrison, Carol E., and Ann Johnson, eds. *National Identity: The Role of Science and Technology*. Vol. 24. Osiris. Chicago: University of Chicago Press, 2009.

Hedayat, Shua' al-din, P. Amirshahy, and B. Khademy. "Frequency of G-6-PD Deficiency Among Some Iranian Ethnic Groups." *Tropical and Geographical Medicine* 19 (1969): 163–68.

Herrick, James B. "Peculiar Elongated and Sickle-Shaped Red Blood Corpuscles in a Case of Severe Anemia. 1910." *Yale Journal of Biology and Medicine* 74, no. 3 (2001): 179–84.

Hilgartner, Stephen. *Reordering Life: Knowledge and Control in the Genomics Revolution*. Cambridge, MA: MIT Press, 2017.

Hirsch, Dafna. "'We Are Here to Bring the West, Not Only to Ourselves': Zionist

Occidentalism and the Discourse of Hygiene in Mandate Palestine." *International Journal of Middle East Studies* 41, no. 04 (2009): 577.

Hirsch, Dafna. "Zionist Eugenics, Mixed Marriage, and the Creation of a 'New Jewish Type.'" *Journal of the Royal Anthropological Institute* 15 (2009): 592–609.

Hirschfeld, Ludwik, and Hanka Hirschfeld. "Serological Differences Between the Blood of Different Races." The *Lancet*, October 18, 1919, 675–79.

Hirszfeld, Ludwik. *Ludwik Hirszfeld: The Story of One Life.* Edited by William H. Schneider. Translated by Marta Aleksandra Balinska. Rochester, NY: University of Rochester Press, 2010.

Hossfeld, Uwe. "Nachruf Prof. Dr. Rer. Nat. Habil. Dr. Med. h. c. Hubert Walter, 14 April 1930–6 Dezember 2008." *Verhandlungen Zur Geschichte Und Theorie Der Biologie* 16 (2010): 281–307.

Houston, Christopher. "An Anti-History of a Non-People: Kurds, Colonialism, and Nationalism in the History of Anthropology." *Journal of the Royal Anthropological Institute* 15 (2009): 19–35.

Hunter, Frederick Mercer. *An Account of the British Settlement of Aden in Arabia.* London: Frank Cass & Co. Ltd., 1877.

Huxley, Henry Minor. "Preliminary Report of Anthropological Expedition to Syria." *American Anthropologist* 4, no. 1 (1902): 47–51.

Huxley, Henry Minor. "Samaritans." In *Jewish Encyclopedia*, 10:674–76, 1906.

Hyun, Jaehwan. "Blood Purity and Scientific Independence: Blood Science and Postcolonial Struggles in Korea, 1926–1975." *Science in Context* 32, no. 3 (2019): 239–60.

Ibrahim, W. N., Karim Kamel, Osaima Selim, Aida Azim, M. F. Gaballah, Fadila Sabry, A. El-Naggar, and Kirk C. Hoerman. "Hereditary Blood Factors and Anthropometry of the Inhabitants of the Egyptian Siwa Oasis." *Human Biology* 46, no. 1 (1974): 57–68.

Iğsız, Aslı. *Humanism in Ruins: Entangled Legacies of the Greek-Turkish Population Exchange.* Stanford: Stanford University Press, 2018.

Ikin, Elizabeth W. "Blood Group Distribution in the Near East." In *Proceedings of the Seventh Congress of the International Society of Blood Transfusion, Rome, September 3–6, 1958*, edited by L. Holländer, 262–65. Basel: S. Karger, 1959.

Ikin, Elizabeth Woodroffe. "The Incidence of the Blood Antigens in Different Populations." PhD diss., University of London, 1963.

Ikin, Elizabeth W., Arthur E. Mourant, and Hermann Lehmann. "The Blood Groups and Haemoglobin of the Assyrians of Iraq." *Man* 65 (July 1965): 110–11. https://doi.org/10.2307/2797446.

İnan, Afet. *L'Anatolie, le pays de la "race" turque: recherches sur les caractères* anthropologiques des populations de la Turquie (enquête sur 64,000 individus). Genève: Imprimerie Albert Kundig, 1939.

İnan, Afet. *L'Anatolie, le pays de la "race" turque: recherches sur les caractères anthropologiques des populations de la Turquie (enquête sur 64,000 individus)*. Genève: Georg, 1941.

İnan, Afet. *Prof. Dr. Afet İnan*. İstanbul: Remzi Kitabevi, 2005.

İnan, Afet. *Türkiye halkının antropolojik karakterleri ve Türkiye tarihi: Türk ırkının vatanı Anadolu (64.000 kisi üzerinde anket)*. Ankara: Türk Tarih Kurumu Basımevi, 1947.

Irmak, Sadi. "Anadolu yürüklerinin kan gruplarına dair araştırmalar." *Tıb Dünyası* 10, no. 11–115 (1937): 3645–48.

Irmak, Sadi. "Die ersten Blutgruppenuntersuchungen bei Nomaden in Kleinasien: Betrachtungen über die türkische-mittelasiatische Serologie." *Zeitschrift für Immunitätsforschung* 92 (1938): 74–78.

Irmak, Sadi. "Türk ırkının biyolojisine dair araştırmalar (kan gruplar ve parmak izleri)." In *İkinci Türk Tarih Kongresi, İstanbul 20–25 Eylül 1937*, 841–45. İstanbul: Kenan Matbaası, 1943.

Izzeddin, Nejla Mustapha. "The Racial Origins of the Druze." PhD diss., University of Chicago, 1934.

Jackson, John P. "'In Ways Unacademical': The Reception of Carleton S. Coon's The Origin of Races." *Journal of the History of Biology* 34, no. 2 (June 2001): 247–85.

Jasanoff, Sheila. "Ordering Knowledge, Ordering Society." In *States of Knowledge: The Co-Production of Science and Social Order*, edited by Sheila Jasanoff, 13–45. London: Routledge, 2004.

Jones, Toby Craig. *Desert Kingdom: How Oil and Water Forged Modern Saudi Arabia*. Cambridge, MA: Harvard University Press, 2010.

Jonxis, J. H. P., and J. F. Delafresnaye, eds. *Abnormal Haemoglobins: A Symposium Organized by the Council for International Organizations of Medical Sciences, Established Under the Joint Auspices of UNESCO and WHO*. Springfield, IL: Charles C. Thomas, 1959.

Kalling, Ken, and Leiu Heapost. "Racial Identity and Physical Anthropology in Estonia, 1800–1945." In *Baltic Eugenics: Bio-Politics, Race and Nation in Interwar Estonia, Latvia and Lithuania, 1918–1940*, edited by Björn M. Felder and Paul J. Weindling, 83–114. Leiden: Brill, 2013. https://doi.org/10.1163/9789401209762_006.

Kamel, Karim, and Ahmed Y. Awny. "Origin of the Sickling Gene." *Nature* 205, no. 4974 (February 1965): 919. https://doi.org/10.1038/205919a0.

Kamel, Karim, and Nadia Moafy. "Some Aspects of Thalassemia and the World

Common Hemoglobinopathies in the Middle East." In *Distribution and Evolution of Hemoglobin and Globin Loci: Proceedings of the Fourth Annual Comprehensive Sickle Cell Center Symposium on the Distribution and Evolution of Hemoglobin and Globin Loci at the University of Chicago, Chicago, Illinois, U.S.A., October 10–12, 1982*, edited by James E. Bowman, 209–20. University of Chicago Sickle Cell Center Hemoglobin Symposia, v. 4. New York: Elsevier, 1983.

Kamel, Karim. "Heterogeneity of Sickle Cell Anaemia in Arabs: Review of Cases with Various Amounts of Fetal Haemoglobin." *Journal of Medical Genetics* 16, no. 6 (December 1, 1979): 428–30. https://doi.org/10.1136/jmg.16.6.428.

[Kansu], Şevket Aziz. "Anadolu ve Rumeli Türklerinin antropometrik tetkikleri, birinci muhtıra: Boy ve gövde nispetleri." *Türk Antropoloji Mecmuası* 12 (1931): 3–39.

[Kansu], Şevket Aziz. "Kan grupları hakkında." *Tıp Fakültesi Mecmuası*, no. 5–6 (1931): 253–61.

Kansu, Şevket Aziz. "Rassengeschichte der Türkei." *Belleten* 40, no. 159 (1976): 353–402.

Kaplan, Sam. "Din-u Devlet All Over Again? The Politics of Military Secularism and Religious Militarism in Turkey Following the 1980 Coup." *International Journal of Middle East Studies* 34 (2002): 113–27.

Kappers, C. U. Ariens. "Anthrūbūlūjiyyat al-sharq al-adanī." *al-Kulliyya* 18, no. 5 (July 1932): 349–65.

Kappers, C. U. Ariens. *An Introduction to the Anthropology of the Near East in Ancient and Recent Times*. Amsterdam: Noord-Hollandsche Uitgeversmaatschappij, 1934.

Kariminezhad, Mohammad Hasan, Houshang Khavari Khorasani, and Mansour Omidi. "Tārīkhchih-i zhinitīk dar Īrān va jahān, bakhsh-i dahhum: zindagīnāmah-i duktur-i Pizishkpūr-i Mustashfī." *Genetics in the 3rd Millennium* 11, no. 3 (2013): 4–8.

Katouzian, Homa. "Alborz and Its Teachers." *Iranian Studies* 44, no. 5 (September 2011): 743–54. https://doi.org/10.1080/00210862.2011.570483.

Katouzian, Homa. "The Campaign Against the Anglo-Iranian Agreement of 1919." *British Journal of Middle Eastern Studies* 25, no. 1 (1998): 5–46.

Katouzian, Homa. *The Persians: Ancient, Medieval, and Modern Iran*. New Haven, CT: Yale University Press, 2009.

Kaufman, Asher. *Reviving Phoenicia: The Search for Identity in Lebanon*. London: I. B. Tauris, 2014.

Kay Ustuvān, Ḥusayn. *Nizhād va zabān*. Tihrān: Sahāmī, 1937.

Kayssi, Ahmed I. "The Rh Blood Groups of the Population of Baghdad." *American Journal of Physical Anthropology* 7, no. 4 (December 1949): 549–51. https://doi.org/10.1002/ajpa.1330070407.

Kayssi, Ahmed I., William C. Boyd, and Lyle G. Boyd. "Blood Groups of the Bedouin near Baghdad." *American Journal of Physical Anthropology* 23, no. 3 (1938): 295–98. https://doi.org/10.1002/ajpa.1330230304.

Keller, Evelyn Fox. *The Century of the Gene*. Cambridge, MA: Harvard University Press, 2002.

Kelly, Melissa. "Transnational Diasporic Identities: Unity and Diversity in Iranian-Focused Organizations in Sweden." *Comparative Studies of South Asia, Africa and the Middle East* 31, no. 2 (2011): 443–54.

Kennedy, James, and Liliana Riga. "Mitteleuropa as Middle America? 'The Inquiry' and the Mapping of East Central Europe in 1919." *Ab Imperio* 7, no. 4 (2006): 271–300.

Kennedy, Walter P., and James MacFarlane. "Blood Groups in Iraq." *American Journal of Physical Anthropology* 21, no. 1 (January 1936): 87–89. https://doi.org/10.1002/ajpa.1330210127.

Kent, Michael, Vivette García-Deister, Carlos López-Beltrán, Ricardo Ventura Santos, Ernesto Schwartz-Marín, and Peter Wade. "Building the Genomic Nation: 'Homo Brasilis' and the 'Genoma Mexicano' in Comparative Cultural Perspective." *Social Studies of Science* 45, no. 6 (December 1, 2015): 839–61. https://doi.org/10.1177/0306312715611262.

Khatib-Shahidi, Rashid Armin. "German Foreign Policy Towards Iran: The Case of the National Bank of Persia." D. Phil., Oxford University, 1999.

Khosrokhavar, Farhad, Shapour Etemad, and Masoud Mehrabi. "Report on Science in Post-Revolutionary Iran—Part II: The Scientific Community's Problems of Identity." *Critique: Critical Middle Eastern Studies* 13, no. 3 (2004): 363–82. https://doi.org/10.1080/1066992042000300693.

Khuri, Fuad I. *An Invitation to Laughter: A Lebanese Anthropologist in the Arab World*. Chicago: University of Chicago Press, 2007.

Kieser, Hans-Lukas. "Türkische Nationalrevolution, Anthropologisch Gekrönt: Kemal Atatürk und Eugène Pittard." *Historische Anthropologie* 1 (2006): 105–18.

Kimel, Sasha Y., Rowell Huesmann, Jonas R. Kunst, and Eran Halperin. "Living in a Genetic World: How Learning About Interethnic Genetic Similarities and Differences Affects Peace and Conflict." *Personality and Social Psychology Bulletin* 42, no. 5 (May 2016): 688–700. https://doi.org/10.1177/0146167216642196.

Kirsh, Nurit. "Genetic Studies of Ethnic Communities in Israel: A Case of Values-Motivated Research Work." *Leo Baeck Institute Yearbook* 72 (2007): 181–94.

Kirsh, Nurit. "Population Genetics in Israel in the 1950s: The Unconscious Internalization of Ideology." *Isis* 94, no. 4 (2003): 631–55.

Koenig, Barbara A., Sandra Soo-Jin Lee, and Sarah S. Richardson, eds. *Revisiting Race in a Genomic Age*. New Brunswick, NJ: Rutgers University Press, 2008.

Koenig, Samuel. "East Meets West in Israel." *Phylon* 17, no. 2 (1956): 167–71.

Kongar, Emre. "Aziz Sancar'ın Düşündürdükleri . . ." *Cumhuriyet*, October 9, 2015. http://www.cumhuriyet.com.tr/koseyazisi/384069/Aziz_Sancar_in_dusundur dukleri....html.

Kossovitch, Nicolas. "Recherches anthropométriques et sérologiques (groupes sanguins) chez les Israélites du Maroc." *Comptes rendus des séances de la Société de Biologie* 109 (1932): 9–11.

Kowal, Emma, Joanna Radin, and Jenny Reardon. "Indigenous Body Parts, Mutating Temporalities, and the Half-Lives of Postcolonial Technoscience." *Social Studies of Science* 43, no. 4 (August 2013): 465–83. https://doi.org/10.1177 /0306312713490843.

Krischner, Harald, and M. Krischner. "The Anthropology of Mesopotamia and Persia A. Armenians, Khaldeans, Suriani (or Aissori) and Christian 'Arabs' from Iraq." *Proceedings of the Royal Academy of Sciences at Amsterdam* 35 (1932): 205–17.

Krischner, Harald, and M. Krischner. "The Anthropology of Mesopotamia and Persia C. The Anthropology of Persia." *Proceedings of the Royal Academy of Sciences at Amsterdam* 35 (1932): 399–410.

Kulozik, A. E., J. S. Wainscoat, G. R. Serjeant, B. C. Kar, B. Al-Awamy, G. J. F. Essan, A. G. Falusi, et al. "Geographical Survey of Bs-Globin Gene Haplotypes: Evidence for an Independent Asian Origin of the Sickle-Cell Mutation." *American Journal of Human Genetics* 39, no. 2 (1986): 239–44.

Kuo, Wen-Hua. "Techno-Politics of Genomic Nationalism: Tracing Genomics and Its Use in Drug Regulation in Japan and Taiwan." *Social Science & Medicine* 73, no. 8 (October 2011): 1200–1207. https://doi.org/10.1016/j.socscimed.2011.06.066.

Kurdi-Haidar, Buran, Philip John Mason, Alain Berrebi, George Ankra-Badu, Amin Al-Ali, Ariella Oppenheim, and Lucio Luzzatto. "Origin and Spread of the Glucose-6-Phosphate Dehydrogenase Variant (G6PD-Mediterranean) in the Middle East." *American Journal of Human Genetics* 47 (1990): 1012–19.

Landau, Jacob M. "Arab and Turkish Universities: Some Characteristics." *Middle*

Eastern Studies 33, no. 1 (January 1997): 1–19. https://doi.org/10.1080/0026320 9708701139.

Latour, Bruno. *Science in Action: How to Follow Scientists and Engineers Through Society.* Cambridge, MA: Harvard University Press, 1987.

Lavie, Smadar. *The Poetics of Military Occupation: Mzeina Allegories of Bedouin Identity Under Israeli and Egyptian Rule.* Berkeley: University of California Press, 1990.

Lederer, Richard. "A New Form of Acute Haemolytic Anaemia: 'Baghdad Spring Anaemia.'" *Transactions of the Royal Society of Tropical Medicine and Hygiene* 34, no. 5 (1941): 387–94.

Lehmann, Hermann. "Distribution of the Sickle Cell Gene: A New Light on the Origin of the East Africans." *Eugenics Review* 46, no. 2 (1954): 101–21.

Lehmann, Hermann. "Distribution of Variations in Human Haemoglobin." In *Abnormal Haemoglobins: A Symposium Organized by the Council for International Organizations of Medical Sciences, Established Under the Joint Auspices of UNESCO and WHO,* edited by J. H. P. Jonxis and J. F. Delafresnaye, 202–15. Oxford: Blackwell Scientific, 1959.

Lehmann, Hermann. "Sickle Cell Anemia 35 Years Ago: Reminiscence of Early African Studies." *American Journal of Pediatric Hematology/Oncology* 6, no. 1 (1984): 72–76.

Lehmann, Hermann. "Sickle-Cell Anaemia in Greece." *The Lancet* 257, no. 6667 (June 9, 1951): 1279.

Lehmann, Hermann. "Sickle-Cell Trait in South Turkey." *The Lancet* 265 (March 26, 1955): 672.

Lehmann, Hermann. "The Sickle-Cell Trait: Not an Essentially Negroid Feature." *Man* 53, no. 4, 5 (1953): 9–10.

Lehmann, Hermann, and Alan B. Raper. "Distribution of the Sickle-Cell Trait in Uganda, and Its Ethnological Significance." *Nature* 164, no. 4168 (September 1949): 494–95. https://doi.org/10.1038/164494a0.

Lehmann, Hermann, and Marie Cutbush. "Sickle-Cell Trait in Southern India." *British Medical Journal* 1, no. 4755 (February 23, 1952): 404–5.

Lehmann, Hermann, and Richard G. Huntsman. *Man's Haemoglobins: Including the Haemoglobinopathies and Their Investigation.* Amsterdam; Oxford: North-Holland Publishing Company, 1974.

Lehmann, Hermann, Fereydoun Ala, S. Hedeyat, K. Montazemi, H. Karini Nejad, S. Lightman, A. C. Kopec, A. E. Mourant, P. Teesdale, and D. Tills. "The

Hereditary Blood Factors of the Kurds of Iran." *Philosophical Transactions of the Royal Society of London* 266 (1973): 195–205.

Lehmann, Hermann, G. Maranjian, and A. E. Mourant. "Distribution of Sickle-Cell Hæmoglobin in Saudi Arabia." *Nature* 198, no. 4879 (May 1963): 492–93. https://doi.org/10.1038/198492b0.

"Les turcs promus 'aryens.'" *Le Temps*, June 14, 1936.

Lightman, Stafford, D. L. Carr-Locke, and Hilary G. Pickles. "The Frequency of PTC Tasters and Males Defective in Colour Vision in a Kurdish Population in Iran." *Human Biology* 42, no. 4 (1970): 665–69.

Lindee, Susan, and Ricardo Ventura Santos. "The Biological Anthropology of Living Human Populations: World Histories, National Styles, and International Networks: An Introduction to Supplement 5." *Current Anthropology* 53, no. S5 (April 2012): S3–16. https://doi.org/10.1086/663335.

Lipphardt, Veronika. "The Jewish Community of Rome: An Isolated Population? Sampling Procedures and Bio-Historical Narratives in Genetic Analysis in the 1950s." *BioSocieties* 5, no. 3 (2010): 306–29.

Lister, Richard W., Neil W. M. Orr, Douglas Botting, Elizabeth W. Ikin, Arthur E. Mourant, and Hermann Lehmann. "The Blood Groups and Haemoglobin of the Bedouin of Socotra." *Man* 1, no. 1 (1966): 82–86.

Little, Michael A. "Human Population Biology in the Second Half of the Twentieth Century." *Current Anthropology* 53, no. S5 (2012): S126–38. https://doi.org/10.1086/662440.

Liu, Jennifer A. "Making Taiwanese (Stem Cells): Identity, Genetics, and Hybridity." In *Asian Biotech: Ethics and Communities of Fate*, edited by Aihwa Ong and Nancy N. Chen, 239–62. Duke University Press, 2010. https://doi.org/10.1215/9780822393207.

Livingstone, Frank B. "Who Gave Whom Hemoglobin S: The Use of Restriction Site Haplotype Variation for the Interpretation of the Evolution of the BS-Globin Gene." *American Journal of Human Biology* 1, no. 3 (1989): 289–302. https://doi.org/10.1002/ajhb.1310010309.

Lorcin, Patricia M. E. *Imperial Identities: Stereotyping, Prejudice and Race in Colonial Algeria*. London; New York: I. B. Tauris, 1995.

Lord Rennell, Bernard Reilly, A. E. Mourant, and O. G. S. Crawford. "The Oxford University Expedition to Socotra: Discussion." *The Geographical Journal* 124, no. 2 (1958): 207–09.

Lotfalian, Mazyar. "The Iranian Scientific Community and Its Diaspora After the Islamic Revolution." *Anthropological Quarterly* 82, no. 1 (2009): 229–50.

Luschan, Felix von. "The Early Inhabitants of Western Asia." *Journal of the Royal Anthropological Institute of Great Britain and Ireland* 41, no. 2 (1911): 221–44.

Luschan, Felix von. *Völker, rassen, sprachen.* Berlin: Welt-Verlag, 1922.

Luzzatto, Lucio, and G. Battistuzzi. "Glucose-6-Phosphate Dehydrogenase." In *Advances in Human Genetics 14*, edited by Harry Harris and Kurt Hirschhorn, 217–329. Boston: Springer, 1985. https://link.springer.com/chapter/10.1007/978-1-4615-9400-0_4.

M'charek, Amade. *The Human Genome Diversity Project: An Ethnography of Scientific Practice.* Cambridge, UK: Cambridge University Press, 2005.

Maarif Vekilliği Neşriyet Müdürlüğü. *Lise Kitapları: Biyoloji II.* Istanbul: Maarif Matbaası, 1939.

"Mabnā-yi nizhādhā-yi insānī." In *Duvvumīn Kungrih-i zhinitīk-i īrān: 26–28 shahrīvar māh 1349*, 1–17. Tihrān: Dānishgāh-i Tihrān, 1970.

Mackintosh, J. Stewart. "Blood Group Percentages." *British Medical Journal* 1, no. 3512 (April 28, 1928): 732.

Mackintosh, J. Stewart. "Evolution of Racial Types of Europe: Its Bearing on the Racial Factor in Disease." *British Medical Journal* 2, no. 2597 (1910): 1049–50.

Mackintosh, J. Stewart. "The Migratory Factor in Eugenics." *British Medical Journal* 2, no. 2744 (1913): 227–29.

Madmoni-Gerber, Shoshana. *Israeli Media and the Framing of Internal Conflict: The Yemenite Babies Affair.* New York: Palgrave Macmillan, 2009.

Magnarella, Paul. "The Development of Turkish Social Anthropology." *Current Anthropology* 17, no. 2 (1976): 263–74.

Mahluji, Mohsen, and Bruce Livingston. "Muṭālaʻah-i baʻẓī az khuṣūṣiyāt-i mardum-shināsī dar panj dihkadah-i fārs." In *Duvvumīn Kungrih-i zhinitīk-i īrān: 26–28 shahrīvar māh 1349*, 92–96. Tihrān: Dānishgāh-i Tihrān, 1970.

Maksudyan, Nazan. "The *Turkish Review of Anthropology* and the Racist Face of Turkish Nationalism." *Cultural Dynamics* 17, no. 3 (2005): 291–322. https://doi.org/10.1177/0921374005061992.

Maksudyan, Nazan. *Türklüğü ölçmek: bilimkurgusal antropoloji ve Türk milliyetçiliğinin ırkçı çehresi, 1925–1939.* İstanbul: Metis, 2005.

Maltzan, Henri de. "Notes de Voyage Sur Les Régions Du Sud de l'Arabie." *Le Globe: Revue Genevoise de Géographie* 10 (1871): 125–56.

Maranjian, George, Elizabeth W. Ikin, Arthur E. Mourant, and Hermann Lehmann.

"The Blood Groups and Haemoglobins of the Saudi Arabians." *Human Biology* 38, no. 4 (1966): 394–420.

Margolis, Emmanuel, Joseph Gurevitch, and David Hermoni. "Blood Groups in Ashkenazi Jews." *American Journal of Physical Anthropology* 18 (1960): 201–3.

Margolis, Emmanuel, Joseph Gurevitch, and David Hermoni. "Blood Groups in Sephardic Jews." *American Journal of Physical Anthropology* 18 (1960): 197–99.

Margolis, Emmanuel, Joseph Gurevitch, and Elijah Hasson. "Blood Groups in Jews from Morocco and Tunisia." *Annals of Human Genetics, London* 22 (1957): 65–68.

Marks, Jonathan. "The Legacy of Serological Studies in American Physical Anthropology." *History and Philosophy of the Life Sciences* 18, no. 3 (1996): 345–62.

Marks, Jonathan. "The Origins of Anthropological Genetics." *Current Anthropology* 53, no. S5 (April 2012): S161–72. https://doi.org/10.1086/662333.

Marks, Paul A., and Ruth T. Gross. "Erythrocyte Glucose-6-Phosphate Dehydrogenase Deficiency: Evidence of Differences Between Negroes and Caucasians with Respect to This Genetically Determined Trait." *Journal of Clinical Investigation* 38, no. 12 (December 1, 1959): 2253–62. https://doi.org/10.1172/JCI104006.

Maroon, Habib. "A Geneticist with a Unifying Message." *Nature Middle East*, March 31, 2013. https://doi.org/10.1038/nmiddleeast.2013.46.

Marshall, Eliot. "Scientific Journals Adapt to New U.S. Trade Sanctions on Iran." *ScienceMagazine*, May 3, 2013. https://www.sciencemag.org/news/2013/05/scientific-journals-adapt-new-us-trade-sanctions-iran.

Masalha, Nur. *Expulsion of the Palestinians: The Concept of "Transfer" in Zionist Political Thought, 1882–1948.* Washington, D.C.: Institute for Palestine Studies, 1992.

Mattson, Greggor. "Nation-State Science: Lappology and Sweden's Ethnoracial Purity." *Comparative Studies in Society and History* 56, no. 02 (2014): 320–50.

Mazower, Mark. *Salonica, City of Ghosts: Christians, Muslims and Jews, 1430–1950.* 1. Vintage Books ed. New York: Vintage, 2006.

McMahon, Richard. "Anthropological Race Psychology 1820–1945: A Common European System of Ethnic Identity Narratives." *Nations and Nationalism* 15, no. 4 (October 2009): 575–96. https://doi.org/10.1111/j.1469-8129.2009.00393.x.

McMahon, Richard. *The Races of Europe: Construction of National Identities in the Social Sciences, 1839–1939.* London: Palgrave Macmillan, 2016.

McMahon, Richard, ed. *National Races: Transnational Power Struggles in the Sciences and Politics of Human Diversity, 1840–1945.* Lincoln: University of Nebraska Press, 2019.

Meneley, Anne. *Tournaments of Value: Sociability and Hierarchy in a Yemeni Town.* Toronto: University of Toronto Press, 2016.

Milani, Abbas. *Eminent Persians: The Men and Women Who Made Modern Iran, 1941–1979.* 1st ed. Syracuse, NY: Syracuse University Press, 2008.

Milani, Abbas. *The Shah.* New York: Palgrave Macmillan, 2012.

Mirdamadi, Mohammad, Ali Fotuhi, and Mehdi Sonbolestan. "Nisbat-i dar ṣad gurūhhā-yi khūnī-i ABO va RH dar 9753 nafar zan az marājiʿīn bih markaz-i pizishkī-i Riẓā Shāh-i Kabīr, Dānishgāh-i Iṣfahān." *Iranian Journal of Public Health* 7, no. 3 (1978): 149–57.

Mizuno, Hiromi. *Science for the Empire: Scientific Nationalism in Modern Japan.* Stanford: Stanford University Press, 2009.

Mohagheghpour, Nahid, Hamideh Tabatabai, and K. Mohammad. "Distribution of HLA Antigens in Zoroastrians." *Tissue Antigens* 17 (1981): 257–60.

Mokhtari, Fariborz. *In the Lion's Shadow: The Iranian Schindler and His Homeland in the Second World War.* Stroud, UK: The History Press, 2011.

Montazemi, Kambiz. "Barʿrasī-i mīzān-i farāvānī-i gurūhhā-yi khūnī dar īrān." *Iranian Journal of Public Health* 7, no. 1 (1978).

Mooreville, Anat. "Eyeing Africa: The Politics of Israeli Ocular Expertise and International Aid, 1959–1973." *Jewish Social Studies* 21, no. 3 (2016): 31–71. https://doi.org/10.2979/jewisocistud.21.3.02.

Morris-Reich, Amos. "Jews Between Volk and Rasse." In *National Races: Transnational Power Struggles in the Sciences and Politics of Human Diversity, 1840–1945*, edited by Richard McMahon, 175–203. Lincoln: University of Nebraska Press, 2019.

Motabar, M., A. Reiss-Sadat, and A. Tabatabai. "Prevalence of High Blood Pressure in Qashqai Tribe, Southern Iran, 1973." *Acta Medica Iranica* 20 (1977): 9–17.

Motadel, David. "Iran and the Aryan Myth." In *Perceptions of Iran: History, Myths and Nationalism from Medieval Persia to the Islamic Republic*, edited by Ali M. Ansari, 119–46. London: I. B. Tauris, 2014.

Motulsky, Arno G. "Metabolic Polymorphisms and the Role of Infectious Diseases in Human Evolution." *Human Biology* 32 (1960): 28–62.

Mourant, Arthur E. "The Blood Groups of the Jews." *Jewish Journal of Sociology* 1 (1959): 153–75.

Mourant, Arthur E. "The Blood Groups of the Peoples of the Mediterranean Area." *Cold Spring Harbor Symposia on Quantitative Biology* 15 (1950): 221–31.

Mourant, Arthur E. *Blood and Stones: An Autobiography*. La Haule, Jersey: La Haule Books, 1995.

Mourant, Arthur E. *The Distribution of the Human Blood Groups, and Other Polymorphisms*. 2nd ed. London: Oxford University Press, 1976.

Mourant, Arthur E. *The Distribution of the Human Blood Groups*. 1st ed. Oxford: Blackwell Scientific, 1954.

Mourant, Arthur E. *The Genetics of the Jews*. Oxford; New York: Clarendon Press, 1978.

Mourant, Arthur E., and Donald Tills. "Phosphoglucomutase Frequencies in Habbanite Jews and Icelanders." *Nature* 214, no. 5090 (May 1967): 810. https://doi.org/10.1038/214810a0.

Mukharji, Projit Bihari. "From Serosocial to Sanguinary Identities: Caste, Transnational Race Science and the Shifting Metonymies of Blood Group B, India c. 1918–1960." *Indian Economic & Social History Review* 51, no. 2 (2014): 143–76. https://doi.org/10.1177/0019464614525711.

Mukharji, Projit Bihari. "Profiling the Profiloscope: Facialization of Race Technologies and the Rise of Biometric Nationalism in Inter-War British India." *History and Technology* 31, no. 4 (October 2, 2015): 376–96. https://doi.org/10.1080/07341512.2015.1127459.

Mukharji, Projit Bihari. "The Bengali Pharaoh: Upper-Caste Aryanism, Pan-Egyptianism, and the Contested History of Biometric Nationalism in Twentieth-Century Bengal." *Comparative Studies in Society and History* 59, no. 2 (April 2017): 446–76. https://doi.org/10.1017/S001041751700010X.

Mukharji, Projit Bihari. *Doctoring Traditions: Ayurveda, Small Technologies, and Braided Sciences*. Chicago; London: University of Chicago Press, 2016.

Myres, J. L. "Shorter Notes: Anthropological Studies in Turkey." *Man* 47 (1947): 29–30.

Naar, Devin E. "The 'Mother of Israel' or the 'Sephardi Metropolis'? Sephardim, Ashkenazim, and Romaniotes in Salonica." *Jewish Social Studies* 22, no. 1 (2016): 81. https://doi.org/10.2979/jewisocistud.22.1.03.

Najjar, Abdallah E., Julius S. Prince, and Russell E. Fontaine. "The 1958 Malaria Epidemic in Ethiopia." *The American Journal of Tropical Medicine and Hygiene* 10, no. 6 (November 1, 1961): 795–803. https://doi.org/10.4269/ajtmh.1961.10.795.

Najmabadi, Hossein, Maryam Neishabury, Farhad Sahebjam, Kimia Kahrizi, Yousef Shafaghati, Nushin Nikzat, Maryam Jalalvand, et al. "The Iranian Human Mutation Gene Bank: A Data and Sample Resource for Worldwide Collaborative

Genetics Research." *Human Mutation* 21, no. 2 (February 2003): 146–50. https://doi.org/10.1002/humu.10164.

Nash, Catherine. "The Politics of Genealogical Incorporation: Ethnic Difference, Genetic Relatedness and National Belonging." *Ethnic and Racial Studies* 40, no. 14 (November 14, 2017): 2539–57. https://doi.org/10.1080/01419870.2016.1242763.

Nash, Catherine. *Genetic Geographies: The Trouble with Ancestry*. Minneapolis: University of Minnesota Press, 2015.

Nelkin, Dorothy, and M. Susan Lindee. *The DNA Mystique: The Gene as a Cultural Icon*. 2nd ed. Ann Arbor: University of Michigan Press, 2004.

Nelson, Alondra. *Body and Soul: The Black Panther Party and the Fight Against Medical Discrimination*. Minneapolis; London: University of Minnesota Press, 2011.

Nelson, Alondra. *The Social Life of DNA: Race, Reparations, and Reconciliation after the Genome*. Boston: Beacon Press, 2016.

Nevo, Sarah. "Genetic Blood Markers in Arab Druze of Israel." *American Journal of Physical Anthropology* 77, no. 2 (October 1988): 183–90. https://doi.org/10.1002/ajpa.1330770206.

Nevo, Sarah, and Hartwig Cleve. "Gc Subtypes in the Middle East: Report on an Arab Moslem Population from Israel." *American Journal of Physical Anthropology* 60, no. 1 (January 1983): 49–52. https://doi.org/10.1002/ajpa.1330600108.

Nevo, Sarah, H. Cleve, A. Koller, E. Eigel, W. Patutschnick, H. Kanaaneh, and A. Joel. "Serum Protein Polymorphisms in Arab Moslems and Druze of Israel: BF, F13B, AHSG, GC, PLG, PI, and TF." *Human Biology* 64, no. 4 (August 1992): 587–603.

Newman, Marissa. "When Israeli Doctors Allegedly Tested Yemenites for 'Negro Blood.'" *Times of Israel*, June 16, 2017. https://www.timesofisrael.com/when-israeli-doctors-allegedly-tested-yemenites-for-negro-blood/.

Nijenhuis, L. E. "Blood Group Frequencies in Iran." *Vox Sanguinis* 9 (1964): 723–40.

Nkhoma, Ella T., Charles Poole, Vani Vannappagari, Susan A. Hall, and Ernest Beutler. "The Global Prevalence of Glucose-6-Phosphate Dehydrogenase Deficiency: A Systematic Review and Meta-Analysis." *Blood Cells, Molecules, and Diseases* 42, no. 3 (May 2009): 267–78. https://doi.org/10.1016/j.bcmd.2008.12.005.

"Nobel Ödüllü Sancar'ı Ailesi Anlattı." *Anadolu Agency*, October 8, 2015. http://www.trthaber.com/haber/turkiye/nobel-odullu-sancari-ailesi-anlatti-207941.html.

Oberling, Pierre. *The Qashqā'ī Nomads of Fārs*. Reprint 2017. Berlin; Boston: De Gruyter Mouton, 1974.

Oikkonen, Venla. *Population Genetics and Belonging*. Cham: Springer International Publishing, 2018. https://doi.org/10.1007/978-3-319-62881-3.

Öktem, Emre. "The Legal Notion of Nationality in the Turkish Republic: From Ottoman Legacy to Modern Aberrations." *Middle Eastern Studies* 53, no. 4 (July 4, 2017): 638–55. https://doi.org/10.1080/00263206.2017.1281125.

Önde, Sertaç, and Aykut Kence. "Distribution of ABO and Rh Gene Frequencies over 67 Provinces of Turkey." *Turkish Journal of Biology* 18, no. 2 (1994): 133–39.

Onur, Nureddin. "Kan grupları bakımından Türk ırkının menşei hakkında bir etüd." In *İkinci Türk Tarih Kongresi, İstanbul 20–25 Eylül 1937*, 845–51. İstanbul: Kenan Matbaası, 1943.

Onur, Nureddin. "Les groupes sanguins chez les Turcs." *Comptes rendus des séances de la Société de Biologie* 89, no. 124 (1937): 521–22.

Onur, Nureddin. "Türklerde kan grupları." *Pratik Doktor* 12 (1936).

Onur, Nureddin. *İnsan ve hayvanlarda kan grupları*. İstanbul: Kader Basımevi, 1941.

Oppenheim, Ariella, Corrine L. Jury, Deborah Rund, Tom J. Vulliamy, and Lucio Luzzatto. "G6PD Mediterranean Accounts for the High Prevalence of G6PD Deficiency in Kurdish Jews." *Human Genetics* 91, no. 3 (April 1993): 293–94. https://doi.org/10.1007/BF00218277.

Ottenberg, Reuben. "A Classification of Human Races Based on Geographic Distribution of the Blood Groups." *Journal of the American Medical Association* 84, no. 19 (May 9, 1925): 1393.

Özek, Ömer. "Kan grupları üzerinde araştırmalar." *Türk Tıb Cemiyeti Mecmuası* 8, no. 6 (1942).

Öztuncay, Bahattin, and Özge Ertem, eds. *Ottoman Arcadia: The Hamidian Expedition to the Land of Tribal Roots (1886)*. İstanbul: Anamed, 2018.

Packard, Randall M. *The Making of a Tropical Disease: A Short History of Malaria*. Baltimore, MD: Johns Hopkins University Press, 2007.

Papiha, Surinder Singh, I. White, Mohammad Taghi Akbari, and Dariush Daneshvar Farhud. "Isoelectric Focusing on Vitamin D Binding Protein (Gc): Genetic Diversity in the Populations of Iran." *Japanese Journal of Human Genetics* 30 (1985): 69–73.

Papiha, Surinder Singh, Y. Seyedna, and Eric Sunderland. "Phosphoglucomutase (PGM) and Group-Specific Component Isoelectric Focusing Among Zoroastrians of Iran." *Annals of Human Biology* 9, no. 6 (1982): 571–74.

Pappé, Ilan. *The Forgotten Palestinians: A History of the Palestinians in Israel*. New Haven, CT: Yale University Press, 2011.

Parr, Leland W. "Blood Studies on Peoples of Western Asia and North Africa." *American Journal of Physical Anthropology* 16, no. 1 (1931): 15–29.

Pauling, Linus, Harvey A. Itano, S. J. Singer, and Ibert C. Wells. "Sickle Cell Anemia, a Molecular Disease." *Science* 110 (November 25, 1949): 543–48.

Pelt, Mogens. *Military Intervention and a Crisis of Democracy in Turkey: The Menderes Era and Its Demise.* London; New York: I. B. Tauris, 2014.

Perry, Tom. "In Lebanon DNA May Yet Heal Rifts." *Reuters*, September 9, 2007. www.reuters.com/article/us-phoenicians-dna/in-lebanon-dna-may-yet-heal-rifts-idUSL0559096520070910.

Picard, Avi. "Yaḥaso shel 'iton ha-Arets la-'aliyatam shel yehude tsefon afriḳah." *Yisra'el* 10 (2006): 117–43.

Pınar, Mehmet. "Türk Tarih Tezi Bağlamında Cumhuriyet Döneminde Nusayriler." *Turkish Studies* 10, no. 9 (2015): 485–485. https://doi.org/10.7827/Turkish Studies.8717.

Pittard, Eugène. "Contribution à l'étude anthropologique des Turcs d'Asie Mineure." *Türk Antropoloji Mecmuası* 8 (1929): 3, 10, 19.

Pittard, Eugène. "Découverte de la civilisation paléolithique en Asie Mineure." *Archives suisses d'anthropologie générale* 2, no. 5 (1928): 135–65.

Pittard, Eugène. *Race and History: An Ethnological Introduction to History.* New York: Alfred A. Knopf, 1926.

Playfair, Robert L. *A History of Arabia Felix or Yemen: From the Commencement of the Christian Era to the Present Time; Including an Account of the British Settlement of Aden.* Bombay: Government at the Education Society's Press, Byculla, 1859.

Podliachouk, Luba, André Eyquem, R. Choaripour, and M. Eftekhari. "Les facteurs sériques Gm(a), Gm(b), Gm(x) et Gm-like chez les Iraniens." *Vox Sanguinis* 7 (1962): 496–99.

Podliachouk, Luba, André Eyquem, R. Choaripour, and M. Eftekhari. "Serum Factor Gma Among the Iranians." *Nature* 191, no. 4789 (1961): 717–18.

Powell, Eve Troutt. *A Different Shade of Colonialism: Egypt, Great Britain, and the Mastery of the Sudan.* Berkeley: University of California Press, 2003.

Prager, Laila. "'Dangerous Liaisons': Modern Biomedical Discourses and Changing Practices of Cousin Marriage in Southeastern Turkey." In *Cousin Marriages: Between Tradition, Genetic Risk and Cultural Change,* edited by Alison Shaw and Aviad E. Raz. New York: Berghahn Books, 2015.

Prakash, Gyan. *Another Reason: Science and the Imagination of Modern India.* Princeton: Princeton University Press, 1999.

"The Problem of Race." *British Medical Journal* 1, no. 3933 (1936): 1060.

Procházka-Eisl, Gisela. *The Plain of Saints and Prophets: The Nusayri-Alawi*

Community of Cilicia (Southern Turkey) and Its Sacred Places. Wiesbaden: Harrassowitz Verlag, 2010.

"Races Congress: The concluding meetings; future work." The *Manchester Guardian*, July 31, 1911.

Radin, Joanna. "Latent Life: Concepts and Practices of Human Tissue Preservation in the International Biological Program." *Social Studies of Science* 43, no. 4 (August 1, 2013): 484–508. https://doi.org/10.1177/0306312713476131.

Radin, Joanna. "Unfolding Epidemiological Stories: How the WHO Made Frozen Blood into a Flexible Resource for the Future." *Studies in History and Philosophy of Science Part C: Studies in History and Philosophy of Biological and Biomedical Sciences* 47 (2014): 62–73. https://doi.org/10.1016/j.shpsc.2014.05.007.

Ragab, A. H., O. S. El-Alfi, and M. A. Abboud. "Incidence of Glucose-6-Phosphate Dehydrogenase Deficiency in Egypt." *American Journal of Human Genetics* 18, no. 1 (1966): 21–25.

Reardon, Jenny. "Race Without Salvation: Beyond the Science/Society Divide in Genomic Studies of Human Diversity." In *Revisiting Race in a Genomic Age*, edited by Barbara A. Koenig, Sandra Soo-Jin Lee, and Sarah S. Richardson, 304–19. New Brunswick, NJ: Rutgers University Press, 2008.

Reardon, Jenny. "The Democratic, Anti-Racist Genome? Technoscience at the Limits of Liberalism." *Science as Culture* 21, no. 1 (March 2012): 25–47. https://doi.org/10.1080/09505431.2011.565322.

Reardon, Jenny. *Race to the Finish: Identity and Governance in an Age of Genomics*. Princeton: Princeton University Press, 2005.

Record of the Proceedings of the First Universal Races Congress: Held at the University of London, July 26–29, 1911. London: P. S. King & Son, 1911.

Redhead, Grace Olivia. "Histories of Sickle Cell Anaemia in Postcolonial Britain, 1948–1997." PhD diss., University College London, 2020.

Reid, Donald M. *Whose Pharaohs? Archaeology, Museums, and Egyptian National Identity from Napoleon to World War I*. Berkeley: University of California Press, 2002.

Rejwan, Nissim. *The Last Jews in Baghdad: Remembering a Lost Homeland*. Austin, TX: University of Texas Press, 2010.

Rezaee-Zavareh, Mohammad Saeid, Hamidreza Karimi-Sari, and Seyed Moayed Alavian. "Iran, Sanctions, and Research Collaborations." The *Lancet* 387, no. 10013 (January 2016): 28–29. https://doi.org/10.1016/S0140-6736(15)01295-7.

Rezaee, Ali Reza, Mohammad Mehdi Banoei, Elham Khalili, and Massoud

Houshmand. "Beta-Thalassemia in Iran: New Insight into the Role of Genetic Admixture and Migration." *The Scientific World Journal* 2012 (2012): 1–7. https://doi.org/10.1100/2012/635183.

Rhode, Maria. "A Matter of Place, Space, and People: Cracow Anthropology, 1870–1920." In *National Races: Transnational Power Struggles in the Sciences and Politics of Human Diversity, 1840–1945*, edited by Richard McMahon, 105–40. Lincoln: University of Nebraska Press, 2019.

Riza Tevfık. "Turkey." In *Papers on Inter-Racial Problems Communicated to the First Universal Races Congress, Held at the University of London, July 26–29, 1911*, edited by Gustav Spiller, 454–61. London; Boston: P. S. King & Son; World's Peace Foundation, 1911.

Roberts, Dorothy. *Fatal Invention: How Science, Politics, and Big Business Re-Create Race in the Twenty-First Century*. New York: New Press, 2011.

Robertson, Jennifer. "Blood Talks: Eugenic Modernity and the Creation of New Japanese." *History and Anthropology* 13, no. 3 (January 2002): 191–216. https://doi.org/10.1080/0275720022000025547.

Robinson, Ronald. "Non-European Foundations of European Imperialism: Sketch for a Theory of Collaboration." In *Studies in the Theory of Imperialism*, edited by Roger Owen and Bob Sutcliffe, 117–40. London: Longman, 1972.

Robson, Laura. *States of Separation: Transfer, Partition, and the Making of the Modern Middle East*. Oakland, CA: University of California Press, 2017.

Roshwald, Mordecai. "Marginal Jewish Sects in Israel II." *International Journal of Middle East Studies* 4, no. 3 (1973): 328–54.

Ruffié, Jacques, and Nagib Taleb. *Étude hémotypologique des ethnies libanaises*. Paris: Hermann, 1965.

Saatçioğlu, Armağan. "An Analysis of the ABO Gene Frequencies in Turkey." *Journal of Human Evolution* 8 (1979): 367–73.

Saatçioğlu, Armağan. *ABO genleri yönünden Türkiye'nin yeri ve bu ülkedeki gensel çeşitlilik üzerine biyometrik bir inceleme*. Ankara: Ankara Üniversitesi Basımevi, 1978.

Saini, Angela. *Superior: The Return of Race Science*. London: 4th Estate, 2019.

Salgırlı, Sanem Güvenç. "Eugenics for the Doctors: Medicine and Social Control in 1930s Turkey." *Journal of the History of Medicine and Allied Sciences* 66, no. 3 (2010): 281–312. https://doi.org/10.1093/jhmas/jrq040.

Samii, A. William. "The Nation and Its Minorities: Ethnicity, Unity, and State Policy in Iran." *Comparative Studies of South Asia, Africa and the Middle East* 20, no. 1 (2005): 128–37.

Sanasarian, Eliz. *Religious Minorities in Iran*. Cambridge, UK: Cambridge University Press, 2000.

"Sancar: Ben Türk'üm o Kadar." *Milliyet*, October 8, 2015. http://www.milliyet.com.tr/nobel-odulunu-kazanan-sancar-in-gundem-2128721.

"Sanctions on Scientific Publication." *Nature Neuroscience* 7, no. 11 (November 2004): 1163–1163. https://doi.org/10.1038/nn1104-1163.

Saran, Nephan. "Kan grupları ve Türkiye'deki dağılımı." *Sosyal antropoloji ve etnoloji dergisi* 2 (1975): 51–60.

Savary, Luzia. *Evolution, Race and Public Spheres in India: Vernacular Concepts and Sciences (1860–1930)*. Abingdon, Oxon; New York: Routledge, 2019.

Sawhney, Kanwarjit Singh, Eric Sunderland, and Dariush Farhud. "Study of Red Cell Enzyme Systems in Tehran and Isfahan Iranians." *Japanese Journal of Human Genetics* 26 (1981): 289–94.

Schaffer, Gavin. *Racial Science and British Society, 1930–62*. Basingstoke; New York: Palgrave Macmillan, 2008.

Schaffer, Simon, Lissa Roberts, Kapil Raj, and James Delbourgo, eds. *The Brokered World: Go-Betweens and Global Intelligence, 1770–1820*. Sagamore Beach, MA: Science History Publications, 2009.

Schayegh, Cyrus. "Hygiene, Eugenics, Genetics, and the Perception of Demographic Crisis in Iran, 1910s–1940s." *Critique: Critical Middle Eastern Studies* 13, no. 3 (2004): 335–61. https://doi.org/10.1080/1066992042000300684.

Schayegh, Cyrus. *Who Is Knowledgeable Is Strong: Science, Class, and the Formation of Modern Iranian Society, 1900–1950*. Berkeley: University of California Press, 2009.

Schieber, Chaim. "Target-Cell Anaemia: Two Cases in Bucharan Jews." The *Lancet* 246, no. 6383 (1945): 851–52.

Schiebinger, Londa L. *Secret Cures of Slaves: People, Plants, and Medicine in the Eighteenth-Century Atlantic World*. Stanford: Stanford University Press, 2017.

Schine, Rachel. "Conceiving the Pre-Modern Black-Arab Hero: On the Gendered Production of Racial Difference in *Sīrat al-Amīrah Dhāt al-Himmah*." *Journal of Arabic Literature* 48, no. 3 (November 27, 2017): 298–326. https://doi.org/10.1163/1570064x-12341346.

Schneider, William H. "La Recherche sur les Groupes Sanguins avant la Deuxième Guerre Mondiale." In *Les sciences biologiques et médicales en France, 1920–1950*, edited by Claude Debru, Jean Gayon, and Jean-François Picard. Paris: CNRS Editions, 1994.

Schneider, William H. "The History of Research on Blood Group Genetics: Initial

Discovery and Diffusion." *History and Philosophy of the Life Sciences* 18, no. 3 (1996): 277–303.

Schreiber, Monika. *The Comfort of Kin: Samaritan Community, Kinship, and Marriage.* Leiden: Brill, 2014.

Segev, Tom. *1949: ha-Yiśre'elim ha-rishonim.* Yerushalayim: Domino, 1984.

Segev, Tom. *1949: The First Israelis.* Edited by Arlen Neal Weinstein. New York: Henry Holt, 1998.

Seidelman, Rhona D., S. Ilan Troen, and Shifra Shvarts. "'Healing' the Bodies and Souls of Immigrant Children: The Ringworm and Trachoma Institute, Sha'ar Ha-Aliyah, 1952–1960." *Journal of Israeli History* 29, no. 2 (September 2010): 191–211. https://doi.org/10.1080/13531042.2010.508956.

Selim, Osaima, Karim Kamel, Aida A. Azim, F. Gaballah, Fadila H. Sabry, W. Ibrahim, Nadia Moafy, and K. Hoerman. "Genetic Markers and Anthropometry in the Populations of the Egyptian Oases of El-Kharga and El-Dakhla." *Human Heredity* 24, no. 3 (1974): 259–72. https://doi.org/10.1159/000152659.

Selim, Samah. "Languages of Civilization: Nation, Translation and the Politics of Race in Colonial Egypt." *The Translator* 15, no. 1 (April 2009): 139–56. https://doi.org/10.1080/13556509.2009.10799274.

Şentuna, Can. "Rh Gen Frekansları Yönünden Türkiye'nin Yeri." *Ankara Üniversitesi Dil ve Tarih-Coğrafya Fakültesi Dergisi* 30, no. 1–2 (1982): 153–79.

Seyedna, S. Y. "A Genetic and Demographic Investigation of the Zoroastrians of Iran." PhD diss., Durham University, 1982.

Shahid, Munib, and Najib Abu Haydar. "Sickle Cell Disease in Syria and Lebanon." *Acta Haematologica* 27 (1962): 268–73.

Shaker, Yehia, A. Onsi, and R. Aziz. "The Frequency of Glucose-6-Phosphate Dehydrogenase Deficiency in the Newborns and Adults in Kuwait." *American Journal of Human Genetics* 18, no. 6 (1966): 609–13.

Shanklin, William M. "Anthropometry of Syrian Males." *The Journal of the Royal Anthropological Institute of Great Britain and Ireland* 68 (July 1938): 379–414. https://doi.org/10.2307/2844134.

Shanklin, William M. "Blood Grouping of Rwala Bedouin." *Journal of Immunology* 29, no. 6 (1935): 427–33.

Shanklin, William M. "The Anthropology of Transjordanians." *Al-Kulliyyah* 21, no. 1 (November 1934): 9–15.

Shanklin, William M., and Nejla Izzeddin. "Anthropology of the Near East Female." *American Journal of Physical Anthropology* 22, no. 3 (April 1937): 381–415. https://doi.org/10.1002/ajpa.1330220303.

Sharim, Yehuda. "The Arch of a Sephardic-Mizrahi Ethnic Autonomy in Palestine, 1926 to 1929." *UCLA Historical Journal* 24, no. 1 (2013): 29–43.

Shaw, Alexander Nicholas. "'Strong, United and Independent': The British Foreign Office, Anglo-Iranian Oil Company and the Internationalization of Iranian Politics at the Dawn of the Cold War, 1945–46." *Middle Eastern Studies* 52, no. 3 (May 3, 2016): 505–24. https://doi.org/10.1080/00263206.2015.1124417.

Sheba, Chaim. "Geneṭiḳah shel shevaṭe-yiśra'el." *Madaʿ* 13 (1968): 94–99.

Sheba, Chaim. "Jewish Migration in Its Historical Perspective." *Israel Journal of Medical Sciences* 7, no. 12 (1971): 1333–41.

Sheba, Chaim. "Nisayon le-shiḥzur nedidat bene-yiśra'el be-'ezrat tavḥinim biyokhimim." *Madaʿ* 4, no. 3–4 (1960): 34–39.

Sheba, Chaim. "Reconstructing Jewish Migration with the Aid of Biochemical Tests: A Working Hypothesis." *Proceedings of the Tel-Hashomer Hospital* 7, no. 3/4 (1968): 91–98.

Sheba, Chaim, Aryeh Szeinberg, Avinoam Adam, and Bracha Ramot. "Distribution of Glucose-6-Phosphate-Dehydrogenase Deficiency Among Various Communities in Israel." In *Proceedings of the Second International Congress of Human Genetics (Rome, September 6–12, 1961)*, edited by Luigi Gedda, 1:633–34. Rome: Instituto G. Mendel, 1963.

Sheba, Chaim, Aryeh Szeinberg, Bracha Ramot, Avinoam Adam, and Israel Ashkenazi. "Epidemiologic Surveys of Deleterious Genes in Different Population Groups in Israel." *American Journal of Public Health* 52, no. 7 (1962): 1101–6.

Sheba, Chaim, Avinoam Adam, and Mariassa Bat-Miriam. "A Survey of Some Genetical Characters in Ethiopian Tribes: Introduction." *American Journal of Physical Anthropology* 20, no. 2 (1962): 168–71.

Shefer-Mossensohn, Miri. *Science Among the Ottomans: The Cultural Creation and Exchange of Knowledge*. Austin: University of Texas Press, 2015.

Sheriff, Abdul. "The Zanj Rebellion and the Transition from Plantation to Military Slavery." *Comparative Studies of South Asia, Africa and the Middle East* 38, no. 2 (2018): 246–60. https://doi.org/10.1215/1089201x-6982029.

Shields, Sarah D. *Fezzes in the River: Identity Politics and European Diplomacy in the Middle East on the Eve of World War II*. Oxford; New York: Oxford University Press, 2011.

Shimazu, Naoko. *Japan, Race, and Equality: The Racial Equality Proposal of 1919*. London; New York: Routledge, 1998.

Shohat, Ella. "Sephardim in Israel: Zionism from the Standpoint of Its Jewish Victims." *Social Text* 19/20 (1988): 1–35.

Shohat, Ella. "The Invention of the Mizrahim." *Journal of Palestine Studies* 29, no. 1 (October 1999): 5–20. https://doi.org/10.2307/2676427.

Shostak, Sara, and Jason Beckfield. "Making a Case for Genetics: Interdisciplinary Visions and Practices in the Contemporary Social Sciences." In *Genetics, Health and Society*, edited by Brea L. Perry, 97–126. Bingley, UK: Emerald, 2015.

Shousha, Ali Tawfic. "On the Biochemical Race-Index of the Egyptians." *Journal of the Egyptian Medical Association* 11, no. 1 (January 1928): 4–11.

Shrum, Wesley. "Collaborationism." In *Collaboration in the New Life Sciences*, edited by John N. Parker, Niki Vermeulen, and Bart Penders, 247–58. Farnham, UK: Ashgate, 2010.

Shvarts, Shifra, Nadav Davidovitch, Rhona Seidelman, and Avishay Goldberg. "Medical Selection and the Debate over Mass Immigration in the New State of Israel (1948–1951)." *Canadian Bulletin of Medical History* 22, no. 1 (2005): 5–34.

Silverman, Rachel. "The Blood Group 'Fad' in Post-War Racial Anthropology." *Kroeber Anthropological Society Papers* 84 (2000): 11–27.

Simpson, Bob. "Imagined Genetic Communities: Ethnicity and Essentialism in the Twenty-First Century." *Anthropology Today* 16, no. 3 (June 2000): 3–6. https://doi.org/10.1111/1467-8322.00023.

Sleeboom-Faulkner, Margaret. "How to Define a Population: Cultural Politics and Population Genetics in the People's Republic of China and the Republic of China." *BioSocieties* 1, no. 4 (December 2006): 399–419. https://doi.org/10.1017/S1745855206004030.

Smith, Harry M., John G. Shiber, Helen M. Hawa, and Fares A. Ghareeb. "Geographic Variation and Human Population Genetics Among the Indigenous Peoples of Lebanon." In *Proceedings of the XII International Congress of Genetics, Tokyo, Japan, August 19–28, 1968*, 1:295. Tokyo: Science Council of Japan, 1968.

Snyder, Laurence H. "Human Blood Groups: Their Inheritance and Racial Significance." *American Journal of Physical Anthropology* 9, no. 2 (1926): 233–63.

Snyder, Laurence H. "The 'Laws' of Serologic Race-Classification Studies in Human Inheritance IV." *Human Biology* 2, no. 1 (1930): 128–33.

Sofer, Sasson. *Zionism and the Foundations of Israeli Diplomacy*. Cambridge, UK: Cambridge University Press, 2007.

Sommer, Marianne. *History Within: The Science, Culture, and Politics of Bones, Organisms, and Molecules*. Chicago: University of Chicago Press, 2016.

Southgate, Minoo. "The Negative Images of Blacks in Some Medieval Iranian Writings." *Iranian Studies* 17, no. 1 (March 1984): 3–36. https://doi.org/10.1080/00210868408701620.

Spiller, Gustav, ed. *Papers on Inter-Racial Problems Communicated to the First Universal Races Congress, Held at the University of London, July 26–29, 1911.* London; Boston: P. S. King & Son; World's Peace Foundation, 1911.

Spivak, Gayatri Chakravorty. *A Critique of Postcolonial Reason: Toward a History of the Vanishing Present.* Cambridge, MA: Harvard University Press, 1999.

Stark, G. H., and Yaninah Henner. "Bediḳot hemaṭologyot etsel torme dam ba-tel-aviv ye-ḥaluḳat ha-tormim li-ḳevutsot dam A, B, O." *Harefuah* 45 (1953): 175–76.

Steffan, Paul. *Handbuch der Blutgruppenkunde.* Munich: J. F. Lehmann, 1932.

Steinweis, Alan E. *Studying the Jew: Scholarly Antisemitism in Nazi Germany.* Cambridge, MA: Harvard University Press, 2006.

Sternfeld, Lior. "The Revolution's Forgotten Sons and Daughters: The Jewish Community in Tehran During the 1979 Revolution." *Iranian Studies* 47, no. 6 (November 2, 2014): 857–69. https://doi.org/10.1080/00210862.2014.948744.

Stocking, George, ed. "Colonial Situations." In *Colonial Situations: Essays on the Contextualization of Ethnographic Knowledge*, 3–8. Madison, WI: University of Wisconsin Press, 1991.

Stone, Richard. "Unsanctioned Science." *Science* 349, no. 6252 (September 4, 2015): 1038–43. https://doi.org/10.1126/science.349.6252.1038.

Strasser, Bruno J. "Linus Pauling's 'Molecular Diseases': Between History and Memory." *American Journal of Medical Genetics* 115, no. 2 (2002): 83–93.

Suárez-Díaz, Edna. "Blood Diseases in the Backyard: Mexican 'Indígenas' as a Population of Cognition in the Mid-1960s." *Perspectives on Science* 25, no. 5 (2017): 606–30.

Suárez-Díaz, Edna. "Indigenous Populations in Mexico: Medical Anthropology in the Work of Ruben Lisker in the 1960s." *Studies in History and Philosophy of Science Part C: Studies in History and Philosophy of Biological and Biomedical Sciences* 47 (September 2014): 108–17. https://doi.org/10.1016/j.shpsc.2014.05.011.

Subramaniam, Banu. *Holy Science: The Biopolitics of Hindu Nationalism.* Seattle: University of Washington Press, 2019.

Summerfield, Daniel. *From Falashas to Ethiopian Jews: The External Influences for Change, c. 1860–1960.* London; New York: Routledge Curzon, 2003.

Sunderland, Eric, and Harry M. Smith. "The Blood Groups of the Shi'a in Yazd, Central Iran." *Human Biology* 38, no. 1 (1966): 50–59.

Sung, Wen-Ching. "Chinese DNA: Genomics and Bionation." In *Asian Biotech: Ethics and Communities of Fate*, edited by Aihwa Ong and Nancy N. Chen, 263–92. Duke University Press, 2010. https://doi.org/10.1215/9780822393207.

Szeinberg, Aryeh. "Investigation of Genetic Polymorphic Traits in Jews. A

Contribution to the Study of Population Genetics." *Israel Journal of Medical Sciences* 9, no. 9–10 (1973): 1171–80.

Szeinberg, Aryeh, and Chaim Sheba. "Hemolytic Trait in Oriental Jews Connected with an Hereditary Enzymatic Abnormality of Erythrocytes." *Israel Medical Journal* 17 (1958): 158–68.

Szeinberg, Aryeh, Chaim Sheba, and Avinoam Adam. "Enzymatic Abnormality in Erythrocytes of a Population Sensitive to Vicia Faba or Haemolytic Anaemia Induced by Drugs." *Nature* 181 (1958): 1256.

Szeinberg, Aryeh, Chaim Sheba, and Avinoam Adam. "Selective Occurrence of Glutathione Instability in Red Blood Corpuscles of the Various Jewish Tribes." *Blood* 13 (1958): 1043–53.

Szeinberg, Aryeh, Chaim Sheba, Nina Hirshorn, and Eva Bodonyi. "Studies on Erythrocytes in Cases with Past History of Favism and Drug-Induced Acute Hemolytic Anemia." *Blood* 12 (1957): 603–13.

Szeinberg, Aryeh, Y. Asher, and Chaim Sheba. "Studies on Glutathione Stability in Erythrocytes of Cases with Past History of Favism or Sulfa-Drug-Induced Hemolysis." *Blood* 13 (1958): 348–58.

Szpidbaum, Henryk. "Die Samaritaner: Anthropobiologische Studien." *Mitteilungen der Anthropologischen Gesellschaft in Wien* 57 (1927): 139–58.

Tabatabai, Hamideh, K. Mohammad, and Nahid Mohagheghpour. "HLA Antigens in Two Iranian Populations: The Armenians and The Jews." *Tissue Antigens* 12, no. 5 (1978): 309–14. https://doi.org/10.1111/j.1399-0039.1978.tb01338.x.

Tabatabai, Hamideh. "Bar'rasī-i muqāyasah'ī-i zhinitīk-i īrāniyān-i armanī va yahūdī." MSc thesis, University of Tehran, 1977.

Taleb, Nagib, and Jacques Ruffié. "Hémotypologie des populations jordaniennes." *Bulletins et Mémoires de la Société d'Anthropologie de Paris* 3, no. 3 (1968): 269–82. https://doi.org/10.3406/bmsap.1968.1420.

Taleb, Nagib, Jacques Loiselet, Fouad Ghorra, and Hodda Sfeir. "Sur La Déficience En Glucose-6-Phosphate-Déshydrogénase Dans Les Populations Autochtones Du Liban." *Comptes Rendus Des Séances de La Société de Biologie* 258 (1964): 5749–51.

TallBear, Kim. "Genomic Articulations of Indigeneity." *Social Studies of Science* 43, no. 4 (August 2013): 509–33. https://doi.org/10.1177/0306312713483893.

TallBear, Kim. *Native American DNA: Tribal Belonging and the False Promise of Genetic Science*. Minneapolis: University of Minnesota Press, 2013.

Tankut, Hasan Reşit. *Nusayriler ve Nusayrilik hakkında*. Ankara: Ulus Basımevi, 1938.

Tanış, Tolga. "Nobel Panteri Aziz Sancar." *Hürriyet Kelebek Magazine*, October 10, 2015. http://www.hurriyet.com.tr/nbel-panteri-aziz-sancar-kizlar-hep-beni-terk-etti-30282793.

Tapper, Melbourne. *In the Blood: Sickle Cell Anemia and the Politics of Race*. Philadelphia: University of Pennsylvania Press, 1999.

Tekiner, Roselle. "Race and the Issue of National Identity in Israel." *International Journal of Middle East Studies* 23, no. 1 (1991): 39–55.

The Race Concept: Results of an Inquiry. Paris: UNESCO, 1952.

Tilley, Helen. "Racial Science, Geopolitics, and Empires: Paradoxes of Power." *Isis* 105, no. 4 (December 2014): 773–81. https://doi.org/10.1086/679424.

Trubeta, Sevasti. "The 'Strong Nucleus of the Greek Race': Racial Nationalism and Anthropological Science." *Focaal* 2010, no. 58 (December 1, 2010): 63–78. https://doi.org/10.3167/fcl.2010.580105.

Trubeta, Sevasti. *Physical Anthropology, Race and Eugenics in Greece (1880s–1970s)*. Boston: Brill, 2013.

Tupasela, Aaro. "Genetic Romanticism—Constructing the *Corpus* in Finnish Folklore and Rare Diseases." *Configurations* 24, no. 2 (2016): 121–43. https://doi.org/10.1353/con.2016.0011.

Turda, Marius. "From Craniology to Serology: Racial Anthropology in Interwar Hungary and Romania." *Journal of the History of the Behavioral Sciences* 43, no. 4 (2007): 361–77. https://doi.org/10.1002/jhbs.20274.

Turda, Marius. "Race, Politics and Nationalist Darwinism in Hungary, 1880–1918." *Ab Imperio* 8, no. 1 (2007): 139–64.

Türkmen, Ahmet Faik. *Mufassal Hatay tarihi*. 4 vols. Istanbul: Cumhuriyet Matbaası, 1937.

"Türklerin genetik şifresi çözüldü." *Cumhuriyet*, February 21, 2014. http://www.cumhuriyet.com.tr/haber/bilim-teknik/43695/Turklerin_genetik_sifresi_co zuldu.html.

Unat, Yavuz, and Ayla Sevim Erol. "Doç. Dr. Armağan Saatçioğlu (1944–1990): unutulmuş bir antropolog." In *Türkiye'de Bilim ve Kadın Kongresi bildirileri, 27–29 Nisan 2009*, edited by Günseli Naymansoy, 271–90. Eskişehir, Turkey: Eskişehir Osmangazi Üniversitesi, 2009.

Ünlü, Barış. "İsmail Beşikçi as a Discomforting Intellectual." *Borderlands* 11, no. 2 (2012).

Vejdani, Farzin. *Making History in Iran: Education, Nationalism, and Print Culture*. Stanford: Stanford University Press, 2014.

Wade, Peter, Carlos López Beltrán, Eduardo Restrepo, and Ricardo Ventura Santos, eds. *Mestizo Genomics: Race Mixture, Nation, and Science in Latin America*. Durham, NC: Duke University Press, 2014.

Wailoo, Keith. *Dying in the City of the Blues: Sickle Cell Anemia and the Politics of Race and Health*. Chapel Hill, NC: University of North Carolina Press, 2001.

Wailoo, Keith, Alondra Nelson, and Catherine Lee, eds. *Genetics and the Unsettled Past: The Collision of DNA, Race, and History*. New Brunswick, NJ: Rutgers University Press, 2012.

Wailoo, Keith, and Stephen Gregory Pemberton. *The Troubled Dream of Genetic Medicine: Ethnicity and Innovation in Tay-Sachs, Cystic Fibrosis, and Sickle Cell Disease*. Baltimore: Johns Hopkins University Press, 2006.

Walker, Deryck G., and James E. Bowman. "Glutathione Stability of the Erythrocytes of Iranians." *Nature* 184, no. 4695 (1959): 1325.

Walter, Hubert, and Mansur Bajatzadeh. "Studies on the Distribution of the Human Red Cell Acid Phosphatase Polymorphism in Iranians and Other Populations." *Acta Genetica* 18 (1968): 421–28.

Walter, Hubert, Dariush Daneshvar Farhud, Heidi Danker-Hopfe, and Pariwash Amirshahi. "Investigations on the Ethnic Variability of the ABO Blood Group Polymorphism in Iran." *Zeitschrift Für Morphologie Und Anthropologie* 78 (1991).

Walter, Hubert, Farideh Azimian, and Dariush Daneshvar Farhud. "Red Cell Enzyme Polymorphisms in Three Iranian Population Samples: Tabriz, Yazd, Mashhad." *International Journal of Anthropology* 6, no. 2 (1991): 119–26.

Walter, Hubert, Mükaddes Gölge, Muzaffer Aksoy, E. Bermik, and A. Sivasli. "Genetic Serum Protein Markers (HP, GC, TF, PI) in Four Turkish Population Samples." *International Journal of Anthropology* 7, no. 4 (1992): 27–32.

Walters, Delores M. "Perceptions of Social Inequality in the Yemen Arab Republic." PhD diss., New York University, 1987.

Watenpaugh, Keith D. "'Creating Phantoms': Zaki al-Arsuzi, the Alexandretta Crisis, and the Formation of Modern Arab Nationalism in Syria." *International Journal of Middle East Studies* 28, no. 3 (1996): 363–89.

Weissenberg, Samuel. "Die autochthone Bevölkerung Palästinas in anthropologischer Beziehung." *Zeitschrift für Demographie und statistik der Juden* 5 (1909): 129–39.

Wiener, Alexander S., Ruth B. Belkin, and Eve B. Sonn. "Distribution of the A1-A2-B-O, M-N, and Rh Blood Factors among Negroes in New York City." *American Journal of Physical Anthropology* 2, no. 2 (June 1944): 187–94. https://doi.org/10.1002/ajpa.1330020206.

"William Haas, 72, Near East Expert, Professor at Columbia Dies." *New York Times*, January 4, 1956.

Wimmer, Andreas, and Nina Glick Schiller. "Methodological Nationalism and Beyond: Nation-State Building, Migration and the Social Sciences." *Global Networks* 2, no. 4 (October 2002): 301–34. https://doi.org/10.1111/1471-0374.00043.

Winter, Stefan. *A History of the 'Alawis: From Medieval Aleppo to the Turkish Republic*. Princeton: Princeton University Press, 2016.

Woodd-Walker, Robert B., Harry M. Smith, and Victor Alan Clarke. "The Blood Groups of the Timuri and Related Tribes in Afghanistan." *American Journal of Physical Anthropology* 27 (1967): 195–204.

World Health Organization, Global Malaria Programme, University of California, San Francisco, Global Health Sciences, Global Health Group, Turkey, and Sağlık Bakanlığı. *The Long Road to Malaria Elimination in Turkey*. Geneva: World Health Organization, 2013.

World Health Organization, Regional Office for the Eastern Mediterranean. "Biographical Note: Sir Aly Tewfik Shousha, Pasha, Chairman of the WHO Executive Board and Director-designate of the WHO Regional bureau in Alexandria." February 10, 1949. World Health Organization Institutional Repository for Information Sharing. EMRO Regional Committee for the Eastern Mediterranean. https://apps.who.int/iris/handle/10665/121216.

Wyman, Leland C., and William C. Boyd. "Human Blood Groups and Anthropology." *American Anthropologist* 37, no. 2 (1935): 181–200.

Yahil, Chaim. "Israel's Immigration Policy." *International Labour Review* 66 (1952): 444–60.

Yanow, Dvora. "From What *Edah* Are You? Israeli and American Meanings of 'Race-Ethnicity' in Social Policy Practices." *Israel Affairs* 5, no. 2–3 (December 1998): 183–99. https://doi.org/10.1080/13537129908719518.

Yāsamī, Rashīd. *Kurd va payvastagī-i nizhādī va tārīkhī-i ū*. Tihrān: Kitābfurūshī-i Ibn Sīnā, 1940.

Yavuz, M. Hakan. "Political Islam and the Welfare (Refah) Party in Turkey." *Comparative Politics* 30, no. 1 (1997): 63–82.

Yiftachel, Oren. *Ethnocracy: Land and Identity Politics in Israel/Palestine*. Philadelphia: University of Pennsylvania Press, 2006.

Yosmaoğlu, İpek. *Blood Ties: Religion, Violence, and the Politics of Nationhood in Ottoman Macedonia, 1878–1908*. Ithaca, NY: Cornell University Press, 2014.

Younovitch, Rina. "Contribution à l'étude sérologique des Juifs de Yémen." *Comptes rendus des séances de la Société de Biologie* 84, no. 111 (1932): 929–31.

Younovitch, Rina. "Étude sérologique des juifs samaritains." *Comptes rendus des séances de la Société de Biologie* 85, no. 112 (1933): 970–71.

Younovitch, Rina. "Les caractères sérologiques des juifs asiatiques." *Comptes rendus des séances de la Société de Biologie* 85, no. 113 (1933): 1101–3.

Zaidman, J. L., H. Leiba, S. Scharf, and I. Steinman. "Red Cell Glucose-6-Phosphate Dehydrogenase Deficiency in Ethnic Groups in Israel." *Clinical Genetics* 9, no. 2 (1976): 131–33. https://doi.org/10.1111/j.1399-0004.1976.tb01558.x.

Zalloua, Pierre A., Daniel E. Platt, Mirvat El Sibai, Jade Khalife, Nadine Makhoul, Marc Haber, Yali Xue, et al. "Identifying Genetic Traces of Historical Expansions: Phoenician Footprints in the Mediterranean." *American Journal of Human Genetics* 83, no. 5 (November 2008): 633–42. https://doi.org/10.1016/j.ajhg.2008.10.012.

Zalloua, Pierre A., Yali Xue, Jade Khalife, Nadine Makhoul, Labib Debiane, Daniel E. Platt, Ajay K. Royyuru, et al. "Y-Chromosomal Diversity in Lebanon Is Structured by Recent Historical Events." *American Journal of Human Genetics* 82, no. 4 (April 2008): 873–82. https://doi.org/10.1016/j.ajhg.2008.01.020.

Zeidman, Lawrence A., and Jaap Cohen. "Walking a Fine Scientific Line: The Extraordinary Deeds of Dutch Neuroscientist C. U. Ariëns Kappers Before and During World War II." *Journal of the History of the Neurosciences* 23, no. 3 (July 3, 2014): 252–75. https://doi.org/10.1080/0964704X.2013.835109.

Zia-Ebrahimi, Reza. *The Emergence of Iranian Nationalism: Race and the Politics of Dislocation.* New York: Columbia University Press, 2016.

Zilberstein, V., and N. Goldstein. "Suge ha-dam etsel nashim yehudiyot mi-'edot ha-mizraḥ." *Harefuah* 54 (1958): 295–96.

Zurayq, Qusṭanṭīn. *al-Aʿmāl al-fikriyya al-ʿāmma li-l-Duktūr Qusṭanṭīn Zurayq.* Vol. 1. Bayrūt: Markaz Dirāsāt al-Waḥdat al-ʿArabīya, 1994.

Index

Page numbers in italic indicate figures.

Abadan oil refinery, 119–20, 123
ABGL (Anthropological Blood Grouping Laboratory), 187, 189, 193–94. *See also* AUB; Smith, Harry Madison
ABO blood groups, *78*, 290n29; alleles in, 72, 85, 87, 121, 236; in apes, 85, 88; compared to anthropometry, 71, 73; failures to find predicted results, 89, 95–96; frequencies among Armenians, 78; frequencies among Bedouins, 76; frequencies among Jews, 119; frequencies in Anatolia, 87, 235; frequencies in Levant, 73, 77; frequencies in Mandate Palestine, 79; frequencies in Turkey, 234; group A, 72, 83–84, 86–87, 236; group AB, 68; group B, 72, 86–87, 93, 221; group O, 72, 76, 86; inheritance patterns of, 68; limitations of as indicators of race, 105; medical genetics and, 126–27; phenotype frequency of, 86; pre-racial evolution theory of, 90; sampling errors in studies, 109; Shanklin fieldwork in, 77; and transfusions, 67. *See also* blood supply chains, sero-anthropology

Abraha's army, 135
Abu El-Haj, Nadia, 115, 117, 161
Abu Haydar, Najib, 146
Adam, Avinoam, 162, 201
Aden, *196*; British withdrawal from, 26, 184, 187, 198–99; Lehmann and Ikin in, 131, 134, 136, 285n20; Marengo-Rowe's hospital in, 198–99; Mourant obtaining samples from, 195–97; RAF deployments to, 183; Socotra

353

Aden (*continued*)
expedition, 101–4, *102*, 110; Veddoid sickle cell carriers in, 141–42
admixture: from Africa, 109, 131, 142, 177, 194–95, 286n48; Aksoy on, 140; Allison on, 135; Altounyan on "Turkomans" and Kurds, 76; among Yemenites, 115; among Zoroastrians, 122; of Bandaris, 230; connotations of term, 5–6, 18; "cyclic", 251; debates over, 73; Hirszfelds on Slavs and Turks, 69; how to evaluate, 71; Ikin on African and Saudi, 112; İnan on Alpine and Dinaric groups, 51–52; of Iranian Plateau, 229; of Jews from Europe, 80; Kurds, 237; Mongol, 235; sickle cell trait, 145–46, 148–49, 151–52; Turks, 251–52; Turks, Kurds, Arabs, 237
Afghans, 60, 121, 193
Africans: admixtures, 109, 112, 131, 140–42, 177, 194–95, 286n48; alleged ancestry of *akhdām*/"Zabidi" class in Yemen, 134–38, 141–43, 146, 150–52, 285n20, 24; Aryeh Gelblum on, 282n48; "Asio-African biochemical race", 72–73; British colonial troops, 67, *68*; expectations regarding Socotrans, 102–3, 105, *102*; G6PD deficiency in, 155–56, 160, 162, 171, 177; as "major race", 103; and Mizrahim, 19, 113–19, 129, 152–53, 161; of North Africa, 11; North Africans in Israel, 114–15; Out of Africa model, 248, 251–52, 255; Parr on, 77; *Rho/cDe* type, 104–5, 109; and Yemenite Jews, 79, 114, 130, 134–35, 138, 206. *See also* Ethiopia; favism; G6PD deficiency; "Negro"/"Negroid"; sickle cell disease

African Americans: Black Panthers, 148; James E. Bowman, 167, 171; G6PD deficiency in, 156, 160; health disparities and activism, 148, 284n6; *Rho/cDe* type, 104; and sickle cell disease, 131, 133, 148. *See also* sickle cell disease

agendas for research, 22, 46, 155–57; of blood banks, 113; in Iranian medicine, 173–77, 220–21, 231

AGHP (Anatolian Genetic History Project), 250–54

Akbari, Mohammad Taghi, 226–30

akhdām/"Zabidi" class in Yemen, 134–38, 141–43, 146, 150–52, 285nn20, 24

Aksoy, Muammer, 240

Aksoy, Muzaffer, xv, *139*, 299n19; on Alawites, 143–44, 150; blood bank donor data, 238–39; brother Muammer's assassination, 240; collaboration with Lehmann, 140–42; critiques of, 145, 149, 286–87n48; early life and background, 138, 152; on "Eti-Turks", 139–44, 149–52, 233; funding sources for, 161, 233; mentoring Sancar, 250; motivations of, 145, 152; on sickle cell anemia in Yemenite Jews, 131, 138; on "Veddoids", 140–42

Ala, Fereydoun, 220–21, 225, 240

Alawites, 55; Aksoy on, 142–44, 150, 152; Arab Alawites, 53; "Eti-Turk"

campaign, 18, 54, 138–41, 144, 149–52, 233, 250; in Lebanon, 144; Nusayri, 53–54, 138; origins of, 54–55, 142; political rifts among, 54; as "proto-Hittites"/"Hittite Turks", 42, 53–55, 139; responding to genetic objectification, 7; sickle-cell anemia among, 138, 150; Turkey's "assimilation" of, 34, 138–39
Alborz College (Tehran), 37
Alevis, 53, 253, 271n94
Algeria, 145
al-Hilāl journal, 38
Allawi, A. A., 125
alleles in blood groups, 72, 85, 87, 121, 236
Allison, Anthony C., 17, 135, 137, 140, 145, 171
al-Muqtaṭaf journal, 38
"Alpine" racial type, 37, 47, 51, 54, 64, 86, 95
Altman, Russ, 258
Altounyan, Aram Assadour, 74
Altounyan, Ernest, 73–76, 275n18
Altounyan Hospital (Aleppo), 73–74
American Indians, 72, 76, 90, 244
Amirshahi, Parivash, 226–29, *228*
Ammar, Abbas Mustafa, 95
Anatolia, 31; ABO frequencies in, 85, 87, 235; Allied occupation of, 12, 70; Anatolian Genetic History Project (AGHP), 250–54; Armenians in, 51; Aziz Sancar and, 249; Babacan research on, 84–85, 93; Greek invasion of, 31; Hittites in, 53–54; National Campaign and population exchange, 70; Ömer Gökçümen research on, 251–52; Osman Çataloluk on, 254; relationship of Kurds to, 51–52, 75, 84; relationship of Turks to, 33–34, 47–54, 70, 84–85, 93, 254; Saatçioğlu research on, 235–37; study of Yürüks in, 86–87
ancestry testing, 63, 264n14
Anderson, Benedict, 8
Andree, Richard, 79
Anglo-Iranian Oil Company, 58, 120
Anglo-Persian Agreement (1919), 56
anonymization of research subjects, 253
Anthropological Blood Grouping Laboratory (ABGL), 187, 189, 193–94. *See also* AUB; Smith, Harry Madison
"anthropological derivation", 137
anthropological genetics, 4, 23, 69, 130, 211, 215–19, 245. *See also* Bonné/Bonné-Tamir, Batsheva
anthropometric studies: Aryan race concept, 36–38, 48, 75, 274n140; AUB promoting, 40–42, 63, 79; compared to sero-anthropology, 71–76; difficulties for field researchers, 58–60, 215; and Iranians, 76, 222; and Iraq, 76, 222; and Lebanon, *41*, 42; manipulation of, 32, 56–57; in Pahlavi dynasty, 24, 34, 57, 60; political significance of, 63; to prove Turks as European, 52; of Samaritans, 45. *See also* race science
antisera, 68; availability of, 109, 113, 116–17, 126, 198; concerns over reliability

antisera (*continued*)
of, 109; in field conditions, 90, 105, 108

Arabs: Arabic speakers, 39, 41, 54, 145; Alawites as "Arab servants", 138, 139, 144, 152; Beckett on, 121; control over blood samples from, 185–86; Kappers on, 41, 79; nationalism and sickle cell disease, 132, 145–46; Parr on, 77–78, 80; use of term, 75–76, 123; Younovitch on, 82–83. *See also* scientific territorialism

Aramco: advising AUB, 193; Armand Phillip Gelpi, 110, 148–49, 171–72; controlling sampling, research, 110–13, 193; George Maranjian, 110–12, 148, 195; malaria eradication campaign by, 111. *See also* Ikin, Elizabeth W.

Arap uşağı/uşakları ("Arab servant/s"), 138, 139, 144, 152

Armenians: Akbari study on, 226, 229–30; Altounyan on, 73–75, 275n18; Bayatzadeh on, 214; Bonné's studies, interactions with, 207, 209–11; Bowman on, 168–69, 171; de Graaf, Nijenhuis on, 123; endogamy among, 7, 168; Farhud on, 231–32; genocide against, 31, 51, 74; George Maranjian, 110; as Hittites, 42, 75; in Iran, 168–69, 175, 227, 257; in Jerusalem and West Bank, 207, 209–10; Kappers on, 42, 79; Mackintosh on, 74–75; nationalism among, 30; origins of, 42, 75, 77, 79, 231–32; Parr on, 77–78; Saatçioğlu on, 235; Tabatabai on, 223–25; in Turkey, 33, 233

Aryan race concept: craniometry and, 57, 270n72; Farhud on Armenians and Azeris, 231–32; and IHGP, 305n38; in Iran, 34, 55–57, 61–66, 122, 169, 223–24, 229, 231; "kindred blood" concept, 66; and Kurds, 75; language families as racial categories, 36; and nationalism, 38; Nazism and, 65–66, 96, 270n72; theories on original Aryans, 37, 42, 57; in Turkey, 63, 65–66; and white supremacy, 36. *See also* anthropometry; race science

Ashkenazim, 155; and Bonné-Tamir studies, 210–12, 246; as "European", 170; and favism, 155–60, 166, 179; Gurevitch serology study, 115–17, 119; Mourant serology study, 118–19; in New Yishuv, 81–82, 84, 91, 93–94; Polish, 44; prejudice of against other Jews, 114–15, 159, 206; serology of in New Yishuv, 79–82; and Yemenite Children Affair, 128–29. *See also* Jews

Asian classification: Asiatic Jews, 79–81, 83, 114; Asio-African racial type, 72. *See also* Mongoloid/yellow/Asian classification

Assyrians, 125–26, 128, 224; Akbari blood group study of, 226, 229–30; Christian communities, 125–26, 188, 226–30; de Graaf/Nijenhuis blood group study of, 123; Farhud on, 232; Gurevitch blood group study of "Assyrian" Jews, 115; as Iranian religious minority, 227; Kappers on, 42, 79; Mourant blood group study of,

125; origins of, 42, 79; and Samaritans, 166; usefulness of research to, 241
Atatürk. *See* Mustafa Kemal and Kemalists
AUB (American University of Beirut): and ABGL, 187, 193–94; Adib Tayyar, 90; Aramco-sponsored field study, 193; controversy over work with Israelis, 192; criticism over findings, ethics, 193–94; formerly Syrian Protestant College, 37; opposing Phoenicianism, 40; promoting anthropometry, 40–42, 63, 79; research in Kuwait, 193; studying ABO frequencies, 73; and Veddoid hypothesis, 146. *See also* Kappers, Cornelius; Parr, Leland W.; Shanklin, William M.; Smith, Harry Madison; Zurayk, Constantine
Aulaqi, Adel, 285n20
Australian racial type, 72, 76
Aygen, Nermin (Nermin Aygen Erdentuğ), 95–96, 234
Azerbaijan/Azeris, 56, 120, 232
Azhir, Ahmad, 119, 221

Babacan, Ethem, 84–86, 91, 93–94
Babylon and Babylonian Jews, 43, 80, 91, 115, 166, 224
Baghdadis, 91, 108, 117
Baghdad Pact, 145
Baghdad spring anemia, 154–55, 177
Baha'i, 259
Bakhtiyari peoples, 120, 123
Baluchis, 216, 226–27, 229

Banihashemi, Kambiz, 256, 305n38
Banse, Ewald, 53
Bantu-speaking tribal groups, 132
Bat-Miriam, Mariassa, 162
Bayatzadeh, Mansur, 213–15, 221–22, 230
B blood. *See* ABO blood groups
Beaconsfield, Peter, 175–76
Beckett, Philip H. T., 121
Bedouins: Altounyan on, 74, 76; as biological, racial, or social group, 18, 39–42, 74; blood groups in, 91; Bonné on South Sinai Bedouins, 185, 200–206, 209, 211; Boyds on, 90–91; Coon on, 136; endogamy among, 7; grouped racially with Native Americans, 76, 90; Hadhrami Bedouin Legion, 197; Ikin on, 112; Israeli research on, 185; Kamel on, 212; Kappers on, 269n50; Maranjian on, 112; no sickle cell disease in, 112–13; as Phoenicians, "true Arabs", 39, 42, 76; Saudi government settlement of, 111; sero-anthropology of, 71; Shanklin and Kappers on, 41–42, 76, 90, 95; Smith and Ghareeb study, 189; Taleb and Ruffié on, 194; tribes of, 76, 90–91. *See also* Jebeliya Bedouins
Beirut universities, 37, 39–40. *See also* AUB
Bekpen, Cemalettin, 250–52, 255
ben Amram, Yitzhak, *46*
Ben-Zvi, Yitzhak, 45, *46*, 83, 163
Berbers, 145
Beşikçi, İsmail, 236–37
Bet Salim, Nadia, *117*

INDEX

BGRL (Blood Group Reference Laboratory): Assyrian blood group testing, 125–26; categorization of communities, 112; as "center of calculation", 104; and Egyptian State Serum Institute, 108; Rh blood typing, 105, 107–8; samples from Fisher, Ibrahim, 108; samples from Maranjian at Aramco, 110–12; samples from Smith, 188–89; samples to Gurevitch, 116–17; sickle cell disease research, 112, 117; Socotrans blood group testing, 102, 105, 110; WHO recognition for, 105. *See also* Ikin, Elizabeth; Mourant, Arthur E.; Smith, Harry Madison

Bier, Jess, 184

Bilim ve Teknik magazine, 252

bin Yahya, Abdullah, 136

"biochemical index", 72–73, 79, 83–86

"biochemical races", 72–73, 76

biocolonialism, 18, 244

"biological principles," Rıza Tevfik on, 30–31

black/African/Negroid classification, 103, 222

Black Panthers sickle cell campaign, 148

blackwater fever, 154, 158. *See also* hemolytic anemia

"blending the exiles" plan (*mizug galuyot*), 114

blood banks, 103–4, 113; Iranian, 119–24, 214; Israeli, 114–19; Turkish, 233, 235, 238. *See also* "traffic in blood"

blood collection issues: blood group typing, 102; criticisms between labs, 109; food or cash for sample, 103; labeling of samples, 125–26, 239; lying to donors, 197; paying for blood, 90; refrigeration and transport, 101–2, 104, 116, 183–84, 202; refusals to donate, 196–97; Tel-Hashomer Government Hospital, 189–90, 192, 195, 198, 200–202, 204, 206

blood groups, 95, 105; MN/MNS blood groups, 88, 95, 107–8, 119, 143, 290n29. *See also* ABO blood groups; Rh antigen system

blood sample sources, 187; Aramco, 110–11; blood banks, 103–4; "blood senders", 298n100; control over, 185–86; Egypt and Iraq, 107–10; Koubba Blood Transfusion Centre, 109; postwar, 163; Saudi Arabia, 110–13; soldiers, 104. *See also* BGRL; blood collection issues

Blumberg, Baruch, 164

Blumenbach, Johann Friedrich, 35

Boahmatlı tribe, 86

Boas, Franz, 268n20

Bodmer, Walter and Julia, 17, 205–6, 208–11, 239

Bonné, Alfred Abraham, 294n15

Bonné/Bonné-Tamir, Batsheva, 187, 216; attitude of toward Armenians, 209; background of, 185, 189–90; collaborations with Boyd, Mourant, Till, Smith, 185, 188–92, 195, 197–205, 296n69; faculty position at Tel Aviv University, 194; Habbanite study,

195, 197–98; having local knowledge, connections, 197, 204–5; HLA collaborations with Bodmers, Nevo, 205–8, 210–11; responding to racism, exploitation charges, 245–46; study of Samaritans, 190–92, 199, 211; study of South Sinai Bedouins, 185, 200–202

Botting, Douglas, 101–4, 110, 121, 124, 195

Bowman, James E., 167–78, 223–24

Boyd, Lyle, 88–95, *89*, 104–5

Boyd, William C., 17, 103, 215; on ABO in apes, 88; ABO/MN/PTC study (1935), 189–91; defending serological genetics, 216–17; field studies of Copts, Bedouins, 88–92, *89*, 104–5; Samaritan study in Nablus, 189–91; serological research with Haldane, 92–94; using threats to obtain samples, 91; views on Ashkenazi origins, 93–94; work with Batsheva Bonné, 189–92; work with Mourant, 108, 189; worldwide blood group data compilation, 91, 106

brachycephaly (broad-headedness), 36–37, 48, 51, 55. *See also* cephalic index

Braun, Hugo, 84–85

Brautbar, Chaim, 210

Britain: and Aden, 134, 184; and Aksoy, 250; Bonné, Bodmers HLA project, 205–6, 208–11; British disagreements with Aksoy, 149–50; Burton on Bedouins, 39; colonialism in India, 36; colonial supply chains, 107–8, 110; and Farhud, 225, 231, 255, 260; favism researchers, 168,

175–76; favoring Christians in Middle East, 38; genetic scientists' expectations, 145; in India, 272n109; in Iran, 120; in Iraq, 125; joint Israeli-British research, 205; Levantine mandates, 46, 63–64, 67; Mandate for Palestine, 12, 63, 70, 73; and Nasserism, 145; opposing Aryanism, 96; race, gender bias in research, 185–88; Sheba in medical services, 158–59, 163; sickle-cell screening by, 131–32; and Sudan, 95; views of sero-anthropology, 74–75, 91; and Yemeni *akhdām*/"Zabidi" class, 134–38, 141–43, 146, 150–52, 285nn20, 24. *See also* Aden; AUB; Mourant, Arthur E.

Broca, Paul, 64

Brooks, Patricia, 189

Bulgaria, 51, 69, 114, 118

Burton, Sir Richard Francis, 39

Canaanites, 39, 43–44

Çatal Höyük, 254

Çataloluk, Osman, 254–55

Caucasoid, 22, 103, 221

Cavalli-Sforza, Luigi Luca, 17, 164, 243–44, 254

cDe chromosome (Rh blood group), 104–5, 109, 137, 194

Cemal Pasha, 74

Center for Iranian Anthropology, 62, 217

cephalic index, 33; applied to Arabs, 39, 42, 79; applied to Jews, Samaritans, 43–44, 79; applied to Turks, Kurds,

360 INDEX

cephalic index (continued)
51, 54–55, 64; versus "biochemical index", 72, 79; creation and application of, 35–36, 75; Franz Boas' critique of, 268n20; in high school textbooks, 49; Kapper and, 42, 79, 274n141; political significance of, 63; popular use of, 63. See also brachycephaly, dolichocephaly
Chehab, Maurice, 40
Chernoff, Amoz I., 287n48
Chinese, origin of modern, 255
chloroquine, 233
CHP (Republican People's Party), 54, 138
Christians: Armenians in Iran, 168, 227, 230–32; Assyrian, 125–26, 188, 226–30; Baghdadi, 91; Coptic, 108; European favoritism toward, 38; Greek Orthodox, 146, 194; Iranian, 226; Israeli, 189, 207–8; Lebanese, 42, 146–47, 193, 245; Maronite, 39, 247; missionary-founded institutions, 37, 103; Syrian, 43, 77–79, 90; Turkish, 70
CISNU (Confederation of Iranian Students–National Union), 214
Clarke, Victor Alan, 188, 193, 220, 225
Cochin, India, 116
Cohen, Amram, 82
Cohen, Jakub, 45
Cohen-Kedmi Commission, 128, 130
collaboration/scientific collaborators: 1, 9, 19–21 26, 108–10, 122, 124, 131–31, 140–42, 145, 149, 152, 157, 160, 175–76, 178, 184–85, 187–88, 192–95, 199–200, 202–7, 211, 231, 238–39, 246, 250, 252, 257–58, 260–61; as knowledge-control regime, 21, 107, 124; political collaborationism, 21, 187, 207, 212, 226
"cold chain" of blood supplies, 104, 107
colonialism: in Aden, 199; anti-colonial ideologies of Turkish Republic and Pahlavi Iran, 46; "biocolonialism", 244; and blood supply chains, 107–10; "comprehensive scientific colonialism", 10; "internal", 212, 223, 264–65n27; Nazi propaganda on, 64; in race science, 35–38; Rıza Tevfik on, 30; Turkish Republic as colonial state, 218; Zionist settler-colonialism, 83
color vision defects, 176, 192, 290n29
Committee of Union and Progress (Young Turks), 30–31
"complexion"/skin color, 14, 55, 89, 92, 135, 155, 159–60, 166
consumer genetic ancestry industry, 244
Coon, Carleton S., 133, 136–37, 151
Copts: endogamy among, 77; pharaonicism regarding, 38, 95, 125; seroanthropology of, 71, 89, 108–9; skin color of, 89
craniometry, 41; at AUB, 40; cranial indices, 33, 37, 42, 75; "craniological science", 32; of deceased poets, scholars, statesmen, 57; laypeople's fears of, 41; in Ottoman Empire, 39; and "racial index", 72; Rıza Tevfik on, 29–30, 32. See also cephalic index

Crusaders, intermarriage with, 78
Cultural Iran (Farhud), 231
Cyrus (the Great), 170, 224

Dabbous, Ibrahim, 146–47
Dameshek, William, 129, 138, 160
Daneshjou, Roxana, 258
data interpretation: separatist fears regarding, 120; sources of error, 93, 124; Rıza Tevfık criticisms of, 29–30
data sharing among researchers, 164, 204–5
Dawlatabadi, Hajji Mirza Yahya, 34
de Graaf, W., 123–24
Deniker, Joseph, 15
Department of School Hygiene (Hadassah), 82
"development towns" (*'ayarot pituaḥ*), 114, 116
Diego blood antigen, 105
Dilaçar, Agop, 54
Dimen, Ahmed Şükrü, 84
Dinaric racial type, 51, 87, 95
disease resistance, 62, 137, 156, 171, 221, 256
Distribution of the Human Blood Groups, The (Mourant), 106, 109, 119, 126
DNA, 24; direct sequencing of, 243–44, 260; DNA mystique, 3–4, 63; era of, 232; GP (Genographic Project), 244–45, 247; HGDP (Human Genome Diversity Project), 243–44; "Phoenician" genetic markers, 245, 247–48
dolichocephaly (long-headedness): associated with race, 36–37; İnan on, 51; "Indo-Persian race", 58; Kurds as, 271n89; Samaritans as, 43. *See also* cephalic index
donors. *See* research subjects
Dreyfuss, Fritz, 129, 133–35, 160
Druze, 40, 42, 78–79, 147, 188, 206–9, 212
DTCF (Dil ve Tarih-Coğrafya Fakültesi) at Ankara University, 49–50, 234–35, 238
Duara, Prasenjit, 8
Duffy blood antigen, 105

Eftekhari, Mirza, 120
Egypt: favism in, 172–73; as part of blood supply chain, 107–10; peace treaty with Israel (1979), 209, 212; pharaonicism, 33, 38, 95, 125; seroanthropology in, 71, 73, 77, 88–89, 95, 107–10; sickle-cell disease in, 140, 145, 147–49; Six-Day War and aftermath, 184, *186*, 194, 198, 201–2, 212; and Sudan, 95; and Syria, 145. *See also* Nasser, Gamal Abdel
Egyptians: Arabness of, 95; "as both colonizer and colonized", 18; interests in race science, 38–39; Mourant disputing findings of, 109–10, 118, 124; as not Aryan, 66; *See also* nationalism (Arab)
El-Dewi, Salah, 109
Elling, Rasmus, 122
El Shakry, Omnia S., 18, 38
endogamy/inbreeding, 5–9, 17–18, 25, 132; among Armenians, 168; among Copts, 77; among Druze, 79; among Jebeliya Bedouins, 201; among Jews,

endogamy/inbreeding (*continued*) 80, 115, 160; among Samaritans, 42–43, 77, 83; among Zoroastrians, 121, 169; within Iran, 256; Lehmann on sickle cell disease, 137

environmental determinism, 138, 151, 180

Erdentuğ, Nermin Aygen, 96, 234

erythrocytes, 280n12

Eseli tribe, 86

Ethiopia, 73, 135–36, 162–63, 170

"ethnic commonsense", 122

"ethnic group" versus "race", 96–97

"Ethno Composition" map of Iran, *228*, 229

"Eti-Turk" Alawites, 18, 54, 138–41, 144, 149–52, 233, 250. *See also* Alawites

eugenics, 14, 23, 96–97, 136

European Ethical Societies movement, 29

European imperialism in Asia, 29–30

European race science, 35–39, 93

evolutionary theory: ABO blood groups and, 72, 88, 90–92, 95; compared to medieval Islamic hierarchy of civilization, 15; evolutionary anthropology, 177–80; favism and malaria hypothesis, 155–57, 164, 172–73, 175–80; and genetic/genomic nationalism, 8; HLA antigens and, 205–6; human evolution, 72, 131, 243–44, 248, 255; Middle East's contribution to, 15–18, 130; "molecular disease" and, 130; and race science, 36; and sickle cell disease, 131–38, 150–52, 157; value of endogamy for, 5–7. *See also* gene flow; natural selection

Eyquem, André, 120–21

Faisal II, 125, 154

Fakhrai, Habibollah (Habib), 220–21, 225, 239–40

Falasha, 162–64, 170, 290n31

falciparum malaria, 156, 171, 172, 176

Family Tree DNA, 244

family trees, 131, 244–45, 248, 255, 258–59

Farhud, Dariush Daneshvar, 1–3, 221–25, 230–32, 255, 260

favism (G6PD deficiency), 127, 153, 290n29; Bowman research on, 167–71, 177–78, 223; "Favism in Iran" seminar, 173–75; as G6PD deficiency, 155, 159–67, 178; Gelpi research in Saudi Arabia, 171–73; in Iran, 167–71, 173–75; in Iranian Kurds, 176; in Israel, 157–59, 178; Jews and, 155–61, 163, 166, 179; Lederer research on, 154–55; and malaria hypothesis, 155–57, 164, 172, 175–80; in Mexico, 177; and "multicentric" origins hypothesis, 160; prevalence of, 155; sex-linked, 164; Sheba's research and views on, 158, 179, 162–66; in Turkey, 233

fellahin (peasant farmers), 39, 43, 139, 269n55

Fellahs, 142, 145

Ferit, Nejat, 47

Fertile Crescent, 79, 155

Field, Henry, 58–62, 71–72, 76

INDEX 363

Field, Marshall, 58
Findlay, H. T., 108
fingerprint patterns, 87, 290n29
Finns, 36
Firdawsi, 57
Firouz, Narcy, 167
Firzli, Salim, 146–47
Fischer, Eugen, 270n79
Fisher, Edgar Jacob, 47
Fisher, J. H., 108
foregone conclusions in research, 56
founder effects, 47, 144, 166, 179
France: and "Alpine" racial type, 37, 51; anthropological genetics, 69, 71; Aryanism in, 37–38, 65; Ashkenazim in, 118; favoring Maronites, 39–40; IBP project, 194; and Iranian anthropology, 62; Lapeyssonnie, Léon, 174–75; Levantine mandates, 46, 53, 63–64, 67; Nicolas Kossovitch, 79, 276n39; Pasteur Institute (Paris), 119–21, 276n39; suppressing pro-Arab groups, 54; tracing sickle cell disease, 131; and William (Wilhelm) Haas, 61
Frank, Erich, 84, 138

G6PD deficiency, 155, 159–67, 178. *See also* favism
Gannett, Lisa, 6, 69
Gelblum, Aryeh, 282n48
Gelpi, Armand Phillip, 110, 148–49, 171–72
"gene drain", 2–3, 260
gene flow, 238

genetic anthropology, 4, 215–22; Bayatzadeh and, 215, 221; Bonné-Tamir and, 281n33; exploitation of Arab communities, 212; Farhud and, 221–22, 225, 231–32; Gökçümen on, 251–52; and medical genetics, 130; as progressive, anti-racist discourse, 244–45, 247–48; relationship to violence, 240–41; Sheba and, 161, 200; in Turkey, 233; Zalloua and, 247
genetic drift, 6, 25, 102, 172, 179
genetic nationalism, 8–9, 260–61
"genetic romanticism", 264n15
Genetics Society of Iran, 219–22, 225
Genna, Giuseppe, 44
Genographic Project (GP), 244–45, 247, 257
Georgians, 47, 115
Ghareeb, Fares Aftimos, 188–89, 193
Gini, Corrado, 44–45, 82–83
Gm^a immunoglobulin, 120–21
Gobineau, Arthur de, 50, 62
Goitein, Shelomo Dov, 133
Gökçümen, Ömer, 251–53, 255
Gölge, Mükaddes, 239
Goodman, Morris, 216
GP (Genographic Project), 244–45, 247
"Greater Lebanon", 40
Greek Orthodox Christians, 54, 146, 194
Greeks: blood types among, 78; as Dinaric type, 51; "European Turks" as, 49; favism among, 160, 166; on İmroz (Imbros) island, 233; invasion of Anatolia, 31; nationalism of, 30–31; "population exchange" with Turkey, 51, 53, 70; race science and,

Greeks (*continued*)
69–70; sickle cell disease among, 131, 133, 137; Trebizond Empire, 236; Turkish ethnic cleansing of, 51
Grenada, 131
Griesemer, James, 6, 69
Gross, Ruth, 160
Gültekin, Timur, 252
Gurevitch, Joseph, 114–19

Haas, William S. (Wilhelm Haas), 61–62
Habbanite Jews, 194–99, *196*, 201–5, 211
Hadassah Department of School Hygiene, 82
Hadassah hospitals, researchers, 73, 79, 113–17, 128–30
Hadhramaut (Hadhramis), 136, 195–99
Haeckel, Ernst, 30
Hafız, 57
Haksever, Oğuz, 252
Haldane, J. B. S., 17, 91–94, 164
Haldun, Zıhni, 47
Hamitic-speaking tribal groups, 60, 95, 132
haptoglobin, 105, 280n12, 290n29
Haqiqi, Fazlallah, 62
Hardy-Weinberg equilibrium, 121
Hatay Crisis, 53–55, 64, 138, 144–45
HbS (hemoglobin S) gene, 130, 143–44, 146, 149, 287–88n64. *See also* sickle cell disease
Hekmat, Ali Asghar, 58
hemoglobin, 105, 130, 216, 280n12, 290n29
hemolytic anemia, 155–56, 158–60, 175
Hermoni, David, 116

Herzfeld, Ernst, 60
HGDP (Human Genome Diversity Project), 5, 243–45, 247, 256
HGDPI (Human Genome Diversity Project of Iran), 255
Higher Education Council (*Yükseköğretim Kurulu*, or YÖK), 238
Hilgartner, Stephen, 21
Hirszfeld, Ludwik and Hanna, 67–73, 78, 84–85, 87, 93, 241
Hirszfeld biochemical index system, 73, 83, 96, 103
Hittites: Alawites as, 42, 53–55, 139, 141–42; Anatolian, 47, 51, 53–54; and Armenians, 75; "proto-Hittites", 42; and Turkish origins, 38, 63. *See also* "Eti-Turk" Alawites, Turkish History Thesis
HLA (human leukocyte antigen) system: Bonné, Nevo, Bodmers project, 205–6, 208, 210–11; Bodmer, Saruhan project, 239; difficulties assigning ethnicity to samples, 239; Mohagheghpour on Zoroastrians, 223–24; Tabatabai thesis on, 223–24
"Hobday, G. R.", 125
Homo sapiens: arrival in Anatolia, 235; versus *erectus*, 255; Linnaeus' racial division of, 35
Hooton, Earnest A., 60, 270n79
Hormozdiari, Mr. H., 224
Hulla valley, 179
human family tree, 131, 245, 248, 258–59
Human Genome Diversity Project (HGDP), 5, 243–45, 247, 256

Human Genome Diversity Project of Iran (HGDPI), 255
Hurrians, 42
Hussein, Saddam, 217
Huxley, Henry Minor, 42–43
Huxley, Julian S., 96
Hygiene Laboratory (University of Istanbul), 85

Ibn Sina, 57
IBP (International Biological Program), 187, 193–95, 200
Ibrahim, Karima A., 108
ideology affecting methodology, 56
IGP (Iranian Genome Project), 257–60
IHGP (Iranian Human Genome Project), 255–57, 305n38
IHMGB (Iranian Human Mutation Gene Bank), 256–57
Ikin, Elizabeth W.: in Aden, 131, 134, 136, 285n20; at BGRL, 189; experimental method and findings, 112; on Levant, 143; on Socotran blood groups, 102; research on Aramco employees, 110; studies of Assyrians from Iraq, 112; on "Zabidi Arabs" (*akhdām*), 143, 150. *See also* BGRL
Illumina, 258
immunoglobulins, 105, 120, 280n12
İmroz (Imbros) island, 233
İnan, Afet, 50–54, 232, 271n83
inbreeding. *See* endogamy
INBTS (Iranian National Blood Transfusion Service), 219–20, 225
India: British colonialism in, 36; caste system in, 272n109; Hirszfelds'
serology, 67, 69; Jews in, 170; Lehmann's research in, 132–37, 40; Mizrahim from, 115; "scientific colonialism" of, 10; sickle cell disease in, 137, 140, 142, 147–49, 151; "Veddoids of", 151
Indians: as Aryans, 36; Parsis, 227; Sheba on hemolysis among, 158; "Veddoids", 136
informed consent from sample donors, 241
Institute for Anthropology (Mainz), 216, 221
Intermediate racial type, 72–73, 78, 84
"internal colonialisms", 13, 18, 35, 212, 223, 264–65n27
internationalism, 261
interpreters, 45, 59, 89, 201, 241, 246
Iranian Oil Refining Company, 119, 123
Iran: Anglo-Iranian Oil Company, 58, 120; Anglo-Persian Agreement (1919), 56; anthropometric studies in, 76; Aryanism in, 34, 55, 57, 61, 64, 223–24; blood banks in, 119–24, 214, 220; "brain drain" from, 225; Center for Iranian Anthropology, 62; conclusions from blood work, 214–15, 221; craniometry research in, 57–61; *Cultural Iran* (Farhud), 231; distrust of Britain, 120; "Ethno Composition" map, *228*, 229; Fars region of, 222; favism in, 167–71; genetic diversity in, 121–24, 222, 229–30; IHGP/HGDPI genome project, 255–57; international sanctions against, 257; Iran-Iraq War, 217, 239; Islamic

Iran (continued)
    Republic of, 216; and James E. Bowman, 167–69; Jews in, 116, 224–25, 259; Kurds in, 234; as "Land of the Aryans", 57; name change from "Persia", 64; relations with Nazi Germany, 64–66, 273n138; salvage genetics on tribes in, 60; shifting borders of, 231; state emphasis on racial unity, 60–61; Wilhelm Haas in, 61–62. *See also* Pahlavi dynasty
Iranians, 123–24; ABO frequencies of, 230; as Aryans, 33–34, 37, 56–57, 61–66, 122, 223–24, 229, 231–32, 305n38; as Europeans, 34, 57; genetic research by, 219–21; IGP study of Iranian Americans, 257–58; inconsistent population definitions of, 119–22; "Iranian Plateau Race", 60; as "white," 57, 64, 121, 258. *See also* Aryan race concept; Persian language; Persian race/ethnicity
Iraq: anthropometric studies, 76; Assyrians in, 125–26, 241; Baghdadis, 91, 108; Baghdad Pact, 145; Baghdad spring anemia, 154–55, 177; Bedouins in, 91; Boyd research in, 88; Iran-Iraq War, 217, 225, 239, 241; Jews in/from, 91, 114, 117, 155, 158–59, 162, 165, 177, 207; Kurds in, 234, 237; sickle cell disease in, 147; Yazidis in, 241
Ireland, 88
Irmak, Sadi, 86–87, 91, 96
Israel: control over subjects, data, 118, 205, 210–11; interest in favism research, 164–67; Israeli-Arab conflict effects on science, 188, 206–7; land expropriation from Palestinians, 207–8; meaning of "ethnicity" in, 16, 206–7; Mizrahi immigrants to, 114–19; post-1967 research in, 199–205, 211; Yom Kippur War, 207. *See also* Palestine; Yishuv
Israelite tribes, 43; Benjamin, 166; Ephraim, 42, 166; Manasseh, 42
Istanbul University Medical Faculty, 84–85
Italians with sickle cell disease, 131
Itano, Harvey, 130, 141
'Izz al-Din, Najla Abu, 269n45

Jacob (biblical patriarch), 167
"Japhethite" women, 166
JDC (Joint Distribution Committee), 161–62
Jebeliya Bedouins, 201–2, 205
Jews: in Baghdad, 117; in Canada, 119; endogamy among, 7; enmity toward Samaritans, 42–43; and favism, 155–61, 163, 166, 179; Habbanites, 194–99; intermarriage with Gentiles, 117–18; Iranian (or Persian) Jews, 60, 80, 170, 223–25, 259; Iraqi Jews, 91, 114, 117, 155, 162, 165, 177, 207; Kurdish, 117, 165–66, *165*, 178–79; Middle Eastern (Mizrahi) Jews, 24, 80–81, 91; narratives of common ancestry for, 161, 179; "New Jews" in Palestine, 94; "non-Ashkenazi" (*lo-ashkenazi*), 159; population categories for, 119; prejudice against

INDEX   367

Mizrahim, 114–19; and race science, 43, 69–70, 74–75, 79–81; refugee emigration to Mandatory Palestine, 70; as religion not race, 79, 279–80n6; in Salonika, 70; Sheba theories on, 160–61, 163, 165–66; Younovitch theories on, 91; Zionists on the "Jewish race", 30–31. *See also* Ashkenazim; Israelite tribes; Mizrahi Jews; Sephardim; Yemenite Jews

Johnson, Eric, 197

Jordan, 145, 184, 189–91, 194, 209–10

Jordan, Charles, 162

Joseph, Dena, 45

June War (1967), 184–85, 199, 211

Justinian I, 201

Kamel, Karim, 147, 212

Kansu, Şevket Aziz, 48–50, 52, 54, 84, 95, 234–35

Kappers, Cornelius Ariens, 40–42, 44, 58, 63, 79, 269n50, 274n141

Karaites, 205

Karami, Ehsan, 1–3

Kasravi, Ahmad, 56, 232

Katznelson, Reuben, 45

Kayssi, Ahmed I., 91, 108

Kay Ustuvan, Husayn, 64

Keane, A. H., 15

Kebedgy, Michel, 31

Kell blood antigen, 95, 105

Kerman, 58, 121, 226–27

Kermanshah, 123

Keshavarz, Mr. K., 224

Khomeini, Ruhollah, 217, 226–27

Khuri, Fuad, 63

"kindred blood" concept, 65–66

Kirsh, Nurit, 115, 117, 161

knowledge-control regime, 21, 106, 124, 205

Kopeć, Ada, 113, 118

Kossovitch, Nicolas, 79, 276n39

Koubba Blood Transfusion Centre, 109

Krischner, Harald, 40, 58, 60

Kurdistan Workers' Party (PKK), 218–19

Kurds, *165*; and AUB, 40; Beşikçi on, 236–37; blood group sampling of, 115, 123; disputes over cephalic index results for, 55–56; favism among, 165, *165*, 175–77; in IHMGB, 257; and "internal colonialism", 264–65n27; in Iran, 55–56, 120–21, 215–16, 226–27, 229, 234; in Iraq, 234; Jews of Kurdistan, 80, 117, 158, 165–66, 176, 178–79; in Northern Iran, 214–17, 226; PTC-tasting and color blindness tests, 176; as racial group, 74–76, 234, 237; reclassified as Turks, 33–34, 51–52, 55, 179, 250; recognized by League of Nations, 52; Republic of Mahabad, 120; Saatçioğlu on, 235–38; security of research subjects, 253; in Turkey, 218–19, 238

Kuwait, 172, 193

Land Day protests, 208

"Land of the Aryans," Iran as, 57

language as determining ancestry, 150

L'anthropologie (French journal), 69

Lapeyssonnie, Léon, 174–75

"Lapps" (Saami), 35–36

Law of Return (Israel), 45, 163
Lawrence, T. E., 74
League of Nations, 12, 23, 38, 52–55, 63, 138
Lebanon, 24; anthropometry research in, *41*, *42*; favism in, 172; Lebanese Maronites, 38; Mount Lebanon communities, 39–42, *78*, 147, 189; Phoenician identity in, 38, 247, 260; sickle cell research in, 132, 143–44, 146–47, 149. *See also* AUB; Levant; Smith, Harry Madison; Taleb, Najib; Zalloua, Pierre
Lederer, Richard, 154–55, 158, 177
Lehmann, Hermann, 17, 285n20, 296n69; background of, 132; defending serological genetics, 215–17; postwar serological research, 149, 163–64; rejecting environmental determinism, 151–52; research in Aden, 131, 134, 136, 285n20; research in India, 132–34; research on Assyrians, 125; research on Kurds, 176, 237; sickle-cell research and theories, 113, 131–38, 140–52, 176–77; use of term "Eti-Turks", 140–41
Levant: ABO frequencies in, 73, 77; Aksoy comparative study in, 143–45, 150; and Eti-Turks, 142; Harry Smith blood grouping studies, 188, 194; and Hatay Crisis, 53; Ikin on, 143; Jewish "gene pool" from, 119, 179; mandates in, 24, 46; racial "living monuments" in, 39–47, 63; Sheba on, 166; Taleb and Ruffié research, 194. *See also* sickle cell disease

Levene, Cyril, 207
Lewis blood antigen, 95, 105
Libya, 114, 116; Libyans, 73, 207
Linnaeus, Carl, 35
Lisker, Rubén, 177
Lister, Richard, 101–2, *102*
Livingstone, Frank, 287–88n64
Los Angeles, Iranian "subgroups" in, 259
Lurs, 229
Lutheran blood antigen, 105
Luzzatto, Lucio, 178–79

"Macedonian Mohammedans", 70, 93
Mackintosh, James Stewart, 74–75
Mainz Institute of Anthropology, 213–14, 216, 221
malaria, 25, 129; among Greeks, 133; blackwater fever from, 154, 158; eradication campaigns, 111, 176, 233; and evolutionary anthropology, 177, 180; falciparum malaria parasites, 156, 172; and favism, 25, 155–57, 163–66, 171–73, 175–80; and hemolytic anemia, 155–56, 158–60, 162, 175; and Kurdish Jews, 165–66, 179; migration patterns and natural selection, 112–13, 153; and sickle cell disease, 25, 129, 132, 135, 137, 147–48, 151, 156–57; in Turkey, 233
Mandatory Palestine, 70
Maranjian, George, 110–12, 148, 195
Marengo-Rowe, Alain, 196–99, 211
Margolis, Emmanuel, 116
Marhiv, Nur, *46*
Marks, Paul, 160

Maronite Christians, 38–40, 78, 188, 247; Maronite Phoenicianism, 41–42
Masud, Dr., 191–92
Matta, Dawood, 89, 95
Mattson, Greggor, 35
Mavendad, Manouchehr, 169
Medes, 58, 224
"medical archaeology"/"medical anthropology", 161–62, 177, 180
medical genetics, 17–18, 23, 96, 126, 130, 163–64, 221–22
Medico-Legal Institute (Baghdad), 108
Mendel, George, 128–30
Mersin, Turkey, 53, 138–44
methodological nationalism, 7, 9, 264n11
Mexico, 171, 177
MHP (*Milliyetçi Hareket Partisi*, Nationalist Action Party), 236, 250
mid-digital hair, 290n29
millet system, 15–16, 69
Minasian, Caro Owen, 168
missing persons identification, 241
Mizrahi Jews, 19, 113–19, 129, 152–53, 157–61
MN/MNS blood groups, 88, 95, 107–8, 119, 143, 290n29
Mohagheghpour, Nahid, 223–24
Mohammad Reza Shah, 167, 213, 216, 223
"molecular disease" concept, 130
Mongoloid/yellow/Asian classification: and Akbari research, 226, 230; considered racially inferior, 33–34; and Farhud research, 232; Hooton and, 60; and Kyrgyz, 235; Mongols, 10; and Şentuna research, 237; Turkish rejection of, 33, 47, 235–36; by UNESCO, WHO, 103, 221–22
"monstrosities" in race science, 35–36
Montazemi, Kambiz, 222
Moroccans, 79–80, 93, 116, 159, 207, 276n39
Mosaddegh, Mohammad, 214
Mostashfi, Pezeshkpour, 220–21, 225
Motamed, Manuchehr, 119
Motulsky, Arno G., 17, 156, 164–65, 171–73, 175–77
"Mountain Turks," Kurds as, 52
Mount Lebanon communities, 39–42, 78, 147, 189
Mourant, Arthur E., 17, 215–16; collaborations with Smith, Bonné, 185, 188–92, 195, 197–205, 296n69; defending serological genetics, 216–17; efforts to control blood traffic, 113; exhibiting race, gender bias, 187–88; favism research, 163–64, 177; forced into retirement, 205; as IBP coordinator, 187–89; inconsistencies in findings, 118–19, 124–26; and Joseph Gurevitch, 116–19; on Kurds, 176, 237; on malaria, inbreeding hypotheses, 176–77; on malaria hypothesis, 176–77; review of Jewish genetic research, 118; on Rh frequencies, 109–10; Socotra research, 102–5, 110, 183; suggesting false positives, 109; views on Assyrian ancestry, 125. *See also* BGRL
Mukharji, Projit, 20
Müller, Friedrich Max, 36

"multicentric" hypothesis, 135, 148–49, 151, 160

Muslims: *akhdām* as, 70; blood group frequencies compared to Copts, 89, 95, 108–9; "Eti-Turks" as, 139; favism frequencies among, 165, 168, 172, 175; and historical race concepts, 15; Iranian, 224, 226; in Israel, 206–8, 212; "lumping" of, 61, 122; Nazi propaganda targeting, 64; relationship of to Copts, 109; relationship of to Zoroastrians, 121–22, 227; as religion not race, 279–80n6; serology results for, 77–78, 90, 121, 193; Shi'i communities, 111–13, 122, 148, 168, 172, 227, 230; sickle cell frequencies among, 112–13, 146–48; in Turkish-Greek population exchange, 70

Mustafa Kemal (Atatürk) and Kemalists, 54; Aksoy and, 138, 140–42, 152, 250; on Alawites, 53; CHP (Republican People's Party), 54, 138; and "Eti-Turks", 250; Muammer Aksoy and, 240; race science under, 12, 46, 47–48, 50, 95, 255; relationship of with Kurds, 52, 218; Saatçioğlu and, 236; Sancar and, 249–50; and Treaty of Sèvres, 70; Turkish History Thesis, 48; on Turks as "Alpine", 64

Nablus, *186*; as example of scientific territorialism, 187–94; Samaritans of, 42, 45, *46*, 185, 189–94, 199, 211

Nader Shah, 57

nasal index, 271n84

Nassar, Gamal Abdel, 145, 184

Nassar, Tamir, 41, *78*

nation, defined, 32

"National Committee for the Protection of Lands", 208

National Consciousness (Zurayk), 40

"national genome", 248, 255–57

nationalism: 26, 31; anticolonial, 14, 18, 21, 35; Arab nationalism, 38, 74, 95, 132, 184; Armenian nationalism 30, 74; ethnic nationalism, 12, 18, 21, 71, 93–94, 219; in Europe, 36–37; genetic nationalism, 8–9, 260–61; Greek nationalism, 30, 49; and historical narratives, 18, 33, 35, 48, 97, 107, 157, 167, 180; Iranian nationalism, 1, 38, 46, 55–57, 61–62, 113, 120, 124, 167–69, 214–17, 219, 231–32, 257, 259–60; Kurdish nationalism, 52, 56, 215, 218, 237; methodological nationalism, 7, 9, 264n11; pharaonicism, 33, 38, 95; Phoenicianism, 33, 38–42, 63, 146, 245; scientific nationalism, 3, 263n2; and territorial claims/competition, 13, 33–34, 52, 232, 240, 247–48; Turkish nationalism, 16, 38, 46, 49–50, 52–53, 70, 86, 88, 95, 132, 141, 144, 146, 150, 152, 215, 232, 234, 236, 248–55; as Zionism, 11, 30, 94, 113. *See also* Mustafa Kemal and Kemalists; Pahlavi dynasty; Zionism

Native Americans, 72, 76, 90, 244

"native informant", 19, 266n47

natural selection: favism and malaria, 164–65, 173, 176; versus genetic drift,

6, 25, 102, 172, 179; sickle cell disease and malaria, 113, 135, 137, 156–57
Nazis/Nazi Germany, 64–67, 96, 270n72, 273n138, 274n141
Neel, James V., 141, 144, 164, 286–87n48
"Negro"/"Negroid": Bandaris as, 226, 230; as "black", 222; and *cDe* chromosome, 109; as ethnoracial category, 69; and G6PD deficiency, 156, 160, 177; and Habbanites, 195; linkage of sickle cells to, 131, 134, 140, 145–47, 151; as "major race", 103; "Negro blood", 109, 129–31, 133; Rh blood types as determinant for, 104–5; Yemenites as, 129–30, 133, 136, 150, 160
Nemazee Hospital, 167, 169
"Neolithic-Chalcolithic populations" in Turkey, 51
Nevo, Sarah, 206–9, 212
New Yishuv, 44–45, 71, 81–82, 91
Nigeria, 149, 188
NIH (National Institutes of Health), 187–88
Nijenhuis, Lourens E., 123–24
Nilgiri Hills of India, 134
Nilotic-speaking tribal groups, 132
Nizhād va Zabān/"Race and Language" (Kay Ustuvan), 64
Noah and sons, 161
nomads, 35, 52, 56, 76, 86, 123
"Nordics," Nordicism, 37, 60, 64, 75, 270n72
Nuffield Blood Centre, 118
Nuffield Foundation, 133
Nuremberg race laws, 65–66
Nusayri Alawites. *See* Alawites

Nusayriler ve Nusayrilik hakkında (Tankut), 54

O blood. *See* ABO blood groups
Odessa Jews, 93–94
Oghuz Turks, 230, 235–36
Ohnesorg, Benno, 298n1
Oikkonen, Venla, 264n14
Onur, Nureddin, 85–88, 94, 96
oral histories, xv, 13, 103, 169
Orientalism, 35, 38, 82, 272n109
Orr, Neil, 101–2
Ottenberg, Reuben, 72–73, 76, 81, 85–86, 96
Ottoman Empire, 30; Armenian genocide, 31, 51, 74; Balkan Wars, 67; craniometry in, 39; dissolution of, 11–12, 24, 33, 46, 70; and European race science, 38, 46; and "European Turks", 49; genetic variation within, 232, 237; *millet* system under, 69–70; Rıza Tevfik and race science, 29–35; and Sanjak of Iskenderun (Alexandretta), 53–55, 138–39; science under, 10–11; typological, genealogical vocabulary of, 14–15; Young Turks, 30–31; Yürük tribe, 86–87. *See also* Syria; Turkey; Turkish peoples
Out of Africa model, 248, 251–52, 255
Özören, Nesrin, 250

Pacific American racial type, 72, 76
Padeh, Baruch, 200
Pahlavi dynasty, 34, 60, 62; and adoption of modern science, 10;

Pahlavi dynasty (*continued*)
anthropometry in, 24, 34, 57, 60; and Aryanist internal colonialism, 223, 231; and Bowman's favism research, 167–69; and Center for Iranian Anthropology, 217–18; centralizing of power by, 56; claiming European kinship, 57; disbanding of tribes by, 60; funding human genetics research, 219–20; Iranian Revolution against, 216–17, 219; labor strikes against, 120; military conscription, 56; Mohammad Reza Shah, 167, 213, 216, 223; racial views during, 15–16, 34, 62; Reza Shah, 46, 56, 58, 62, 64–65 (as Reza Khan, 12); standardization of education, 56; suppression of separatism and regional autonomy, 56–57, 60. *See also* Aryan race; Iran

paleoanthropology, 234–35, 255, 270n79

Palestine: 71, 81–82, 158, 166, 194; anthropometric research in, 43–45; Ashkenazi settlers in, 93–94; as British Mandate, 12, 24, 70, 73, 79, 114, 157, 189, 209; Palestine Exploration Fund, 39; Palestine Liberation Organization, 207, 209; Samaritan origins in, 44–46; seroanthropology research in, 77, 79–81; Turkic speakers in, 232; Zionist claims to, 31

Palestinians: and Armenians, 209–10; blood samples from, 189, 194, 207–8, 212; Bonné-Tamir and, 185–87, 246, 294n15; Bowman's comments regarding, 171; Farhud on, 232; genetic resemblance to Jordanians, 194; indigenous populations of, 43; as indistinguishable from Mizrahim, 115; as "Israeli Ethnic Groups", 205–10; Jewish Israeli attitudes toward, 246; Land Day protests, 208

Pallister, M. A., 183–84, 195

Palmyra, Syria, 39, 269n50

pamaquine, hemolytic response to, 158

Paris Peace Conference (1919), 74

Parr, Leland W., 77–80, 83, 88–90

Parsa Community Foundation, 258

Pasteur Institute (Paris), 119–21, 276n39

Pauling, Linus, 130

P blood antigen, 105

Persepolis, 60

Persia: name change to "Iran", 64; Qajar dynasty, 10–12, 30, 34–35, 38, 56, 167; Sassanian Empire of, 169

Persian Gulf, 223, 226

Persian language and speakers, xi, 14–15, 57, 64, 167–68, 217, 221–25

Persian race/ethnicity, 34, 40, 58; Akbari study of, 229–30; and Aryanism, 305n38; conflated with Iranian majority population, 120-23; Farhud study of, 230–32; IHGP on, 305n38; Persian Jews, 80, 115; Zoroastrians and, 169. *See also* Aryan race concept; nationalism (Iranian); Pahlavi dynasty

philology, 36, 64

Phoenicians: Bedouins as, 42; claims of non-Arab genealogy, 38–40; Kappers on, 42, 269n50; Maronites as,

40–41; search for living descendants of, 33, 38, 42, 44, 63, 245; Sheba on, 166; and sickle cell disease, 146; Wells/Zalloua DNA project, 245, 247, 260
Pirniya, Hassan, 56, 231
Pittard, Eugène, 48–50, 54–55, 84
PKK (Kurdistan Workers' Party), 218–19
Podliachouk, Luba, 120–21
population structure, 210, 251
"population transfers", 23, 53, 241
Portuguese, 230
primaquine, 156, 159, 162, 175, 233
prisoners: testing on, 156; of war, 23, 31, 68
"proto-Hittites", 42
PTC (phenylthiocarbamide) tasting ability, 88, 176, 290n29

Qajar dynasty, 10–12, 30, 34–35, 38, 56, 167
Qashqa'i tribes, 120, 123, 222–23
Quest for the Phoenicians documentary, 245

Rabin, Yitzhak, 208
race: defined, 32; "ethnic group" versus "race", 96–97; and genetic disorders, 150–53; geographical versus religious determinants, 42; versus nation, 52, 75; no biological basis for, 252; persistence of concept, 260
Race, Robert R., 106, 108
Race and History (Pittard), 55
"Race History of Turkey" (Kansu), 235

race science, 14, 32; Afet İnan's use of, 50–54; colonial versus nationalist approaches to, 38; correlates of, 12–13; Dawlatabadi and, 34; European, 17, 35–39; in Iran, 61; Jewish, 11, 43; Kemalist funding of, 46; in League of Nations, 38, 53; in Middle East, 16–17, 38, 63–64; and post-WWI boundaries, 31–32; precursors of, 14–15; scholarship on, 18; shifting terminology within, 97; Rıza Tevfik and, 30–32; UNESCO on, 96–97; Yāsamī on, 55; Zurayk and, 40. *See also* anthropometry
Races of Man, The (Deniker), 15
Rachel (biblical matriarch), 167
racial anthropology, 16, 22, 31, 49, 68, 235, 272n109
"racial crossing", 73
"racial index", 72, 85, 103. *See also* "biochemical index"
"racial stocks", 33, 41, 60, 87
Ramot, Bracha, 170
rape in Iran-Iraq War, 239
Red Crescent (Turkey), 237
Red Lion and Sun Society (Iran), 119
relationality of genetic data, 247–48
religion: and favism, 156, 167–71; Jews as religion versus race, 79–80, 224–25, 259; preventing intermarriage and preserving endogamy, 5, 7, 17, 25, 42–43, 77, 80, 83, 115, 121, 133, 160, 168–70, 291n47
Renan, Ernest, 62
restriction-enzyme genotyping, 148–49, 178

Retzius, Anders, 35–36
Reza Shah, 46, 56, 58, 62, 64–65; as Reza Khan, 12
Rh antigen system, 126–27, 290n29; BGRL work on, 107–8; Gurevitch on, 115–16; medical genetics and, 126–27; Mourant on, 109, 118–19; Nijenhuis on, 123; Rh_0/cDe type, 95, 104–5, 137; Şentuna on Turkish findings, 237–38; typing of soldiers, 109; Younovitch on, 115. See also blood groups
Rhodesia, 131
Rıza Tevfik, 29–32, 34
Robert College (Istanbul), 37, 47
Ronaghi, Mostafa, 257–58
Rosh Ha'ayin, 116–17, 128–30, 133–34, 206
Ruffié, Jacques, 193–94
Ruppert, Johannes, 65

Saami ("Lapps"), 35–36
Saatçioğlu, Armağan, 234–38
Sakhnin, land expropriation in, 207–8
Salonika, 67–70, 68, 93, 276n39
salvage anthropology/genetics, 5, 44, 59–60, 234, 243
Samaritans, 46, 294n24; as "ancient Hebrews", 43; as "Asiatic", 83; claims of Israelite descent, 43, 77; interpretations regarding favism, 166–67; issue of payment for blood samples, 189–92; Kappers on, 42; and Law of Return (1950), 45; as mostly endogamous, 24, 42–45, 77, 83–84; Parr on, 77; resembling Sephardic Jews, 44, 82; scientists' access to, 185, 189–95, 198–99, 203–6, 209, 211, 246; serology studies of, 71, 79, 189–92; Sheba on, 166–67; as vanishing tribe, 43–45, 83–84; Younovitch on, 79, 82–83. See also Bonné/Bonné-Tamir, Batsheva

sample collection: by Egyptian military physicians, 109; exploitation of research subjects, 211–12; importance of sample size, 8, 85, 109–10, 112, 123–24, 208; issue of payment for samples, 189–92; lying to sample donors, 197; selection and categorization, 124–25; as "traffic in blood", 103–6, 243
Sancar, Aziz, 249–50
Sancar, Mithat, 249
Sanjak of Iskenderun (Alexandretta), 53–55, 138–39
San Remo Conference, 70
Saran, Nephan, 234, 238
Sardari, Abdolhossein, 65
Sardinia, favism in, 154, 160, 166
Saruhan, Güher, 239
Saudi Arabia, 124, 132, 145, 172. See also Aramco
Sawhney, Kanwarjit Singh, 301n46
Ṣāʿid al-Andalusī, 15
Schayegh, Cyrus, 10, 19
Schiebinger, Londa, 20
Schwidetzky, Ilse, 213
scientific collaboration. See collaboration
scientific territorialism, 187–94

scientific trafficking. *See* blood supply chains; "traffic in blood"
Second Turkish History Congress, 49, 86–87, 95
sectarian/religious categories, 69–70, 112–13
"self-described origins," accuracy of, 253
Semites (sons of Shem), 73, 166
Şentuna, Can, 237–38
Şenyürek, Muzaffer Süleyman, 270n79
Sephardim: definition of, 79–80, 115–16, 159; of Mexico City, 177; Samaritan connection with, 44, 82; serology studies of, 79–80, 115, 118
sero-anthropology: ABO groups found in apes, 88; compared to anthropometry, 71–76; disputes over data interpretation, 93; Haldane's defense of, 91–93; legacy of, 94–97; origins of, 48, 69–71; Parr study of Jews, 77, 79; post-WWII fieldwork methods, 103–7; skepticism toward, 215–16; in Turkish Republic, 84–88; Younovitch study of Yemenites, 133. *See also* ABO blood groups; "traffic in blood"
"Serological Anthropology", 5
Seyedna, Yousuf, 226–27
Shahid, Munib, 146, 193
Shahrzad, Moubed Rustam, 224
Shanklin, William M., 40–41 (*41*), 58, 63, 76–77, 88, 90, 95
Sheba, Chaim: aiding Bonné's research, 200, 202, 205–6; background of, 157; Bowman and, 170–71; on Falasha origins, 162–63, 290n31; favism research and theories, 158–63, 165–67, 171–73, 176–79; at 1961 Jerusalem genetics conference, 164; "medical archaeology" of, 162; on Samaritan origins, 166, 291n39; theories regarding Sephardim, 159, 177, 209, 291n39
Shimony, Shimon, 45
Shi'i communities, 111–13, 122, 148, 168, 172, 227, 230
Shiraz, 58, 123, 155, 167–68, 172, 174
Shusha, Ali Tawfiq (Shusha Pasha, Shousha), 73, 275n17
Sicily, favism in, 154
sickle cell anemia, 129-130, 138, 143, 177
sickle cell disease, 129–38; Aksoy's work on, 138–45; Allison's work on, 135; among Alawites, 138, 150; among Shi'i communities, 148; Black Panthers screening campaign, 148; and Black-white racial dichotomy, 152; carriers, 130, 135, 141, 148; erroneous findings in Yemenites, 117, 129, 133–34; evolution of, 134–35; field research on, 105; frequencies among Sunnis, Shi'is, Bedouins in Saudi Arabia, 112–13; geographic distribution of, 134–35; and health-care activism, 284n6; initially tied to "Negro blood", 129–33, 145, 147–49, 156; Lehmann's work on, 113, 131–38, 140–52, 176–77; malaria and natural selection for, 113, 135, 148–49, 165
Six-Day War (or June War), 184–85, 199, 211

skelic index, 50, 271n84
skin color, 14, 55, 89, 92, 135, 155, 159–60, 166
skull shape, 33, 36–37, 41, 51, 55, 92. *See also* cephalic index
Slavs, 49, 69
smallpox, 221
Smith, Harry Madison, 199, 211; and ABGL, 187–89, 194; academic, research background, 188; and Aramco physicians, 193; Beirut laboratory, 199, 211, 220; collaborations with Mourant, Till, Bonné, 185, 188–92, 195, 197–205, 211, 296n69; favoring foreign over local staff, 188
Snyder, Laurence, 72–73, 81, 85, 96
social anthropology, 61–62, 233
Society for National Heritage (Iran), 57
Socotra, 101–5, *102*, 110, 112, 183–84, 195–96
Somalis, 136
Soviet Union, 22, 88, 120, 145, 168, 232
Speakman, Cummins E., 188
Spencer, Herbert, 30
SPGL (Serological Population Genetics Laboratory), 195, 197–200, 202–4
Spivak, Gayatri Chakravorty, 266n47
Stamm, William P., 183
State Commission of Inquiry into the Disappearance of Yemenite Children (Israel), 128
Stern, Curt, 164
Suárez-Díaz, Edna, 177
"subgroups" designation, 259
Sudan, 18, 95
Sudanese, 136

"Sufis" in Iran, 259
sulfapyridine, hemolytic response to, 158
Sunderland, Eric, 226, 301
"supra-ethnicities", 122
Sweden, 35–36
Syria: Aksoy on Alawites in, 143–44; Altounyan on, 75–76, 275n18; Boyds' work in, 88–91; Christians in, 43, 77–79, 90; defense pact with Egypt, Saudi Arabia, 145; effects of Israeli-Arab conflict on, 184, 194, 198; Egypt, Egyptians and, 145; endogamy among peoples in, 43; Fellahs in, 142; Kappers fieldwork, 40; meaning of "Syrian", 39; Muslims in, 43, 78–79, 90; Palmyra, 39, 269n50; Parr on Armenians in, 78; Parr on Christians in, 79, 90; peoples of, 76–77; Sanjak of Alexandretta issue, 53–54, 138–39; sero-anthropology studies in, 88; Shanklin work on Bedouins in, 41–42, 76–77, 95; sickle-cell disease in, 140, 146; skull collections from, 39; Younovitch on Samaritans, Sephardim in, 82–84. *See also* AUB
Szeinberg, Aryeh, 170, 174–75, 178, 296n69
Szpidbaum, Henryk, 44

Tabatabai, Hamideh, 223–25
Tabulae Biologicae, 91, 93
Taleb, Najib, 193–94
Tankut, Hasan Reşit, 54
Taqizadeh, Hassan, 61

Tayyar, Adib, 90
Tehran: Aryanism revival in, 223; blood banks, transfusion practices, 214; blood samples from, 119–21, 214, 226, 229; Field's testing in, 60–61; IHMGB (Iranian Human Mutation Gene Bank), 256–57; Institute for Public Health Research, 176; race science in, 37, 55; under Reza Shah, 56; Tehran University, 175, 219–22, 225–26; Zoroastrians in, 227
Tel-Hashomer Government Hospital: budget issues, 161, 167; favism research and theories, 159, 161–63, 167; field research by, 162; Israeli geneticists at, 189–90, 192, 195, 198, 200–202, 204, 206; responsibility for Sinai Bedouin health care, 200–202; TB treatments and hemolytic anemia, 162. *See also* Bonné/Bonné-Tamir, Batsheva; Sheba, Chaim
territorial integrity, 13, 26, 52, 55–56, 231–32, 240–41, 257, 260
TGP (Turkey Genome Project), 250–54
thalassemia, 156, 165
Thoumaian, Garabed and Lucy, 31
Tills, Donald, 197, 199–205
tongue-rolling ability, 290n29
Topinard, Paul, 64
"traffic in blood", 103–6, 243. *See also* scientific trafficking
transferrin, 105, 150, 280n12, 290n29
Transjordan, 41
transplant rejection, 205
Treaty of Sèvres, 52, 70
Tsedaka, Yisra'el, 190

TÜBİTAK, 233
Tünakan, Seniha, 234, 270n79
Tunisia, 93, 116
Turkey Genome Project (TGP), 250–54
Turkey: *Arap uşakları* ("Arab servants") in, 138; Armenians in, 233; college student essays on, 47; "European" and "Asian" division, 49, 51; geographical subdivisions, 236; Greek invasion of Anatolia, 31; homogenization goals of Turkish Republic, 52, 232–39; Kurds as "Mountain Turks", 52; Levantine mandates, 46; malaria in, 233; under martial law, 218–19; National Campaign (*milli mücadele*), 70; Turkish "disease", 249; Turkish History Thesis, 47–48, 50, 84, 87; war for independence, 46; Young Turks, 30–31. *See also* Ottoman Empire; Young Turks
Turkish peoples: Anatolians as "pure" Turks, 84–85; Babacan on, 84–86, 93–94; claimed as Type A origin, 84, 87; "culturally Turkish", 279–80n6; disputes over origin of, 31, 33, 254; fingerprint patterns, 87; Gökçümen on, 251–53, 255; in Iran, 229; Onur on, 85–87, 93–94, 96; Pittard on, 48; sero-anthropological studies of, 49, 69, 74, 84–85, 233–34; as white race, 47; Yürük tribes, 86–87. *See also* "Eti-Turks" (Alawites); Young Turks
Turkish Anthropometric Survey (1937), 50, 58
Turkish History Thesis, 47–48, 50, 84, 86–88, 271n95

378 INDEX

Turkish Institute of Anthropology, 48, 84, 270n72
"Turkish-Islamic synthesis", 218
Turkish Malaria Eradication Organization, 233
Turkmen (or Turkoman), 74, 76, 121, 226–27, 230, 271n95

UAR (United Arab Republic), 145
Uganda, 131–32
ülkücüler ("idealists"), 236
"umbrella" Iranian category, 122, 259
UNESCO (United Nations Educational, Scientific and Cultural Organization), 96–97, 103, 106, 141
"unity in diversity" framing, 248, 257
Universal Races Congress, 29–31, 34
Urumiya (Rezaiyeh), Iran, 226, 229

vaccination, 117
"vanishing indigene" narrative, 5
Veddoids, 136–37, 140–43, 146–47, 150–52
Verse of the Elephant (Qur'an), 135
Vidal, Federico, 111–13
von Anrep, Gleb, 89
von Eickstedt, Egon Freiherr, 213
von Luschan, Felix, 29, 39–40, 54–55, 75

Wales, 88
Walker, Deryck G., 168
Walter, Hubert, 213–14, 221, 230, 239
Weissenberg, Samuel, 79, 269n55
Wells, Spencer, 244–45, 254
Wenner-Gren Foundation, 134, 190, 216

whiteness, 39, 57, 64, 146, 247, 258; white/European/Caucasoid classification, 103, 222; "white Jews", 81
WHO (World Health Organization), 105–6, 110, 156, 174–75, 221, 233
Wiener, Alexander S., 104–5
World's Peoples, The (Keane), 15
World War I, 23–24, 31, 241
World War II, 23, 62, 63–64, 102, 104
Wyman, Leland C., 88

Yarsani (religious community), 259
Yāsamī, Rashīd, 55–56, 62
Yazd, 58, 168, 175, 226–27
Y-DNA haplotypes, 247, 254
yellow classification. *See* Mongoloid/yellow/Asian classification
Yemen: *akhdām*/"Zabidi" class, 134–38, 141–43, 146, 150–52, 285nn20, 24; Coon's research in, 133, 136; Jews from, 79, 81, 114, 130, 133–35, 138, 152, 206; sickle cell disease in, 134, 140, 149, 152. *See also* Aden
Yemenite Jews: admixture among, 115; BGRL and, 112; Bonné's blood collection work, 185, 195, 197–98; as having "Negro blood", 129–30, 133, 151, 160; HLA study on, 206; Sheba's findings of favism among, 158–59; and sickle cell disease, 117, 129–31, 133–34, 138, 152; Yemenite Children Affair, 128–30
Yishuv: New Yishuv, 44–45, 71, 81–82, 91; Yishuv period, 16, 114, 116
Yoruba people, 188
Yosmaoğlu, İpek, 23

Young Turks, 30–31
Younovitch (née Goldberg), Rina, 79–83, 91, 93–94, 114–15, 133
Yükseköğretim Kurulu or YÖK (Higher Education Council), 238
Yürük tribes, 86–87

"Zabidi Arabs"/*akhdām*, 134–38, 141–43, 146, 150–52, 285nn20, 24
Zabolis, 226, 229
Zalloua, Pierre, 245, 247
Zangwill, Israel, 30
Zavitzianos, Spyridon, 31
Zaydan, Jurji, 15
Zionism: among Salonika Jews, 70; and anthropological genetics, 211, 246; and Armenian expulsion, 209; and "Asiatic" communities, 81, 83; Bowman's beliefs regarding, 170; historical legitimacy of, 43; and Jewish biology, 116, 118–19; Jewish settlement of Palestine, 31; and medicine, public health, 157; Mourant's and Gurevitch's beliefs regarding, 116–18; nationalism, 11, 30; and Samaritan identity, 43–45, 83–84; Sheba's beliefs regarding, 157, 159, 161, 170; against Young Turks, 30–31; Younovitch beliefs regarding, 79–84

Zollschan, Ignaz, 31
Zoroastrians, 229; Amirshahi study of, 229; at Assembly of Experts, 227; Beckett study of, 121–22; Bowman study of, 169–70, 175; endogamy among, 7, 121, 168–69, 291n47; favism frequencies among, 175; in IGP, 259; in IHMGB, 257; Iranian beliefs regarding, 25; low favism rates among, 168, 170, 175; Mohagheghpour study of, 223–24; origins of, 58, 125; Persians as, 169; Seyedna study of, 226–27; in United States, 259
Zuckerkandl, Emile, 216
Zurayk, Constantine, 40

The authorized representative in the EU for product safety and compliance is:
Mare Nostrum Group
B.V Doelen 72
4831 GR Breda
The Netherlands

www.ingramcontent.com/pod-product-compliance
Lightning Source LLC
Chambersburg PA
CBHW031750220426
43662CB00007B/348